More information about this series at http://www.springer.com/series/4927

Physics of Megathrust Earthquakes

Edited by
Sylvain Barbot

Previously published in *Pure and Applied Geophysics* (PAGEOPH),
Volume 176, No. 9, 2019

Editor
Sylvain Barbot
University of Southern California
Los Angeles, CA, USA

ISSN 2504-3625
Pageoph Topical Volumes
ISBN 978-3-030-43571-4

Cover illustration: Slow Slip Events spatio-temporal distribution along the Cascadia subduction zone (Courtesy of Sylvain Michel, Tristan Michel, Adriano Gualandi and Jean-Philippe Avouac)

This book is published under the imprint Birkhäuser, www.birkhauser-science.com by the registered company Springer Nature Switzerland AG
The registered company address is: Gewerbestrasse 11, 6330 Cham, Switzerland

Contents

Pure Appl. Geophys. 176 (2019), 3813–3814
© 2019 Springer Nature Switzerland AG
https://doi.org/10.1007/s00024-019-02308-y

Physics of Megathrust Earthquakes: Introduction

SYLVAIN BARBOT[1]

Subduction zones represent a key element of plate tectonics, whereby oceanic plates are recycled in the mantle and volatiles are expelled to the atmosphere, contributing to many facets of the Earth system, including seafloor spreading, continental drift, landscape formation, and some important climate forcing factors. Because the subducting plate shaves the bottom of the continental crust at low angle, a large area is available to faulting. Hence, subduction zones create the largest earthquakes on our planet, often associated with devastating tsunamis. The earthquake phenomenon fascinates because of the sudden devastation that it brings and the hidden mysteries of the subsurface. The physics of earthquakes is complicated to the extreme, but a recent realization in the scientific community is the inter-connectedness of the physical processes that contribute to earthquake generation. For example, the long quiescent period between large earthquakes sets the stage for several thermo-mechanical processes that play important roles in controlling the recurrence times and the characteristics of subsequent ruptures. This Physics of Megathrust Earthquakes Topical Issue of Pure and Applied Geophysics presents contributions discussing several important aspects of the seismic cycle at subduction zones.

First, the review by Okal (2018) in this issue presents a recount on how the familiar earthquake magnitude scale has been designed, and the multiple challenges associated with assigning a magnitude to remote earthquakes, based solely on records of the seismic waves that they generate. Lin et al. (2019) discuss a numerical method to calculate the static

stress drop, another important earthquake rupture source property.

The next studies in this Topical Issue document the complexity of fault slip during the mostly quiescent period of the seismic cycle. Michel et al. (2018) describe the emergence of slow slip and slow earthquakes at the Cascadia subduction zone using long time series of geodetic data. Li et al. (2019) combine various geodetic data to document the spatial complementarity of seismic coupling with the rupture area of the 2015 $Mw = 7.9$ Gorkha, Nepal earthquake. Peña et al. (2019) investigate the quasi-static deformation that accrued for many years following one of the five largest instrumented earthquakes, the 2010 $Mw = 8.8$ Maule, Chile earthquake. Liu et al. (2018) provide sophisticated tools to investigate this kind of slow, distributed deformation at subduction zones. Shibazaki et al. (2019) investigate another facet of post-seismic deformation, focusing on the slow afterslip that spreads around the coseismic rupture of the 2011 $Mw = 9.1$ Tohoku-Oki, Japan megathrust earthquake.

The three following papers in this issue focus on the physics of megathrust earthquake ruptures by investigating the role of frictional properties on rupture style (Senatorski 2019), the role of geometrical segment boundaries on the down-dip segmentation of the megathrust (Ong et al. 2019), and the coupling between rupture propagation and tsunami generation (Lotto et al. 2018).

Last but not least, van Dinther et al. (2019) in this issue present sophisticated models of subduction zone dynamics that link the short-term seismic cycle with the long-term tectonic processes to argue for the presence of a secondary zone of uplift that can help understand the rheological structure of subduction zones.

[1] University of Southern California, Los Angeles, USA.
E-mail: sbarbot@usc.edu

I thank all authors for their excellent contributions to a better understanding of the physics of megathrust earthquakes. I extend my gratitude to the many anonymous reviewers, who helped improve the quality of the work that is presented here. I am thankful for the additional editorial work of Dr. Yuning Fu and Dr. Carla F. Braitenberg. Finally, I am grateful for the indefatigable help provided by Dr. Renata Dmowska during the editorial process.

Publisher's Note Springer Nature remains neutral with regard to jurisdictional claims in published maps and institutional affiliations.

References

Li, S., Wang, Q., Chen, G., He, P., Ding, K., Chen, Y., et al. (2019). Interseismic coupling in the Central Nepalese Himalaya: Spatial correlation with the 2015 Mw 7.9 Gorkha Earthquake. *Pure and Applied Geophysics*. https://doi.org/10.1007/s00024-019-02121-7.

Lin, R., Fang, X., Gan, Y., & Zheng, Y. (2019). A damped dynamic finite difference approach for modeling static stress–strain fields. *Pure and Applied Geophysics*. https://doi.org/10.1007/s00024-019-02207-2.

Liu, T., Fu, G., She, Y., & Zhao, C. (2018). Green's functions for post-seismic strain changes in a realistic earth model and their application to the Tohoku-Oki Mw 9.0 earthquake. *Pure and Applied Geophysics*. https://doi.org/10.1007/s00024-018-2054-z.

Lotto, G. C., Jeppson, T. N., & Dunham, E. M. (2018). Fully-coupled simulations of megathrust earthquakes and tsunamis in the Japan Trench, Nankai Trough, and Cascadia Subduction Zone. *Pure and Applied Geophysics*. https://doi.org/10.1007/s00024-018-1990-y.

Michel, S., Gualandi, A., & Avouac, J. P. (2018). Interseismic coupling and slow slip events on the Cascadia Megathrust. *Pure and Applied Geophysics*. https://doi.org/10.1007/s00024-018-1991-x.

Okal, E. (2018). Energy and magnitude: A historical perspective. *Pure and Applied Geophysics*. https://doi.org/10.1007/s00024-018-1994-7.

Ong, S. Q. M., Barbot, S., & Hubbard, J. (2019). Physics-based scenario of earthquake cycles on the ventura thrust system, California: The effect of variable friction and fault geometry. *Pure and Applied Geophysics*. https://doi.org/10.1007/s00024-019-02111-9.

Peña, C., Heidbach, O., Moreno, M., Bedford, J., Ziegler, M., Tassara, A., et al. (2019). Role of lower crust in the postseismic deformation of the 2010 Maule earthquake: Insights from a model with power-law rheology. *Pure and Applied Geophysics*. https://doi.org/10.1007/s00024-018-02090-3.

Senatorski, P. (2019). Effect of slip-weakening distance on seismic-aseismic slip patterns. *Pure and Applied Geophysics*. https://doi.org/10.1007/s00024-019-02094-7.

Shibazaki, B., Noda, H., & Ikari, M. J. (2019). Quasi-dynamic 3D modeling of the generation and afterslip of a Tohoku-oki earthquake considering thermal pressurization and frictional properties of the shallow plate boundary. *Pure and Applied Geophysics*. https://doi.org/10.1007/s00024-018-02089-w.

van Dinther, Y., Preiswerk, L., & Gerya, T. (2019). A secondary zone of uplift due to megathrust earthquakes. *Pure and Applied Geophysics*. https://doi.org/10.1007/s00024-019-02250-z.

(Published online September 2, 2019)

Pure Appl. Geophys. 176 (2019), 3815–3849
© 2018 Springer Nature Switzerland AG
https://doi.org/10.1007/s00024-018-1994-7

Energy and Magnitude: A Historical Perspective

Emile A. Okal[1]

Abstract—We present a detailed historical review of early attempts to quantify seismic sources through a measure of the energy radiated into seismic waves, in connection with the parallel development of the concept of magnitude. In particular, we explore the derivation of the widely quoted "Gutenberg–Richter energy–magnitude relationship"

$$\log_{10} E = 1.5 M_s + 11.8 \qquad (1)$$

(*E* in ergs), and especially the origin of the value 1.5 for the slope. By examining all of the relevant papers by Gutenberg and Richter, we note that estimates of this slope kept decreasing for more than 20 years before Gutenberg's sudden death, and that the value 1.5 was obtained through the complex computation of an estimate of the energy flux above the hypocenter, based on a number of assumptions and models lacking robustness in the context of modern seismological theory. We emphasize that the scaling laws underlying this derivation, as well as previous relations with generally higher values of the slope, suffer violations by several classes of earthquakes, and we stress the significant scientific value of reporting radiated seismic energy independently of seismic moment (or of reporting several types of magnitude), in order to fully document the rich diversity of seismic sources.

Key words: Radiated seismic energy, earthquake magnitudes, historical seismicity, seismic scaling laws.

1. Introduction

This paper presents a historical review of the measurement of the energy of earthquakes, in the framework of the parallel development of the concept of magnitude. In particular, we seek to understand why the classical formula

$$\log_{10} E = 1.5 M_s + 11.8 \qquad (1)$$

referred to as "Gutenberg [and Richter]'s energy–magnitude relation" features a slope of 1.5 which is not predicted a priori by simple physical arguments. We will use Gutenberg and Richter's (1956a) notation, *Q* [their Eq. (16) p. 133], for the slope of $\log_{10} E$ versus magnitude [1.5 in (1)].

We are motivated by the fact that Eq. (1) is to be found nowhere in this exact form in any of the traditional references in its support, which incidentally were most probably copied from one referring publication to the next. They consist of Gutenberg and Richter (1954) (*Seismicity of the Earth*), Gutenberg (1956) [the reference given by Kanamori (1977) in his paper introducing the concept of the "moment magnitude" M_w], and Gutenberg and Richter (1956b). For example, Eq. (1) is not spelt out anywhere in Gutenberg (1956), although it can be obtained by combining the actual formula proposed for *E* [his Eq. (3) p. 3]

$$\log_{10} E = 2.4m + 5.8 \qquad (2)$$

with the relationship between the "unified magnitude" *m* (Gutenberg's own quotes) and the surface-wave magnitude M_s [Eq. (1) p. 3 of Gutenberg (1956)]:

$$m = 0.63 M_s + 2.5, \qquad (3)$$

neither slope (2.4 or 0.63) having a simple physical justification. The same combination is also given by Richter (1958, pp. 365–366), even though he proposes the unexplained constant 11.4 instead of 11.8 in (1), a difference which may appear trivial, but still involves a ratio of 2.5. It is also given in the caption of the nomogram on Fig. 2 of Gutenberg and Richter (1956b), which does provide separate derivations of (2) and (3). As for Gutenberg and Richter (1954) (the third edition, generally regarded as definitive, of

---

[1] Department of Earth and Planetary Sciences, Northwestern University, Evanston, IL 60208, USA. E-mail: emile@earth.northwestern.edu

Seismicity of the Earth), the only mention of energy is found in its Introduction (p. 10)

"In this book, we have assumed for radiated energy the partly empirical equation

$$\log_{10} E = 12 + 1.8M. \quad (4)$$

This seems to give too great energy. At present (1953), the following form is preferred:

$$\log_{10} E = 11 + 1.6M''. \quad (5)$$

While Eq. (4), with $Q = 1.8$, was derived from Gutenberg and Richter (1942), Eq. (5), with $Q = 1.6$, was apparently never formally published or analytically explained.

The fact that none of the three key references to "Gutenberg and Richter's energy–magnitude relation" actually spells it out warrants some research into the origin of the formula, from both historical and theoretical standpoints. In order to shed some light on the origin of (1), and to recast it within modern seismic source theory, this paper explores the development of the concept of earthquake energy and of its measurement, notably in the framework of the introduction of magnitude by Richter (1935). In particular, we examine all of Gutenberg's papers on the subject, using the compilation of his bibliography available from his obituary (Richter 1962).

2. The Modern Context and the Apparent "Energy Paradox"

Understanding the evolution of the concept of magnitude and the attempts to relate it to seismic energy must be based on our present command of seismic source theory. In this respect, this section attempts to provide a modern theoretical forecast of a possible relation between magnitude and energy. We base our discussion on the concept of *double couple* **M** introduced by Vvedenskaya (1956), and later Knopoff and Gilbert (1959) as the system of forces representing a seismic dislocation, its scalar value being the seismic moment M_0 of the earthquake.

Note that we consider here, as the "energy" of an earthquake only the release of elastic energy stored during the interseismic deformation of the Earth, and not the changes in gravitational and rotational kinetic

energy resulting from the redistribution of mass during the earthquake, which may be several orders of magnitude larger (Dahlen 1977).

2.1. The Energy Paradox

We first recall that *magnitude* was introduced by Richter (1935) as a measure of the logarithm of the amplitude of the seismic trace recorded by a torsion instrument at a distance of 100 km, and thus essentially of the ground motion generated by the earthquake. In the absence of source finiteness effects, and given the linearity of the equations of mechanics governing the Earth's response [traceable all the way to Newton's (1687) "$\mathbf{f} = m\,\mathbf{a}$"], that ground motion, A in the notation of most of Gutenberg's papers, should be proportional to M_0, and hence any magnitude M should grow like $\log_{10} M_0$. This is indeed what is predicted theoretically and observed empirically, for example for the surface-wave magnitude M_s below about six (Geller 1976; Ekström and Dziewoński 1988; Okal 1989).

In most early contributions, it was generally assumed that the energy of a seismic source could be computed from the kinetic energy of the ground motion imparted to the Earth by the passage of a seismic wave, which would be expected to grow as the square of the amplitude of ground motion. Since the concept of magnitude measures the logarithm of the latter, this leads naturally to $Q = 2$, as featured by earlier versions of Eq. (1) (Gutenberg and Richter 1936).

By contrast, using the model of a double-couple **M**, the seismic energy E released by the source is simply its scalar product with the strain ε released during the earthquake. The absolute value of the strain should be a characteristic of the rock fracturing during the earthquake, and as such an invariant in the problem, so that E should be proportional to M_0. Again, in the absence of source finiteness effects, the linearity between seismic source and ground motion ("$\mathbf{f} = m\,\mathbf{a}$") will then result in E being directly proportional to ground motion, and hence in a slope $Q = 1$ in Eq. (1).

We thus reach a paradox, in that the two above arguments predict contradictory values of Q. The highly quoted Gutenberg and Richter relationship (1), which uses the intermediate value $Q = 1.5$, may appear as a somewhat acceptable compromise, but satisfies neither interpretation. Thus, it deserves full understanding and discussion.

2.2. A Modern Approach

The origin of this apparent, and well-known, "energy paradox" can be traced to at least three effects:

(i) Most importantly, the proportionality of energy to the square of displacement holds only for a monochromatic harmonic oscillator (with the additional assumption of a frequency not varying with size), while the spectrum of seismic ground motion following an earthquake is distributed over a wide range of frequencies, and thus the resulting time-domain amplitude at any given point (which is what is measured by a magnitude scale) is a complex function of its various spectral components;

(ii) Because of source finiteness ("a large earthquake takes time and space to occur"), destructive interference between individual elements of the source causes ground motion amplitudes measured at any given period to grow with moment slower than linearly, and eventually to saturate for large earthquakes (Geller 1976);

(iii) Ground motion can be measured only through the use of instruments acting as filters; while some of them could in principle be so narrow as to give the seismogram the appearance of a monochromatic oscillator, justifying assumption (i), the inescapable fact remains that most of the energy of the source would then be hidden outside the bandwidth recorded by the instrument. In parallel, ground motion can be measured only at some distance from the source, and anelastic attenuation over the corresponding path will similarly affect the spectrum of the recorded seismogram.

In modern days, the effect of (iii) is vastly reduced by the availability of broadband instrumentation. Using modern theory, it is possible to offer quantitative

models of the concept of finiteness (ii), as first described by Ben-Menahem (1961), and, when integrating it over frequency [which takes care of (i)], to reconcile quantitatively the paradox exposed above.

In practice, seismic magnitudes have been, and still are, measured either on body waves, or on surface waves (exceptionally on normal modes). As discussed by Vassiliou and Kanamori (1982), the energy radiated in P and S wavetrains can be written as

$$E^{\text{Body}} = F^{\text{B}} \cdot \frac{M_0^2}{t_0^3} \qquad (6)$$

where t_0 represents the total *duration* of the source (the inverse of a *corner frequency*), which under seismic scaling laws (Aki 1967) is expected to grow like $M_0^{1/3}$, and F^{B} is a combination of structural parameters (density, seismic velocities) and of the ratio x of rise time to rupture time, which are expected to remain invariant under seismic scaling laws. As a result, E/M_0 is also expected to remain constant, its logarithm being estimated at -4.33 by Vassiliou and Kanamori (1982), and -4.90 by Newman and Okal (1998). Extensive datasets compiled by Choy and Boatwright (1995) and Newman and Okal (1998) have upheld this invariance of $\log_{10}(E/M_0)$ with average values for shallow earthquakes of -4.80 and -4.98, respectively, a result later upheld even for microscopic sources by Ide and Beroza (2001).

In the case of surface waves, we have shown (Okal 2003) that the energy of a Rayleigh wave can be similarly expressed as

$$E^{\text{Rayleigh}} = F^{\text{R}} \cdot \omega_c^3 M_0^2, \qquad (7)$$

where ω_c is a corner frequency, not necessarily equal to its body wave counterpart, but still expected to behave as $M_0^{-1/3}$ [see Eq. (A8) of Okal (2003)], and F^{R} is a combination of structural parameters and Rayleigh group and phase velocities, which can be taken as constant. While we argue in Okal (2003) that the energy carried by Rayleigh waves represents only a small fraction (less than 10 %) of the total energy released by the dislocation, the combined result from (6) and (7) is that energy, when properly measured across the full spectrum of the seismic field, does grow linearly with the seismic moment M_0.

As for the growth of magnitudes, Geller (1976) has used the concept of source finiteness heralded by Ben-Menahem (1961) to explain how any magnitude measured at a constant period T starts by being proportional to $\log_{10} M_0$ for small earthquakes, and then grows slower with moment, as the inverse of the corner frequencies characteristic of fault length, rise time, and fault width become successively comparable to, or longer than, the reference period T. Under Geller's (1976) model, the slope of M versus $\log_{10} M_0$ should decrease to 2/3, then 1/3, and finally 0, as any magnitude measured at a constant period reaches an eventual saturation. The latter is predicted and observed around 8.2 for the 20-s surface-wave magnitude M_s, and predicted around 6.0 for m_b, if consistently measured at 1 s (occasional values beyond this theoretical maximum would reflect measurements taken at periods significantly longer than the 1-s standard). Figure 1 plots this behavior (m_b as a long-dashed blue curve, M_s in short-dotted red), as summarized by the last set of (unnumbered) equations on p. 1520 of Geller (1976). Because of the straight proportionality between E and M_0, the vertical axis also represents $\log_{10} E$, except for an additive constant.

The conclusion of these theoretical remarks is that Q is expected to grow with earthquake size, from its unperturbed value of 1 in the domain of small sources unaffected by finiteness, to 1.5 under moderate finiteness, and to a conceptually infinite value when M has fully saturated. Note that these conclusions will hold experimentally only under two conditions: (1) that the energy should be computed (either from body or surface waves) using an integration over the full wave packet, in either the time or frequency domain, these two approaches being equivalent under Parseval's theorem; and (2) that magnitudes for events of all sizes should be computed using the same algorithms, most importantly at constant periods.

3. The Quest for Earthquake Energy: A Timeline

In this general framework, we present here a timeline of the development of measurements of

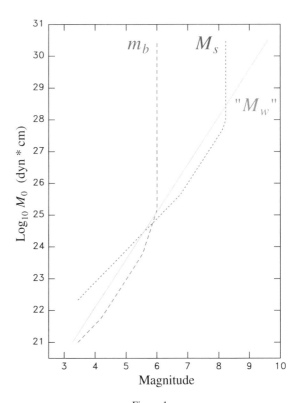

Figure 1
Variation of body-wave magnitude m_b (long dashes, blue) and surface-wave magnitude M_s (short dots, red) with seismic moment M_0, predicted theoretically from Eqs. (16) and (17) (Geller 1976). Superimposed in solid green is the relationship (18), casting values of M_0 into the scale M_w (Kanamori 1977)

earthquake energy, and of the refinement of the concept of magnitude. The discussion of the critical papers by Gutenberg and Richter which underlie Eq. (1) will be reserved for Sect. 4 and further detailed in the Appendix; contributions from other authors will be discussed here.

1. To our knowledge, the first attempt at quantifying the energy released by an earthquake goes back to Mendenhall (1888), who proposed a value of 3.3×10^{21} erg (2.4×10^{14} ft lbs) for the 1886 Charleston earthquake. This figure is absolutely remarkable, given that modern estimates of the moment of the event are around $(1-10) \times 10^{26}$ dyn cm (Johnston 1996; Bakum and Hopper 2004), which would suggest an energy of about $(1.2-12) \times 10^{21}$ erg, according to global scaling laws (Choy and Boatwright 1995; Newman and Okal 1998). Mendenhall's (1888) calculation was based on Lord Kelvin's

description of the energy of a "cubic mile" of vibrating matter moving with the wave front (Thomson 1855), a concept nowadays expressed as an *energy flux* (Wu 1966; Boatwright and Choy 1986). However, Mendenhall uses grossly inadequate estimates of particle velocities (on the order of 15 cm/s), and of the cross section and "thickness" of the wavefront, taken as 10^4 square miles and 1 mile, respectively. Thus the surprising accuracy of his result stems from the fortuitous compensation of fatally erroneous assumptions.

2. By contrast, the first calculation of the energy released by an earthquake whose methodology would be upheld under present standards goes back to Reid (1910). In his comprehensive report on the 1906 San Francisco earthquake, Reid proposed a value of 1.75×10^{24} erg, based on an estimate of the forces necessary to offset the strain accumulated around the fault. This number is about 10 times too large given generally accepted values of the event's seismic moment (Wald et al. 1993), but its computation is nevertheless remarkable.

   A year later, Reid embarked on measuring the energy of 12 additional earthquakes from an analysis of the areas of their Rossi–Forel intensity III isoseismals, scaled to that of the 1906 San Francisco earthquake, an idea originally found in Milne (1898). Reid does comment on the apparent scatter of his results, and in a truly visionary discussion, identifies probable sources of errors which would be described in modern terms as source radiation patterns, departure from earthquake scaling laws, and even the preferential attenuation of shear waves by the asthenosphere (Reid 1912).

3. The next reference to the computation of an earthquake's energy seems to be the work of Golitsyn (1915) on the Sarez earthquake of 18 February 1911[1] in present-day Tajikistan, which was later assigned a magnitude of $7\frac{3}{4}$ by Gutenberg and Richter (1954). Golitsyn's (1915)

computation is important, because it is the first one making use of a recorded seismogram. It was then revised and discussed, first by Klotz (1915) and later by Jeffreys (1923, 1929), and it became part of the small original dataset used by Gutenberg and Richter (1936) in their first attempt to relate magnitude to energy [see (6) in this list]. An extensive study of the Sarez earthquake was recently published by Ambraseys and Bilham (2012), who recomputed a surface-wave magnitude $M_s = 7.7$, but did not perform any waveform analysis. These numbers would suggest a moment of about 5×10^{27} dyn cm, under the assumption that the event follows scaling laws. We were able to obtain an independent, modern, estimate of the moment of the earthquake by applying the M_m algorithm (Okal and Talandier 1989, 1990) to original Love and Rayleigh wavetrains recorded at De Bilt and Uppsala, yielding an average value of $M_0 = 3 \times 10^{27}$ dyn cm. As detailed by Ambraseys and Bilham (2012), the earthquake was accompanied by a catastrophic landslide, later surveyed by the Imperial Russian Army at an estimated volume of 2.5–3.5 km^3, which dammed the Murgab River, creating Sarez Lake, a 17-km^3 reservoir (Shpil'ko 1914, 1915); incidentally, the potential failure of the resulting natural Usoi Dam remains to this day a significant hazard in the region (Lim et al. 1997; Schuster and Alford 2004), especially in view of recent large-scale seismic activity in its neighborhood (Elliott et al. 2017; Negmatullayev et al. 2018).

Golitsyn's purpose in computing the energy of the seismic waves was to address the question of the causality of the landslide, i.e., was it a result of the earthquake or the opposite, the argument being that, if the slide was caused by the earthquake, it should have only a fraction of the seismic energy released. Golitsyn's (1915) work constituted a significant improvement on Mendenhall's (1888) in that he used actual seismograms (in this case Rayleigh waves at Pulkovo) to compute an energy flux, which he then integrated over the observed duration of the wavetrain, and, mistakenly as noticed by Jeffreys

---

[1] In the Gregorian calendar ("new style"). The date is 5 February in the Julian calendar ("old style") used in the Russian Empire at the time.

(1923), over the lower focal hemisphere, to derive the energy released at the source. Golitsyn (1915) obtained a value of 4.3×10^{23} erg, comparable to his estimate of $(2–6) \times 10^{23}$ erg for the energy of the rockslide, and concluded that the slide was "not the consequence, but the cause of the earthquake." Once again, and remarkably, his value of the seismic energy is not unreasonable, being only ten times larger than expected under the assumption of scaling laws (Boatwright and Choy 1986; Newman and Okal 1998) applied to our estimate of the seismic moment (3×10^{27} dyn cm), but as we will show, this remains a coincidence, resulting from a number of compensating errors.

In what amounts to a translation of Golitsyn's paper, Klotz (1915) revised the seismic energy estimate slightly upwards, to 7×10^{23} erg, but offered no definitive comment on the matter of the possible trigger. By contrast, Jeffreys (1923) argued that Golitsyn's (1915) computation was erroneous, since he had not taken into account the decay of Rayleigh wave amplitudes away from the Earth's surface, which invalidates the integration over the focal hemisphere. Having corrected the calculation, and performed a cylindrical, rather than spherical, integration around the focus, Jeffreys (1923) came up with a considerably lower value of the seismic energy of the Rayleigh waves of only 1.8×10^{21} erg. To this he added a much smaller contribution from the S waves, estimated at 9×10^{19} erg.

In the context of modern theories and energy computations, we can point to a number of fatal shortcomings in Jeffreys' (1923) arguments, the most obvious one being that he assigns more energy to Rayleigh waves than to S waves. Indeed, we now know that most of the energy radiated by a seismic source is initially carried by high-frequency S waves, but those attenuate so fast in the far field that their contribution must be calculated by scaling that of the less attenuated P waves (Boatwright and Choy 1986); note that Jeffreys (1923) neglects anelastic attenuation altogether. Furthermore, he uses the model of a harmonic oscillator (despite expressing some reservations in this respect), whereas

modern computations using digital data show that the major contribution to the energy integral at teleseismic distances is usually around 1 Hz. In his surface wave calculation, Jeffreys (1923) uses a single period of 14 s, and by ignoring other spectral contributions, once again underestimates the final energy of the wave, by a factor of about six with respect to the theoretical value predicted for an earthquake of that size [Okal (2003); Eq. (45), p. 2209]. In short, Golitsyn (1915) was grossly overestimating the Rayleigh energies, but we now know that they carry only a fraction of the seismic energy released, so that in the end, his result might have been correct, while Jeffreys (1923) was underestimating both Rayleigh and S energies, by considering only single frequencies. A scientific exchange between Golitsyn and Jeffreys would certainly have been enlightening, but unfortunately, Prince Golitsyn died of natural causes on 17 May (n.s.) 1916, aged only 54.

Incidentally, we now understand that the whole argument about causality was totally flawed, since the two phenomena considered (the earthquake and the rockslide) express the release of two forms of potential energy of a different nature, one elastic and the other gravitational. If one phenomenon simply *triggers* the other, the relative amounts of energy released by the two processes are unrelated, since they come from different energy reservoirs.[2] As such, an earthquake could conceivably trigger a landslide more energetic than itself (this is possibly what happens during "orphan" slides for which the triggering mechanism is simply too small to be detected), and the converse might also be envisioned, i.e., a landslide triggering a more energetic earthquake in a tectonically ripe environment.

Indeed, the interpretation of the 1911 Sarez earthquake by Golitsyn (1915) and Jeffreys (1923) as being due to the rockslide was questioned both by Oldham (1923), who argued that the earthquake source was normal (and in

---

[2] These remarks illustrate scientifically the popular expression "the straw that broke the camel's back."

particular could not be superficial) on account of its isoseismal distribution and of its aftershock sequence, and by Macelwane (1926), based on a comparison with the case of the Frank, Alberta slide of 29 April 1903, for which no evidence of triggered seismicity could be found (admittedly for a significantly smaller, if still massive, rockslide, and during the dawn years of instrumental seismology). Jeffreys (1923) discusses in considerable detail the transfer of energy from a small body (the slide) falling on a big one (the Earth), and proposes a figure of 1/300 for its efficiency. His approach is correct only under his assumption "that the blow to the ground caused by the fall might have been the cause of the seismic disturbance," and thus that the elastic energy carried by the seismic waves was traceable to the deformation of the Earth upon the impact of the slide, and not to the release of tectonic strain independently accumulated during the interseismic cycle. The low efficiency of this mode of triggering was verified in the case of the collapse of the World Trade Center on 11 September 2001, for which Kim et al. (2001) proposed a value of between 10^{-4} and 10^{-3}. Incidentally, those authors also verified that the character of the seismic waves and the geometry of the spreading area differed significantly from those of a traditional earthquake, thus supporting in retrospect Oldham's (1923) criticism of Golitsyn's model for the 1911 Sarez events.

4. A few years later, in the second edition of *The Earth*, Jeffreys (1929) listed several additional values of earthquake energies, apparently all obtained from *S* waves: 10^{21} erg for the Montana earthquake of 28 June 1925, only 10 times smaller than suggested by (1) based on the magnitude of $6\frac{3}{4}$ later assigned by Gutenberg and Richter (1954); 5×10^{16} erg for the much smaller Hereford, England earthquake of 15 February 1926, deficient by a factor of 200 when applying (1) to the earthquake's present magnitude estimate, $M_{\mathrm{L}} = 4.8$ (Musson 2007); and about 10^{19} erg for the Jersey event of 30 July 1926, for which no definitive magnitude is available. He also lists a value of 5×10^{16} erg for the Oppau, Germany explosion of

ammonium-nitrate-based fertilizer on 21 September 1921 (Wrinch and Jeffreys 1923), although we would nowadays question the concept of using *S* waves to quantify the source of an explosion.

5. In 1935, Richter published his landmark paper introducing the concept of magnitude, in which he cautioned that "[its] definition is in part arbitrary; an absolute scale, in which the numbers referred directly to shock energy [...] measured in physical units, would be preferable" (Richter 1935). Notwithstanding this disclaimer, Richter could not resist the temptation of relating his newly defined magnitude scale to physical units, using Jeffreys' (1929) energy estimate for the 1925 Montana earthquake. Since the latter was outside the domain of his original study (limited to California and Nevada), Richter assigned it $M = 7.5$ based on an interpretation of its isoseismals, and proceeded to scale magnitude to energy. While not expressed verbatim, the relation

$$\log_{10} E = 2M + 6 \tag{8}$$

can be inferred from the several examples discussed on pp. 26–27 of Richter (1935). The slope $Q = 2$ in Eq. (8) is not explained, probably because it looked obvious to Richter that energy should scale as the square of amplitude, under the model of a harmonic oscillator.

6. In collaboration with Richter, Gutenberg wasted no time expanding the concept of magnitude, and within one year the two Caltech scientists had published the third in their series of "On Seismic Waves" papers (Gutenberg and Richter 1936), in which they extended the concept of magnitude to teleseismic distances and thus to earthquakes worldwide, using exclusively surface waves. In parallel, they proposed the first formal relation between magnitude and energy [their Eq. (15) p.124]:

$$\log_{10} E = 2M + \log_{10} E_0, \tag{9}$$

E_0 being the energy of a shock of magnitude 0, "the smallest ones recorded" (of course we now know that events of negative magnitude can exist and be recorded). They also laid the foundations for the extension of the magnitude concept to teleseismic distances, in the process revising down the magnitude of the 1925 Montana earthquake to 6.8 [later transcribed as $6\frac{3}{4}$ in Gutenberg and Richter (1954)]. This has the effect of increasing the constant 6 in (8) [or E_0 in (9)], but Gutenberg and Richter (1936) do not publish a definitive value of E_0, indicating rather that it ranges between 10^7 and 3×10^8 erg. On the other hand, they specifically justify the slope $Q = 2$ in (9), stating "Since the magnitude scale is logarithmic in the amplitudes, doubling the magnitudes gives a scale logarithmic in the energies."

Note that, up to this stage, Gutenberg and Richter do not compute their own energies based on any personal analytical approach, but rather use published values, e.g., from Reid (1910) or Jeffreys (1929); while they write "Energies have been found for earthquakes by many investigators using different methods," the modern reader is left at a loss to figure out exactly who these scientists were and what their methods might have been.

7. This situation apparently changes in the next few years, resulting in the first compilation of magnitudes derived from individual measurements of ground motion, as part of the first version of *Seismicity of the Earth*, published as a "Special Paper" of the Geological Society of America (Gutenberg and Richter 1941), although earthquakes remain grouped in "magnitude classes," from a ($M \approx 5$) to e ($M \approx 8$).

8. Gutenberg and Richter's (1942) next paper on the subject makes a number of fundamental breakthroughs. A critical discussion of the analytical parts of this paper will be given in Sect. 4.1. We present here only the general milestones in that contribution. First, the authors work out a detailed expression of the energy of the source based on a model of the vertical energy flux at the epicenter, expressed as a function of the maximum acceleration a_0 observed above a point source buried at depth h [their Eq. (27) p. 178]:

$$\log_{10} E = 14.9 + 2\log_{10} h + \log_{10} t_0 \\ + 2\log_{10} T_0 + 2\log_{10} a_0. \tag{10}$$

where t_0 is the duration of the signal, and T_0 its (dominant) period. In this respect, and as indicated by the title of their paper, the authors are clearly motivated by relating magnitude to maximum intensities in the epicentral area, which are generally related to accelerations, rather than displacements. Similarly, the experimental data available to them in epicentral areas came primarily in the form of strong-motion accelerograms, hence the emphasis on acceleration a_0, rather than displacement A_0, in (10). Second, and perhaps more remarkably, they propose that parameters such as the duration t_0 or the prominent period T_0 of the wavetrains are not constant but rather vary with the size of the earthquake. As such, their study represents the first introduction of the concept of a *scaling law* for the parameters of a seismic source. The authors suggest that $\log_{10} t_0$ varies as $\frac{1}{4}M$ [their Eq. (28)], albeit without much justification. In modern terms, t_0 would be called a *source duration*, controlled by the propagation of rupture along the fault, and hence expected to grow like $M_0^{1/3}$, which would be reconciled with Gutenberg and Richter's (1942) Eq. (28) if $\log_{10} M_0$ were to grow like $\frac{3}{4}M$, an unlikely behavior. The fundamental empirical observation in Gutenberg and Richter (1942) is that the maximum acceleration a_0 at the epicenter grows slower with magnitude than the amplitude of ground motion used to compute M [their Eq. (20) p. 176]:

$$M = 2.2 + 1.8\log_{10} a_0, \tag{11}$$

from which they infer a slope of 0.22 between $\log_{10} T_0$ and M [their Eq. (32), p. 179]. T_0 represents the dominant period of an accelerogram at the epicenter, which in modern terms may be related to the inverse of a *corner frequency*, itself controlled by a combination of source duration and rise time, and thus T_0 would be expected to grow like $M_0^{1/3}$ (Geller 1976), which is reconciled with Eq. (32) of Gutenberg and Richter (1942) if $\log_{10} M_0$ were to grow like $\frac{2}{3}M$, again an unlikely behavior. When substituting the dependence of t_0 and T_0 with M into their

Eq. (27), the authors finally obtain their Eq. (34) p. 179:

$$\log_{10} E = 8.8 + 2 \log_{10} h + 1.8M \qquad (12)$$

as a replacement for the previous version featuring a slope $Q = 2$ (Gutenberg and Richter 1936). Note that this formula would correspond to (10) for $h = 40$ km, which is significantly greater than the typical depth of the Southern California earthquakes that constituted the dataset used by the authors for most of their investigations, and would even be greater than the probable depth of focus of most of their teleseismic events. They correctly assign the reduction in slope to issues nowadays described as evolving from *source finiteness*, even though these effects remain underestimated by modern standards.

Gutenberg and Richter (1942) then proceed to estimate (only to the nearest order of magnitude) energies for 17 large earthquakes, based on intensity reports at their epicenter, or on their radius of perceptibility, both approaches being related to epicentral ground acceleration, and the latter reminiscent of Milne's (1898) methodology. The values proposed are significantly overestimated (reaching 10^{26} to 10^{27} erg), which may result from underestimating the effect of source finiteness, ignoring in particular the existence of several corner frequencies, a phenomenon leading to a saturation of acceleration [ground accelerations in excess of one g are not observed systematically for great earthquakes, but rather in certain tectonic environments in the case of even moderate, but "snappy," shocks (e.g., Fry et al. 2011)]. As a result, the growth of E with M [the slope $Q = 1.8$ in Eq. (12)] remains too strong, and this eventually overpredicts energy values, especially for large events. However, from a *relative* standpoint, it is remarkable that, in addition to the truly great earthquakes in Assam (1897) and San Francisco (1906), the three events given the strongest energies by Gutenberg and Richter (1942) are the shocks of 1926 off Rhodos, 1939 in Chillán, Chile, and 1940 in Vrancea, Romania. The

Chillán event has been shown by Okal and Kirby (2002) to feature an anomalously high energy-to-moment ratio, a property expected to be shared by the other two on account of their location as intermediate-depth intraslab events (Radulian and Popa 1996; Ambraseys and Adams 1998). Finally, and in retrospect, a significant problem with the multiple regression (10) underlying Gutenberg and Richter's (1942) model is that it predicts a logarithmic discontinuity in E as the source reaches the surface ($h \rightarrow 0$). This limitation reflects the simplified model of a point source, which is rendered invalid as soon as the fault's width W and length L become comparable to h.

9. In 1945, Gutenberg published three papers establishing the computation of magnitudes from surface and body waves on a stronger operational basis. First, in Gutenberg (1945a), he formalized the calculation of a surface-wave magnitude M_s by introducing the distance correction $1.656 \log_{10} \Delta$ [his Eq. (4)]. This slope, still of an empirical nature, is significantly less than suggested by Gutenberg and Richter (1936); for example, the 42 points at distances less than $150°$ on their Fig. 6 (p. 120) regress with a slope of $(-2.08 \pm 0.09) \log_{10} \Delta$ (or (-1.94 ± 0.15) for $\Delta < 55°$). The new slope (rounded to 1.66) was to be later inducted (albeit after considerable debate) into the Prague formula for M_s (Vaněk et al. 1962); while never derived theoretically, it was justified as an acceptable empirical fit to a modeled decay of 20-s Rayleigh wave amplitudes with distance (Okal 1989).

In the second paper, Gutenberg (1945b) used the concept of geometrical spreading, initially described by Zöppritz [and written up as Zöppritz et al. (1912) following his untimely death], to extend to teleseismic distances the calculation of an energy flux pioneered at the epicenter by Gutenberg and Richter (1942), thus defining a magnitude from the body-wave phases P, PP and S. The most significant aspects of this paper are (1) the difficulty of the author to obtain both local and distant values of magnitudes for the same event; (2) the necessity to invoke station corrections reflecting site responses; (3)

the relatively large values of the periods involved (up to 7 s even when considering only *P* phases); and (4) the introduction of correction terms for magnitudes larger than seven, clearly related to the effects of source finiteness.

In the third paper, Gutenberg (1945c) attacked the problem of deep earthquakes, and produced (for *P*, *PP*, and *S*) the first versions of the familiar charts for the distance–depth correction, generally referred to as $Q(\Delta, h)$, but as $A(\Delta, h)$ in early papers,[3] to be applied to the logarithm of ground motion amplitude. After this correction was significantly adjusted by Gutenberg and Richter (1956b) [their Fig. 5, as compared with Gutenberg's (1945c) Fig. 2], it was to be retained in the Prague formula for m_b (Vaněk et al. 1962), and has remained to this day the standard for the computation of magnitudes from *P* waves. By imposing that the same earthquake should have the same magnitude when measured at different distances, it was possible, at least in principle, to obtain empirically the variation of $Q(\Delta, h)$ with distance, and in particular to lock the body-wave magnitude scale with Richter's (1935) original one. However, the dependence with depth obviously required a different approach. In the absence of a physical representation of the source by a system of forces, Gutenberg (1945c) elected to impose that shallow and deep earthquakes of similar magnitude should have the same radiated energy, which allowed him to obtain the first version of the $A(\Delta, h)$ chart (his Fig. 2), through a generalization of the concept of geometrical spreading. We now understand that this approach tacitly assumes a constant stress drop $\Delta\sigma$, which may not be realistic for deep sources at the bottom of subduction zones.

In hindsight, Gutenberg's (1945c) approach suffered from ignoring the presence of the low-velocity zone in the asthenosphere, as well as of the seismic discontinuities in the transition zone,

which result in significant distortion of slowness as a function of distance, and hence of its derivative controlling geometrical spreading. In addition, the attenuation structure of the mantle was at the time little if at all understood, which later prompted Veith and Clawson (1972) to propose a new, more streamlined, distance–depth correction, $P(\Delta, h)$, based on a more modern representation of attenuation as a function of depth; however, they elected to use a structure derived from Herrin's (1968) tables, which does not include mantle discontinuities, even though the latter were documented beyond doubt by the end of the 1960s (e.g., Julian and Anderson 1968). Note finally that Gutenberg (1945b, Eq. 18, p. 66; 1945c, Eq. (1), p. 118) introduces the ratio A/T rather than A in the computation of m_B, presumably motivated by the quest for a closer relation between magnitude and energy, the latter involving ground velocity rather than displacement in its kinetic form. This change from A to A/T is potentially very significant, since T is expected to vary with earthquake size.

10. The next decade saw the compilation of the definitive version of *Seismicity of the Earth*, published as a monograph in two subsequent editions (Gutenberg and Richter 1949, 1954), in which earthquakes are assigned individual magnitudes. In their introduction to the final edition (p. 10), the authors argued that the value $Q = 1.8$, derived in Gutenberg and Richter (1942) and used in the first one, overestimated energies, and suggested the lower value $Q = 1.6$; however, they did not revise their discussion of energy, perhaps because they felt that their new formula (5), which they had not yet formally published (and eventually would never publish), was itself not definitive.

11. In the meantime, the scientific value of the concept of magnitude as a quantification of earthquake sources had become obvious to the seismological community, and many investigators developed personal, occasionally competing, algorithms, and more generally offered comments on Gutenberg and Richter's ongoing work. Among them, Bullen (1953) suggested that

---

[3] The correction is unrelated to the slope Q between magnitude and energy, and to the amplitude of ground motion A used to measure magnitudes. It also bears no relation to the quality factor Q later defined as the inverse of anelastic attenuation.

energies derived from (12) were too large for the largest magnitudes, which would go in the direction of the reduction of Q. On the other hand, following Jeffreys' (1923) criticism of Golitsyn's (1915) calculation of the energy of Rayleigh waves, Båth (1955) proposed to correct it by restricting the vertical cross-section of the teleseismic energy flux (which he evaluates at the Earth's surface) to a finite depth, H, which he took as 1.1 times the wavelength Λ. While this approach is indeed more sound than Golitsyn's, it is still limited to Rayleigh waves, and thus does not quantify the contribution of body waves, even though Båth (1955) does state that the latter is probably important. Finally, his empirically derived slope Q reverts to the value 2, presumably because of his exclusive use of narrowband instruments, essentially converting the Earth's motion into that of a harmonic oscillator. Similarly, Di Filippo and Marcelli (1950) obtained $Q = 2.14$ from a dataset of Italian earthquakes using Gutenberg and Richter's (1942) methodology, this higher value reflecting the departure of the dominant period T from the scaling proposed by Gutenberg and Richter. Sagisaka (1954) attempted to reconcile Reid's (1910) approach, based on integrating the energy of the elastic deformation, with Gutenberg and Richter's (1942) values derived from magnitudes in the case of several Japanese earthquakes (both shallow and deep), and noticed that the latter were consistently excessive, e.g., by a factor of at least 100 in the case of the great Kanto earthquake of 1923. We have proposed (Okal 1992) a seismic moment of 3×10^{28} dyn cm for that event, which under modern scaling laws, would suggest an energy of about 4×10^{23} erg, in general agreement with Sagisaka's (1954) figure, and confirming that (12) overestimates energies, at least for large events.

12. The year 1956 sees no fewer than three new publications on the subject by Gutenberg and Richter. Gutenberg (1956) and Gutenberg and Richter (1956b) constitute in particular their last efforts at trying to reconcile the various magnitude scales they had arduously built over the previous 20 years, into a single "unified"

magnitude. In this framework, they proceed to develop empirical relations between magnitude scales (and hence with energy) which become more complex, and in particular involve nonlinear terms. We now understand that, because they were measuring different parts of the seismograms at different periods and the scaling laws underlying the concept of a unified magnitude were distorted differently [different *corner frequencies* apply to different wavetrains (Geller 1976)], it was impossible for them to find a simple relation between magnitudes that would apply for all earthquake sizes. For example, we note that the relation proposed in Eq. (20), p. 134 of Gutenberg and Richter (1956a):

$$\log_{10} E = 9.4 + 2.14M - 0.054M^2 \qquad (13)$$

would regress with a slope $Q = 1.6$ between magnitudes 1 and 9, but only $Q = 1.4$ between the more usual values of 5.5 and 8.5. We note that Gutenberg (1956) elects to align a "unified" magnitude m on his body-wave m_B. This may sound surprising since we now understand that, being higher frequency than a surface-wave magnitude, m_B is bound to suffer stronger distortion from the effect of source finiteness, and saturate earlier (Geller 1976). However, as suggested in Gutenberg and Richter (1956a), Gutenberg may have been motivated by the goal of matching Richter's (1935) initial scale, which would be regarded today as a local magnitude M_L, measured at even higher frequencies, and therefore more closely related to a body-wave scale. Also, Richter's initial magnitude used torsion records with deficient response at longer periods.

In many respects, the third paper (Gutenberg and Richter 1956a) [whose preparation apparently predated Gutenberg and Richter (1956b)] stands in a class by itself, and constitutes a superb swan song of the authors' collaboration on this subject. In particular, it presents a review of previous efforts at extracting energy from seismograms, and revises the power laws relating phase duration t_0 and dominant period T_0 to magnitude. The authors now propose $\log_{10} t_0 = 0.32M - 1.4$

[rather than $0.25M - 0.7$ in Gutenberg and Richter (1942)], which would be in remarkable agreement with modern scaling laws predicting a slope of 1/3, but only in the absence of source finiteness. However, the dominant period T_0 in the seismogram is described as being essentially independent of magnitude. The authors suggest an algorithm for the computation of seismic energy from P waves ("integrating $(A_0/T_0)^2$ over the duration of large motion of the seismogram") which constitutes a blueprint for later and definitive algorithms, such as Boatwright and Choy's (1986), but of course, in the absence of digital data, they can only propose an approximate methodology (integrating the maximum value of velocity squared over an estimated duration of the phase), which by modern standards, keeps too much emphasis on representing the source as a monochromatic oscillator.

In the course of what was to be their last contribution on this topic, Gutenberg and Richter (1956a) make a number of visionary statements, notably concerning the limitations inherent in the model of a point source, which they argue make it uncertain that "there is a one-to-one correspondence between magnitude [...] and the total energy radiation." They conclude by emphasizing the necessity of a complete revision of the magnitude scale, which they claim to be in preparation, based on ground velocity (A/T) rather than amplitude, this statement reflecting once again both the pursuit of Richter's (1935) original goal of relating magnitudes to the physical units of energy, but also the common and erroneous model of a monochromatic oscillator.

Beno Gutenberg died suddenly on 25 January 1960, and Charles Richter, the inventor of the magnitude concept, never published any further contribution on the topic of magnitudes or seismic energy.

13. Developments regarding magnitude scales and the computation of seismic energy following Gutenberg's death are better known and will be only briefly summarized here. The torch was then passed to Eastern Bloc scientists. Following the International Union of Geodesy and Geophysics

(IUGG) meeting in Helsinki in 1960, Czech and Russian scientists met in Prague in 1961, and standardized the calculation of body- and surface-wave magnitudes, respectively, m_b (measured in principle at 1 s) and M_s (at 20 s) (Vaněk et al. 1962).[4] Both algorithms were based on the use of A/T, which the authors specifically justified as linked to energy, thus once again tacitly assuming the model of a monochromatic oscillator. In operational terms, a major difference arises between m_b and M_s: because of the prominence of 20-s waves strongly dispersed along oceanic paths, and suffering little attenuation in the crust, and of the universal availability of narrowband long-period instruments peaked around its reference period, M_s has indeed been measured close to 20 s, while measurements of m_b have often been taken at periods of several seconds, which considerably affects not only its relation to physical source size (i.e., seismic moment and hence energy), but also its saturation for large earthquakes. As a consequence, the Prague-standardized M_s has been more comparable to previous versions of surface-wave magnitudes than its m_b counterpart may have been to previous body-wave scales such as Gutenberg's (1945b) m_B, for which measurements were often taken at more variable periods, apparently as long as 7 s.

14. In the meantime, an interesting development had taken place in the then Soviet Union with the establishment of the so-called K-class scale for regional earthquakes. We refer to Rautian et al. (2007) and Bormann et al. (2012) for detailed reviews, and will only discuss here a number of significant points. That work originated with the need to compile and classify the very abundant seismicity occurring in the area of Garm, Tajikistan, notably in the aftermath of the Khait earthquake of 10 July 1949. It is remarkable that, even though it was conducted under what amounted to material and academic autarky

[4] This landmark paper was published simultaneously in Moscow (Vaněk et al. 1962) and Prague (Kárník et al. 1962). The different listing of coauthors reflects the use of different alphabetical orders in the Cyrillic and Latin alphabets.

(Hamburger et al. 2007), it nevertheless proceeded with great vision, and was rooted, at least initially, in the principles which would later constitute the foundations of modern algorithms using digital data (e.g., Boatwright and Choy 1986).

Initial work on the *K*-class was clearly intent on deriving a physically rigorous measurement of energy from seismic recordings by setting an a priori relation with energy (Bune 1955), later formalized as

$$K = \log_{10} E \qquad (14)$$

with *E* in joules (Rautian 1960). This approach differed fundamentally from Richter's (1935) empirical definition of magnitude, and indeed pursued the work of Golitsyn (1915) by seeking to compute an energy flux at the receiver (Bune 1956). It is interesting to note that Bune clearly mentions the need to work in the Fourier domain, but in the absence of digital data (and of computational infrastructure), he resorted to a time-domain integration over a succession of individual wavetrain oscillations, which may be justifiable under the combination of Parseval's theorem and the presence of strong systematic dispersion during propagation; however, the latter may not be sufficiently developed at the short distances involved. Faced with these challenges, Rautian (1960) reverted to the simplified practice of adding the absolute maximum amplitudes of *P* and *S* traces, in order to define an adequate "amplitude" to be used in the computation of the energy flux. She also pointedly recognized the influence of instrumentation, and later the variability required when exporting the algorithm to geologically different provinces (e.g., Fedotov 1963; Solov'ev and Solov'eva 1967), both of which can be attributed to a filtering effect (by instrument response and regional anelastic attenuation), before interpreting a complex source spectrum through a single number in the time domain.

Attempts to relate *K*-class values to magnitudes (as measured in the Soviet Union) were marred by the fact that the former was built for small events recorded at regional distances, while the latter, based on the application of the Prague formula to Love waves (Rautian et al. 2007), tacitly assumed larger shocks recorded at teleseismic distances. Nevertheless, the initial work of Rautian (1960) and systematic formal regressions later performed by Rautian et al. (2007) indicate an average slope $Q = 1.8 \pm 0.3$ [note, however, that in an application to larger earthquakes, Rautian (1960) suggested the use of a "total" seismic energy E_0, growing slower than E (in fact like $E^{2/3}$; her Eq. (19)), leading to a lower value of $Q = 1.1$]. The origin of the relatively high value $Q = 1.8$ is unclear, but probably stems from the simplified algorithm used by Bune (1956) and Rautian (1960), which may amount to narrow-bandpass filtering, thus approaching the conditions of a monochromatic signal ($Q = 2$).

In retrospect, the clearly missing element in the *K*-class algorithm is the key link between the energy flux at the station and the radiation at the source, which is now understood in the model of ray tubes, and quantified using a formula expressing geometrical spreading, e.g., in the algorithm later developed by Boatwright and Choy (1986). This concept, initially published as Zöppritz et al. (1912), could not be applied to the regional phases (P_n, S_n, L_g) on which the *K*-class was built, which motivated Rautian (1960) to develop other, more empirical algorithms.

15. Further significant progress could be made in the early 1960s on account of several theoretical developments. First, the double-couple was introduced by Vvedenskaya (1956), as the physical representation of a dislocation along a fault in an elastic medium, and later formalized by Knopoff and Gilbert (1959); its amplitude, the scalar moment M_0, materializes the agent, measurable in physical units, to which Richter (1935) had lamented he had no access. In addition, Ben-Menahem (1961) published a landmark investigation of the effect of source finiteness on the spectrum of seismic surface waves, introducing the concept of directivity, which was to prove critical in resolving the "energy paradox" exposed in Sect. 2.1. Based on the representation

theorem, and considering a source of finite dimension, Haskell (1964) gave the first expression for the energy radiated into P and S waves by a finite source of arbitrary geometry, which he found proportional to

$$E \propto \frac{W^2 (\Delta u)^2 L}{\tau^2}, \qquad (15)$$

where we have rewritten his Eq. (39) (p. 1821) with the more usual notation W for fault width, L for fault length, Δu for fault slip, and τ for rise time. Under generally accepted scaling laws (Aki 1967), we anticipate that E would indeed be proportional to M_0, in agreement with Vassiliou and Kanamori's (1982) later work. Haskell's paper is remarkable in that it is the first one to express seismic energy in the frequency domain by means of a Fourier decomposition. By then, efficient Fourier-transform algorithms were becoming available (Cooley and Tukey 1965), but digitizing analog records at time samplings adequate to compute energy fluxes remained a formidable challenge, which would be resolved only with the implementation of analog-to-digital converters in the 1970s (Hutt et al. 2002); it is not fortuitous that the only spectral data published in Haskell (1964) and its statistical sequel (Haskell 1966) are limited to $f < 12$ mHz [his Fig. 2, reproduced from Ben-Menahem and Toksöz (1963), and incidentally relating only to source phase]. The same improvement in computational capabilities in the 1960s had led to the systematic development of ray-tracing techniques which revitalized the estimate of energy fluxes in the far field [whose idea, we recall, can be traced all the way back to Mendenhall (1888)], based in particular on the concept of geometrical spreading defined by Zöppritz et al. (1912). In particular, Wu (1966) laid the bases for the computation of seismic energy from digitized teleseismic body waves, but in practice, it could be applied only to long-period records until digital data became available in the 1970s. The state of affairs in the mid-1960s is detailed in Båth's (1967) review paper, which rather surprisingly does not mention Ben-Menahem's

(1961) work on source finiteness, which was to play a crucial role in any understanding of magnitude and energy for large earthquakes.

16. By contrast, Brune and King (1967) addressed the question of the influence of finiteness for large earthquakes when using conventional magnitudes measured at insufficient periods (e.g., M_s), and embarked on computing "mantle" magnitudes using 100-s Rayleigh waves (their proposed M_M, which we call here M_{100}), thus becoming the first to propose a systematic use of mantle waves to quantify large earthquake sources. They documented a dependence of M_{100} on M_s (actually more precisely on an undefined M) in the shape of a stylized "S", whose intermediate regime corresponds to events with initial corner periods between 20 and 100 s. Brune and Engen (1969) complemented this study with measurements on Love waves, this time taken in the frequency domain in order to separate individual periods in the absence of dispersion. Remarkably, Brune and Engen (1969) recognized two striking outliers in their (M_{100} : M) dataset: they found the 1933 Sanriku event deficient in M_{100} despite a record $M = 8.9$ assigned by Richter (1958); we have documented this shock as a "snappy" intraplate earthquake, which violated the scaling laws tacitly implied by Richter when converting a long-period magnitude into a "unified" one based on a short-period algorithm [see Okal et al. (2016) for a detailed discussion]. On the opposite side, Brune and Engen (1969) noticed that the 1946 Aleutian earthquake, now known as a "tsunami earthquake" of exceptional source slowness (Kanamori 1972; López and Okal 2006), featured a mantle magnitude significantly larger than its standard M_s. In a visionary statement, which unfortunately remained largely unnoticed at the time, they stressed the potential value of a mantle magnitude in the field of tsunami warning (Brune and Engen 1969; p. 933).

17. The connection between directivity, as introduced by Ben-Menahem (1961), and full saturation of magnitude scales measured at constant periods was described in the now classic papers by Kanamori and Anderson (1975) and

Geller (1976). As mentioned in Sect. 2, Geller's last set of (unnumbered) equations on p. 1520 detail the relationship between M_s, measured at 20 s, and $\log_{10} M_0$, and in particular the evolution of the slope between the latter and former from a value of 1 at low magnitudes, through 1.5 over a significant range of "large" earthquakes $(6.76 \leq M_s \leq 8.12)$, a narrow interval where a slope of 3 is predicted, and a final saturation of M_s at a value of 8.22. We reproduce these relations, complete with moment ranges, here:

$$M_s = \log_{10} M_0 - 18.89 \quad \text{for}$$
$$\log_{10} M_0 \leq 25.65 \, (M_s \leq 6.76), \tag{16a}$$

$$M_s = \frac{2}{3} (\log_{10} M_0 - 15.51) \quad \text{for}$$
$$25.65 \leq \log_{10} M_0 \leq 27.69 \, (6.76 \leq M_s \leq 8.12), \tag{16b}$$

$$M_s = \frac{1}{3} (\log_{10} M_0 - 3.33) \quad \text{for} \quad 27.69 \leq \log_{10} M_0$$
$$\leq 28.00 \, (8.12 \leq M_s \leq 8.22), \tag{16c}$$

$$M_s = 8.22 \quad \text{for} \quad \log_{10} M_0 \geq 28.00. \tag{16d}$$

The combination of these equations with the relationships derived between m_b and M_s (Geller 1976; first set of unnumbered equations, p. 1520) leads to the following four-segment expression for the variation of m_b with moment:

$$m_b = \log_{10} M_0 - 17.56 \quad \text{for}$$
$$\log_{10} M_0 \leq 21.75 \, (m_b \leq 4.19), \tag{17a}$$

$$m_b = \frac{2}{3} (\log_{10} M_0 - 15.47) \quad \text{for}$$
$$21.75 \leq \log_{10} M_0 \leq 23.79 \, (4.19 \leq m_b \leq 5.55), \tag{17b}$$

$$m_b = \frac{1}{3} (\log_{10} M_0 - 7.16) \quad \text{for} \quad 23.79 \leq \log_{10} M_0$$
$$\leq 25.16 \, (5.55 \leq m_b \leq 6.00), \tag{17c}$$

$$m_b = 6.00 \quad \text{for} \quad \log_{10} M_0 \geq 25.16. \tag{17d}$$

These relations were obtained assuming a constant stress drop, a constant aspect ratio (L/W) of the fault, and constant particle and rupture velocities. Equations (16) and (17) are plotted in Fig. 1.

18. In his landmark paper, Kanamori (1977) introduced the concept of "moment magnitude" M_w by casting an independently obtained bona fide scientific measurement of the seismic moment M_0 (in dyn cm) into a magnitude scale through

$$M_w = \frac{2}{3} \left[\log_{10} M_0 - 16.1 \right]. \tag{18}$$

This definition of M_w seeks to (1) relate M_w to earthquake energy (hence the subscript "w"); and (2) make its values comparable to those (M) previously published, notably by Gutenberg and Richter. In doing so, it specifically assumes a ratio of 0.5×10^{-4} between energy and moment [Kanamori 1977, Eq. (4$'$) p. 2983], and also "Gutenberg and Richter's energy–magnitude relation" (1). While the former can be derived under scaling laws, we have seen that the latter lacked a satisfactory theoretical derivation. By further seeking to ensure the largest possible continuity between M_w and traditional M, Kanamori (1977) forces the factor 2/3 into (18) and thus tacitly assumes that the dataset of M is always taken in the size range where the magnitude has started to feel the effects of source finiteness, characterized by a slope of 1.5 in Fig. 1. This was further developed in the case of local magnitudes M_L by Hanks and Kanamori (1979) and can also be applied conceptually to the case of the body-wave magnitude m_b. The relationship (18) is then predicted to give an estimate M_w approaching a classical magnitude as long as the latter is computed exclusively in a domain where it has started to be affected by source finiteness, but with only *one* corner frequency (that relating to fault length L) lower than the reference frequency of that magnitude scale, thus resulting in a slope of 1.5 in Fig. 1. In other words, M_w should coincide with a traditional magnitude measurement if and only if that magnitude is M_s for reasonably large earthquakes $(25.65 \leq \log_{10} M_0 \leq 27.69)$, m_b for smaller events $(21.75 \leq \log_{10} M_0 \leq 23.79)$, and presumably M_L at even smaller moments.

Otherwise, and notably for smaller events (e.g., if using M_s for magnitude 5 events), one should expect a discrepancy between M_w and M_s; this is indeed the subject of the "bias" reported by Ekström and Dziewoński (1988). In general terms, Fig. 1, on which we have superimposed (in solid green) the relationship predicted by (18), features such a trend, but the values of M_w are smaller, by about 0.35 logarithmic units, than predicted by the dotted red and dashed blue lines. Similarly, in the regime for smaller-size events (slope of 1), the constant (18.89) relating M_s to $\log_{10} M_0$ in (16a) is smaller than derived theoretically by Okal (1989) (19.46, a difference of 0.57 units) or obtained experimentally by Ekström and Dziewoński (1988) (19.24; 0.35 units). This observation is traceable to the modeling of the $M_s : M_0$ relationship by Geller (1976), copied here as Eq. (16); note in particular that the fit of these relations to the dataset in his Fig. 7 deteriorates significantly for earthquakes in the moment range 10^{27} to 10^{28} dyn cm. A possible explanation is Geller's use of a relatively high $\Delta\sigma = 50$ bar as an average stress drop. For this reason, we prefer to replace Eq. (16) with

$$M_s = \log_{10} M_0 - 19.46 \quad \text{for}$$
$$\log_{10} M_0 \leq 26.22 \, (M_s \leq 6.76), \tag{19a}$$

$$M_s = \frac{2}{3}(\log_{10} M_0 - 16.08) \quad \text{for}$$
$$26.22 \leq \log_{10} M_0 \leq 28.26 \, (6.76 \leq M_s \leq 8.12), \tag{19b}$$

$$M_s = \frac{1}{3}(\log_{10} M_0 - 3.90) \quad \text{for}$$
$$28.26 \leq \log_{10} M_0 \leq 28.56 \, (8.12 \leq M_s \leq 8.22), \tag{19c}$$

$$M_s = 8.22 \quad \text{for} \quad \log_{10} M_0 \geq 28.56, \tag{19d}$$

shown in Fig. 2 as the thick red line superimposed on the background of Geller's (1976) Fig. 7. Note that a better fit is provided to interplate earthquakes (solid dots), especially in the range of moments 10^{27}–10^{28} dyn cm, characterized by the slope of 2/3 in (19b); upon

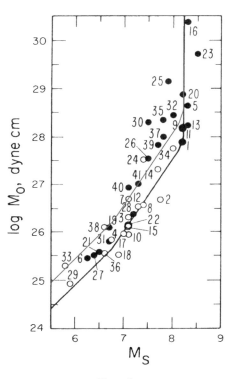

Figure 2
Reproduction of Fig. 7 of Geller (1976), with our preferred relationship (19) between M_s and $\log_{10} M_0$ superimposed in red

reduction of stress drop, corner frequencies are in principle lowered, which results in a slight displacement of saturation effects to higher moments, and a better agreement with (18). We similarly replace (17) with

$$m_b = \log_{10} M_0 - 18.18 \quad \text{for}$$
$$\log_{10} M_0 \leq 22.36 \, (m_b \leq 4.19), \tag{20a}$$

$$m_b = \frac{2}{3}(\log_{10} M_0 - 16.08) \quad \text{for}$$
$$22.36 \leq \log_{10} M_0 \leq 24.41 \, (4.19 \leq m_b \leq 5.55), \tag{20b}$$

$$m_b = \frac{1}{3}(\log_{10} M_0 - 7.76) \quad \text{for}$$
$$24.41 \leq \log_{10} M_0 \leq 25.76 \, (5.55 \leq m_b \leq 6.00), \tag{20c}$$

$$m_b = 6.00 \quad \text{for} \quad \log_{10} M_0 \geq 25.76. \tag{20d}$$

In Fig. 3, it is clear that the fit to M_s and m_b in the ranges where they feature a slope of 2/3 is much improved; we also achieve a small

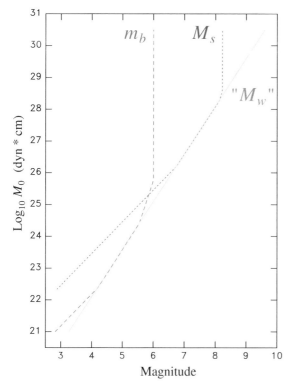

Figure 3
Same as Fig. 1, with our preferred relationships (19) and (20) for m_b and M_s. Note the better agreement of M_w with both, in the ranges where either has started to saturate

reduction in the gap between moments separating regimes for which M_w coincides with either M_s or m_b.

In this context, and in retrospect, it is unfortunate that Brune and Engen (1969) would not (or could not) push their measurements to periods longer than 100 s. This should have been possible at least for the largest earthquakes they studied, and would probably have led them to the explanation of the full saturation of magnitude scales, given that even M_{100} will eventually be affected by finite source dimensions (bringing the second elbow in the stylized "S" in their Fig. 3), and then fully saturate for even larger events. In this respect, note that the dataset plotted on that figure uses as abscissæ "Magnitudes" which cannot be strict M_s values, the latter saturating around 8.2 (Geller 1976, Fig. 7, p. 1517), and that the stylized "S" fails to take into account the effect of subsequent corner frequencies, for both M_s and M_{100}, which is expected to lead to the

saturation of both. This is shown in Fig. 4, where we plot a theoretical version of their relationship, obtained by adapting Eq. (19) to a reference period 5 times longer (which simply amounts to multiplying all elbow moments by a factor of 125). However, the fully saturated value $M_{100} = 9.77$ would occur at $M_0 = 1.2 \times 10^{30}$ dyn cm, which would make it essentially unobservable.

19. Vassiliou and Kanamori (1982) applied Haskell's (1964) concept to derive energy estimates by extracting the seismic moment M_0 and the time-integrated value of the source rate time function squared (I_t in their notation) from hand-digitized analog long-period records of teleseismic body waves. They concluded that E/M_0 was essentially constant for shallow (and even a few deep and intermediate) sources, but suggested a slope $Q = 1.8$ when regressing their estimates of $\log_{10} E$ against published values of M_s. This is probably due to the regression sampling into a range of moments ($\geq 10^{27.9}$) where severe saturation affects M_s and drives it away from its domain of variation as $\frac{2}{3} \log_{10} M_0$ (their Fig. 9a and Table 1). It is noteworthy that Vassiliou and Kanamori (1982) were the first to publish logarithmic plots of energy-to-moment datasets, later used systematically by Choy and Boatwright (1995) and Newman and Okal (1998). Without access to high-frequency digital data, they elected to directly estimate the ratio E/M_0^2 [their Eq. (3) p. 373] from what amounts to modeling the shape of long-period body waves, which were at the time the only ones they could hand-digitize from paper records. However robust the procedure may be with respect to parameters in the simple tapered-boxcar source model they use (their Fig. 1), their approach will perform poorly with complex, jagged sources such as those characteristic of a number of tsunami earthquakes (Tanioka et al. 1997; Polet and Kanamori 2000).

20. Following the deployment of short-period and broadband digital networks in the 1970s, the landmark paper by Boatwright and Choy (1986) finally regrouped all necessary ingredients for the routine computation of radiated energy, including the far-field energy flux approach (Wu 1966),

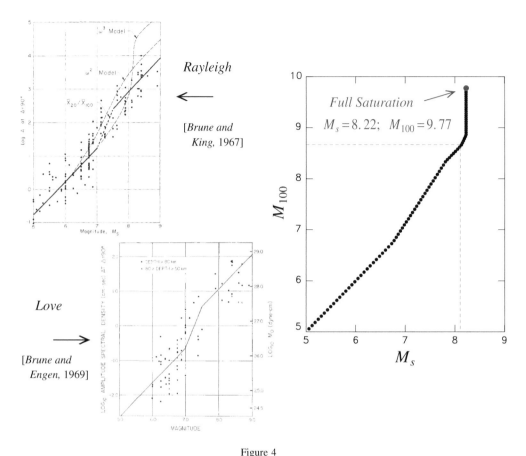

Figure 4

Left: Variation of mantle magnitudes versus conventional ones, reproduced from Brune and King (1967) (Rayleigh; *top*) and Brune and Engen (1969) (Love; *bottom*). Note that the abscissæ are probably not strict M_s values, which are supposed to saturate around 8.2. *Right:* Theoretical behavior of a 100-s magnitude versus 20-s M_s as predicted using Geller's (1976) model. The dashed line reproduces the domain of study of Brune's papers, in agreement with the stylized "S" curves at left. Note the eventual saturation of M_{100} at a value of 9.77 ($M_0 = 1.2 \times 10^{30}$ dyn cm)

a computation in the Fourier domain as suggested by Haskell (1964), and adequate corrections for geometrical spreading, anelastic attenuation, and focal mechanism orientation. It set the stage for the systematic computation of radiated energies, the catalog of Choy and Boatwright (1995) providing the first extensive demonstration of a generally constant ratio E/M_0. Outliers to this trend, few in number but of critical importance in terms of seismic or tsunami risk and providing insight into ancillary problems in plate tectonics, were later identified by Choy et al. (2006). Newman and Okal (1998) developed a simplified alternative to Boatwright and Choy's (1986) algorithm, allowing rapid computation of $\Theta = \log_{10}(E/M_0)$, a robust

estimate of the slowness of a seismic source, which was implemented as part of tsunami warning procedures (Weinstein and Okal 2005). Finally, Convers and Newman (2013) have combined the measurement of radiated energy with that of rupture duration, both obtained from P waves, to identify as rapidly as possible anomalously slow events bearing enhanced tsunami risk; a similar approach can be found in the definition of Okal's (2013) parameter Φ.

21. Later developments in the study of the energy radiated by seismic sources largely transcend the mainly historical scope of the present paper, and we will only review them succinctly.

Using a wide variety of sources from the largest megaearthquakes to microearthquakes induced during shaft excavation in granite (Gibowicz et al. 1991), McGarr (1999) and Ide and Beroza (2001) documented that, remarkably, the general constancy of E/M_0 can be extended over 17 orders of magnitude of seismic moment. However, as summarized, e.g., by Walter et al. (2006), later studies failed to bring a firm consensus for either a constant stress drop across the range of significant earthquake sources, as suggested, e.g., by Prieto et al. (2004) and more recently Ye et al. (2016), or for a detectable increase of "scaled energy" E/M_0 for large earthquakes (e.g., Mayeda et al. 2005). The origin of this disparity of results remains obscure, but may be rooted in the difficulty (or impossibility) to obtain adequate, universal models of attenuation at high frequencies (Sonley and Abercrombie 2006), with Venkataraman and Kanamori (2004) also mentioning the possible effect of source directivity in biasing the computation of radiated energy for large earthquakes.

4. A Discussion of Gutenberg and Richter's Derivations

The timeline in Sect. 3 provides a comprehensive examination of the history of estimates of the energy radiated by seismic sources, as well as of the developments of the concept of magnitudes. We return here to the origin of Eq. (1) and specifically to the derivation of the slope $Q = 1.5$ implicit in Gutenberg and Richter (1956a, b), and of the earlier value $Q = 1.8$ (Gutenberg and Richter 1942). The process by which these values were obtained results from a combination of parameters (σ_i and μ_j, see below) generally expressing power laws controlling the relative growth with earthquake size of various physical quantities; these parameters were usually obtained in an empirical fashion by Gutenberg and Richter, but can now be explored in the context of seismic source scaling laws. Tables 1 and 2 summarize their values and properties, in particular their poor robustness.

As explained in Sect. 2.2, radiated energy is proportional to seismic moment M_0, and this was verified eventually from datasets such as Choy and Boatwright's (1995) or Newman and Okal's (1998). On the other hand, the concept of magnitude, which measures the logarithm of ground displacement, should be proportional to $\log_{10} M_0$, at least in the absence of saturation effects due to source finiteness. The combination of these two remarks should lead to a theoretical value of $Q = 1$.

We start by noting that energy is carried mostly by high-frequency body waves, and as such should be measured on high-frequency seismograms. On a global scale, this restricted early authors to measurements of acceleration by strong-motion instruments, at least initially (in the 1930s and 1940s). It is probable that, partly because of this instrumental restriction and partly because their model to trace back the energy flux to a point source was developed only above the hypocenter, all their energy calculations were performed at the epicenter (with the subscript zero on all relevant variables, such as A_0, a_0, etc.). This remains surprising since short-period instruments providing high performance in the far field had been developed at Caltech (Benioff 1932), and were operating routinely by the late 1930s. Their records were indeed used by Gutenberg (1945b, c) to develop his body-wave magnitude m_B. One can only speculate as to why Gutenberg and Richter did not embark on a teleseismic measurement of radiated energy, especially since Gutenberg was obviously cognizant of the concept of geometrical spreading, having written up (with Geiger) the landmark paper by Zöppritz et al. (1912) following the first author's untimely death in 1908.

As a result, when trying to relate energy and magnitude on a global scale, the authors compute an energy in the near field, from what are primarily measurements of accelerations, but a magnitude from ground motions (or perhaps estimates of velocity) obtained in the regional or far field. This is the source of the complex, occasionally arcane, nature of their derivations; in order to streamline the argument, we have relegated critical details of their calculations to the Appendix, which also presents a discussion of the (generally poor) robustness of the resulting parameters, including the slopes Q.

Table 1

Summary of parameters σ_i used by Gutenberg and Richter (1942)

Parameter	Description	Defining equation	G–R value	Expected modern value	Robustness
σ_1	Slope of M versus $\log_{10}$ of displacement A	(21)	1		Forced
σ_2	Slope of M versus $\log_{10}$ of acceleration a	(22)	1.8	4 to 8	Very poor
σ_3	Inverse slope of M versus $\log_{10}$ of duration t_0	(23)	0.25	1/3 before initiation of saturation; grows thereafter	Poor
σ_4	Power law coefficient of kinetic energy E versus dominant period T_0 (at constant a_0)	(25)	2		Forced
σ_5	Power law coefficient of kinetic energy E versus acceleration a_0 (at constant T_0)	(25)	2		Forced
σ_6	Slope of $\log_{10}$ of dominant period T_0 versus magnitude M $\sigma_6 = \frac{1}{2} - \frac{1}{2\sigma_2}$	(29)	0.22	1/3 before initiation of saturation; grows thereafter	Poor
σ_7	Power law coefficient of dominant period T versus energy E $\sigma_7 = \sigma_6/Q$	(30)	0.12	1/3	Poor
σ_8	Power law coefficient of kinetic energy E versus acceleration a_0 (including effect of T_0) $\sigma_8 = 1 + \sigma_2(1 + \sigma_3)$	(A.2)	3.25	3 to 7, see Fig. 6	Very poor
Q	Slope of $\log_{10} E$ versus M in G–R relation $\boldsymbol{Q = 1 + \sigma_3 + \frac{1}{\sigma_2}}$	(27)	1.8	1 before initiation of saturation	Poor

Table 2

Summary of parameters μ_j used by Gutenberg and Richter (1956b)

Parameter	Description	Defining equation	G–R value	Expected modern value	Robustness
μ_1	Slope of m versus $\log_{10}$ of displacement-to-period ratio $q = \log_{10} A/T$	(31)	1		Forced
μ_2	Power law exponent of energy E versus duration t_0 (at constant q_0)	(32)	1		Forced
μ_3	Power law exponent of energy E versus logarithmic ratio q_0 (at constant t_0)	(32)	2		Forced
$\mu'_4; -\mu''_4$	Regression coefficients of q_0 versus local magnitude M_L and M_L^2	(33)	0.8; 0.01		
μ_4	μ'_4 for a linear regression ($\mu''_4 = 0$). Also, slope of m versus M_L	(34)	0.7 (inferred)	1 before initiation of saturation	Poor
μ_5	Slope of $\log_{10}$ of dominant period T versus logarithmic ratio q	(36)	0.4		
μ_6	Slope of $\log_{10}$ of energy E versus magnitude m $\mu_6 = \mu_3 + \mu_5$	(38)	2.4		Poor
μ_7	Slope of "unified" magnitude m versus surface-wave magnitude M_s	(39)	0.63	1 before initiation of saturation; 2/3 and greater thereafter	Poor
Q	Slope of $\log_{10} E$ versus M in G–R relation $\boldsymbol{Q = \mu_6 \cdot \mu_7}$	(40)	1.5	1 before initiation of saturation	Poor

4.1. The Derivation of $Q = 1.8$ by Gutenberg and Richter (1942)

The derivation proposed in that paper is based on:

(i) The definition of magnitude M from the amplitude of ground motion A_0 recorded by a torsion instrument, which we can write schematically as

$$M = \sigma_1 \log_{10} A + c_1 = \sigma_1 \log_{10} A_0 + c'_1 \qquad (21)$$

with a slope σ_1 identically equal to 1, as imposed by Richter (1935); note that A in (21), being measured at a regional distance (typically up to a few hundred kilometers), is not necessarily an epicentral value (A_0), but is taken as such by the authors, the difference being absorbed into the constant c'_1.

(ii) The correlation (11) between magnitude and epicentral acceleration a_0 [their Eq. (20) p. 176] reproduced here as

$$M = \sigma_2 \log_{10} a_0 + c_2, \qquad (22)$$

the slope $\sigma_2 = 1.8$ being obtained empirically from the comparison of magnitude values measured on torsion instruments in the regional field and accelerations derived from strong-motion seismograms in the vicinity of the epicenter;

(iii) An empirical relation between the duration of "strong ground shaking" t_0 and magnitude [their Eq. (28) p. 178]:

$$\log_{10} t_0 = \sigma_3 M + c_3, \qquad (23)$$

where $\sigma_3 = 0.25$;

(iv) The calculation (10) of seismic energy E from the energy flux radiated vertically at the epicenter above a point source [their Eq. (24) p. 178], which we rewrite as

$$E = C_4 \cdot t_0 \cdot V_0^2, \qquad (24)$$

where V_0 is ground velocity, and all c_i and C_i are constants independent of event size.

We have rewritten (10) as (24) to emphasize that (21), (22), and (24) involve *different physical quantities*, namely displacement [used in Richter's (1935) original definition], acceleration (available as strong-motion data in the epicentral area), and ground velocity (defining the kinetic energy flux). The authors' ensuing combinations of these equations through the use of the "period T_0" of the signal, presumably the dominant one, tacitly imply a harmonic character for the source, an additional complexity being that the concept of signal duration, t_0, is sensu stricto incompatible with this model, since a monochromatic signal is by definition of infinite duration.

Under that ad hoc assumption, they then derive their Eqs. (24) or (27) for E, which we rewrite as

$$\log_{10} E = c_4 + \log_{10} t_0 + \sigma_4 \log_{10} T_0 + \sigma_5 \log_{10} a_0 \qquad (25)$$

with $\sigma_5 = 2$ identically [from (24), i.e., the power of 2 in the kinetic energy] and $\sigma_4 = 2$ identically (two powers of T_0 going from V_0^2 to a_0^2). Similarly, they transform (21) into

$$M = \sigma_1 \log_{10} a_0 + \sigma_4 \log_{10} T_0 + c_5, \qquad (26)$$

equivalent to their Eq. (31). Substituting (22), (23), and (26) into (25) (note that σ_4, in principle equal to 2, and hence all *direct* reference to T_0, are eliminated), they obtain

$$\log_{10} E = QM + C_0 \qquad (27)$$

with

$$Q = \sigma_1 + \sigma_3 + \frac{\sigma_5 - \sigma_1}{\sigma_2} = 1 + \sigma_3 + \frac{1}{\sigma_2} \approx 1.8. \qquad (28)$$

The slope $Q = 1.8$ is thus explained as a combination of the various slopes σ_i. The latter are of a very different nature: $\sigma_1 = 1$ and $\sigma_5 = 2$ were fixed in Gutenberg and Richter's (1942) model, while σ_2 and σ_3 (1.8 and 0.25, respectively) were obtained empirically by the authors, and as such could vary, impacting significantly the value of Q derived from (28). We emphasize this point in the third member of Eq. (28), which leaves only σ_2 and σ_3 as variables.

As summarized in Table 1, the critical examination of this derivation detailed in the Appendix shows that the robustness of $Q = 1.8$ with respect to various assumptions underlying the computations is poor.

Note finally that Gutenberg and Richter (1942) combined their Eqs. (20) and (31) (our Eqs. 22 and 26) to obtain the dependence with magnitude of the dominant period T_0, which had dropped out of their derivation of Q:

$$\log_{10} T_0 = \sigma_6 M + c_6 \qquad \text{with}$$
$$\sigma_6 = \frac{1}{\sigma_4} - \frac{\sigma_1}{\sigma_4 \cdot \sigma_2} = 0.22, \qquad (29)$$

further leading to

$$\log_{10} T_0 = \sigma_7 \log_{10} E + c_7; \qquad \sigma_7 = \frac{\sigma_6}{Q} = 0.12, \tag{30}$$

some limitations of those two relations being further discussed in the Appendix.

4.2. The Final Regressions: the Road to $Q = 1.5$ (Gutenberg and Richter 1956b)

In this section, we paraphrase that paper to underscore the fundamental steps which led the authors to $Q = 1.5$. In the various equations relating E, m, A, etc., we will use formal slopes comparable to the σ_i used above, but with the notation μ_j to avoid confusion; γ_j will be constants substituting for the c_i above, largely irrelevant in the present discussion. Note that indexing of the parameters σ_i and μ_j is in both cases sequential in the derivation, so that σ and μ of identical index ($i = j$) may not describe comparable physical relations. The results are summarized in Table 2, in a format similar to that of Table 1; again, additional details can be found in the Appendix.

The situation differs from the previous derivation [Sect. 4.1 above, after Gutenberg and Richter (1942)] in several respects. First, and most importantly, the authors introduce the variable $q_0 = \log_{10}(A_0/T_0)$; following Gutenberg (1945c), the body-wave magnitudes, in particular m_B and hence m, are now defined from q_0, rather than from the logarithm of A_0:

$$m = \mu_1 \cdot q_0 + \gamma_1 = \mu_1(\log_{10} A_0 - \log_{10} T_0) + \gamma_1 \tag{31}$$

with $\mu_1 = 1$ identically (Gutenberg and Richter 1956b; Eq. 11). Next, the authors now consider as their reference magnitude the unified magnitude m rather than Richter's (1935) original scale (restricted to local records of California earthquakes). Finally, several relationships now feature nonlinear (albeit weak) terms.

The fundamental difference between (31) and (21) stems from an implicit variation of the dominant period T_0 with earthquake size, which would be expected from scaling laws. However, the combined influence of anelastic attenuation and instrument response can act as a bandpass filter, reducing variations in the dominant period, which may then significantly violate scaling laws, an effect enhanced at the short periods considered by the authors. For this reason, it may not be possible to define an expected value of the parameters μ under modern theories, and we leave blank several entries in the relevant column in Table 2.

Gutenberg and Richter's (1956b) next step is their Eq. (8) p. 10, equivalent to (24), and leading to

$$\begin{aligned} \log_{10} E &= \gamma_2 + \mu_2 \log_{10} t_0 + \mu_3(\log_{10} A_0 - \log_{10} T_0) \\ &= \gamma_2 + \mu_2 \log_{10} t_0 + \mu_3 q_0 \end{aligned} \tag{32}$$

with $\mu_2 = 1$ and $\mu_3 = 2$ identically. The introduction of q_0 rather than A_0 (with a nonconstant period T_0) then results in their new Eq. (9):

$$q_0 = \gamma_3 + \mu_4' M_L - \mu_4'' M_L^2 \tag{33}$$

with $\mu_4' = 0.8$ and $\mu_4'' = 0.01$ (empirical values). Note that (33) expresses the relationship between m and M_L (Gutenberg and Richter 1956b; Eq. 14), and under the assumption that its curvature can be neglected, μ_4 is also the slope of m versus M_L:

$$m = \mu_4 M_L + \gamma_4. \tag{34}$$

Note that, by comparison with (29), and assuming again that μ_4'' can be neglected, Eq. (33) leads to

$$\sigma_6 = \frac{1 - \mu_4}{\sigma_1} = 0.2. \tag{35}$$

Next, the authors study empirically the relation between the duration t_0 in the epicentral area and the newly defined parameter q_0:

$$\log_{10} t_0 = \mu_5 q_0 + \gamma_5. \tag{36}$$

They obtain a slope $\mu_5 = 0.4$ [their Eq. (10)], which we have confirmed (0.40 ± 0.06) by regressing the 19 points on their Fig. 6a [which are probably a subset of the previously mentioned dataset in Fig. 1 of Gutenberg and Richter (1956a)].

Note that the combination of (21), (23), (35), and (36) leads to

$$\mu_5 = \frac{\sigma_3 \sigma_1}{1 - \sigma_6} = \frac{\sigma_3}{\mu_4}, \tag{37}$$

which takes the observed value $\mu_5 = 0.4$ for $\mu_4 = 0.8$ and $\sigma_3 = 0.32$ as favored by Gutenberg and Richter (1956a), but $\mu_5 = 0.31$ for $\sigma_3 = 0.25$ under the first combination. Although not specifically spelt out in Gutenberg and Richter (1956b), the combination of (31), (32), and (37) then leads, for $\mu_5 = 0.4$, to

$$\log_{10} E = \mu_6 m + \gamma_6 \text{ with } \mu_6 = \frac{\mu_3 + \mu_2 \mu_5}{\mu_1} \quad (38)$$
$$= \mu_3 + \mu_5 = 2.4,$$

and then to (2), which as discussed in the Appendix, remains dependent on σ_3 through (37), and hence on the dataset in Fig. 1 of Gutenberg and Richter (1956a).

Using this slope of 2.4 in (2), and in order to finalize $Q = 1.5$ in (1), there remains to justify (3). That relation traces its origin to Fig. 9 of Gutenberg and Richter (1956a, p. 138), which examines the slope μ_7 regressing m versus M_s:

$$m = \mu_7 M_s + \gamma_7 \quad (39)$$

[note, however, that instead of m in (39), Gutenberg and Richter (1956b) apparently use the uncorrected body-wave magnitude (but with the notation M_B, as opposed to m_B)]. They describe an average slope of 0.4 between $(M_s - M_B)$ and M_s, equivalent to $\mu_7 = 0.6$, later revised to $\mu_7 = 0.63$ by Gutenberg and Richter (1956b). We have verified that a modern regression of the 105 points in their Fig. 9 does yield $1 - \mu_7 = 0.37 \pm 0.03$. Then, the slope Q is just the product

$$Q = \mu_6 \cdot \mu_7, \quad (40)$$

which takes the value 1.51, justifying (1).

4.3. Discussion

The detailed examination of the derivations of the slopes $Q = 1.8$ in Gutenberg and Richter (1942) and later $Q = 1.5$ inferred from Gutenberg and Richter (1956b), and the critical analysis of their underlying assumptions carried out in the Appendix, should severely limit the confidence of the modern reader in the resulting slopes Q and in particular in the value 1.5, proposed in their last contributions and now universally enshrined into seismological dogma. We recap here some of the most serious limitations documented in the present study:

- *First, it is not clear that they correctly measured the energy flux at the epicenter.*

As we have stated several times, the algorithm used by Gutenberg and Richter (1942) to evaluate energy flux near the epicenter consists of squaring the product of the maximum ground acceleration and the dominant period, and multiplying the result by the duration of sustained maximum ground motion. This is only a gross approximation to the integral defining energy flux, as defined later, e.g., by Haskell (1964), and suffers from an inherent flaw, the concept of a single (or dominant) frequency being incompatible with that of a finite signal duration.

- *Next, the authors were using a simplified analytical model.*

The latter essentially predated any knowledge of anelastic attenuation, for which models would start being available (albeit at much lower frequencies) in the late 1950s and early 1960s (Ewing and Press 1954; Satô 1958; Anderson and Archambeau 1964); of an adequate physical representation of the earthquake source as a double-couple (Knopoff and Gilbert 1959); or even of the concept of a spatially extended source, an idea pioneered by Lamb (1916) and heralded analytically by Ben-Menahem (1961). Feeling constrained to work in the near field, Gutenberg and Richter were using the model of a point source, whereas most earthquakes in their datasets would have had fault lengths comparable to, or greater than, their hypocentral depth. It would take six years following Gutenberg's death for Wu (1966) to regroup these later developments into a blueprint for the definitive computation of radiated energy from teleseismic datasets, which would be put to fruition only when digital data became available in the 1970s.

- *In addition, the robustness of their results is seriously cast in doubt.*

The final slopes Q are obtained from the parameters σ_i and μ_j, which relate observables of a high-frequency nature, strongly affected if not fully controlled by small-scale heterogeneity within the seismic source, as now universally documented by source tomography studies of recent megathrust events (Ishii et al. 2007; Lay et al. 2011). Such

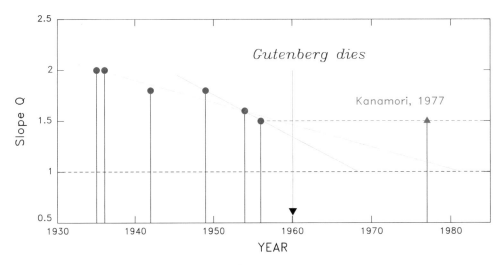

Figure 5

Evolution with time of the value of the slope Q in the various relations proposed by Gutenberg and Richter between seismic energy and magnitude. The dashed line is the value ($Q = 1$) derived theoretically for a point source double-couple. The vertical downpointing arrow indicates Gutenberg's death on 25 January 1960. The green line shows a linear regression through the six published values (1935–1956) and the pink one through the last three (1949–1956). The dashed blue line expresses the value $Q = 1.5$, frozen into perpetuity as the last one published before B. Gutenberg's death

source structure acts to defy the scaling laws which are inherent in the derivations of Q proposed by Gutenberg and Richter (1942, 1956b). Not surprisingly then, Eqs. (22) and (23) [or (31) and (36)] formalizing these scaling laws have not been upheld by modern strong-motion studies in the near field.

One can only speculate as to how Gutenberg and Richter may have pursued this line of work, but for the former's sudden death in 1960. Despite the promise of a full revision of magnitude scales (Gutenberg and Richter 1956a; abstract p. 165), no such work came forth during the next three years, a time when Gutenberg apparently shifted his main activity to studies in structural seismology, especially regarding the Earth's core (Gutenberg 1958). However, it is highly probable that he would have been, at the time of his death, cognizant of the new representation of the seismic source as a double-couple quantified by a seismic moment M_0, introduced in the West a year earlier by Knopoff and Gilbert (1959). One can only assume that, eventually, Gutenberg would have sought to relate magnitude to M_0, even though the first measurement of seismic moment from long-period waves had to wait until Aki's (1966) study of the 1964 Niigata earthquake, and a sufficient dataset of values until the early 1970s,

allowing the landmark studies by Kanamori and Anderson (1975) and Geller (1976).

Figure 5 summarizes the evolution of the slope factor Q in Gutenberg and Richter's papers from 1935 until Gutenberg's death. A further step in speculation would address how this inescapable, regular *decrease* would have continued, had the two authors been able to keep a proactive collaboration. A linear regression of the three points in the 1950s, when the authors' work was most supported analytically, predicts that they would have reached the more justifiable value $Q = 1$ in 1968. A full regression of the six values published since 1935 would suggest 1981. However, we note that, by then, Gutenberg would have been 92 years old.

5. Conclusion and Perspective

A detailed, occasionally forensic examination of the algorithms used by Gutenberg and Richter (1942; 1956b) suggests that they may have been legitimate under what was the state of the art of seismology in the 1940s and 1950s, but are nowadays difficult to reconcile with modern advances in source theory and Earth structure, including anelasticity, as well as with the advent of high-frequency digital data. In this context, the physical models underlying the derivation of the successive values of the slope Q, including the empirical relations used by Gutenberg and Richter, can now appear simplistic if not outright inaccurate, e.g., the model of a point source, and the implicit assumption of a monochromatic signal. As a result, the derived values of Q clearly lack robustness, and in particular the last one (1.5) hardly deserves the apparently sacred character which led to its being enshrined into the definition of moment magnitude M_w by Kanamori (1977).

We recall that, in their last paper on the subject, Gutenberg and Richter (1956a) commented that there may not necessarily be a single relationship between seismic energy and any magnitude scale. In retrospect, this statement appears visionary, as we now understand that the concept of radiated energy is fundamentally anchored in the high-frequency part of the source spectrum, whereas individual magnitude scales target specific periods which can belong to significantly different parts of the spectrum. Note that Kanamori (1983) echoed this statement, asserting that "It is impossible to represent all [...] parameters by a single number, the magnitude."

Since the advent of digital seismology in the late 1970s, it has become clear that many seismic sources do indeed follow laws of similitude allowing the description of many, if not all, of an event's source properties on the basis of a single number, namely its seismic moment M_0, which Kanamori (1977) has proposed to cast into the "magnitude" M_w, thus introducing a parameter combining the rigor of the underlying quantification of a physical quantity with the practicality of empirically derived magnitude scales. Such source properties generally include radiated energy E, as documented by extensive catalogs such as Choy and Boatwright's (1995).

At the same time, a number of violations to these scaling laws have been regularly documented, both in the form of "slow" events, whose red-shifted spectrum leads to a deficiency in radiated energy, and of "snappy" ones featuring a blue-shifted spectrum. Such rogue events bear crucial societal relevance, since the former include the so-called tsunami earthquakes (Kanamori 1972), whose real-time identification remains a challenge in operational

tsunami warning, while the latter have proven ominously destructive due to exceptional levels of ground acceleration, e.g., reaching $2.2g$ during the relatively small $(M_w = 6.2)$ 2011 Christchurch earthquake (Holden 2011).

However, little progress has been made in the understanding of parameters possibly controlling their occurrence. For example, it is not clear whether all megaquakes (defined as having moments greater than 10^{29} dyn cm) feature source slowness (Okal 2013, Fig. 10), or whether there exists a regional trend controlled by simple tectonic patterns to the distribution (and hence to the predictability) of tsunami earthquakes, an idea hinted at by Okal and Newman (2001), and further supported by the 2006 Java and 2012 El Salvador events. In this context, the availability of catalogs asserting a diversity of source parameters is of crucial necessity for further research into such topics.

Yet, such catalogs are not (or no longer) made readily available. For example, an updated extension of Choy and Boatwright's (1995) extremely valuable dataset is not widely distributed, and to our best knowledge, their dataset is no longer being routinely updated. Similarly, for many years, the National Earthquake Information Center of the United States Geological Survey had distributed an electronic catalog of epicenters, listing for each event a set of *various* and occasionally different conventional magnitudes (m_b, M_s, M_L), in addition to moment estimates transcribed as M_w. This allowed the easy identification of anomalous events, e.g., through a simple search for an $m_b : M_s$ disparity (a predecessor to the E/M_0 ratio in the pre-digital age). As of 2015, this practice has been discontinued and replaced by the issuance of a single magnitude, whose nature (body, surface, local, moment, etc.) is not clearly specified, with the probable goal of making the catalog less confusing to the lay user, admittedly a legitimate concern. As a result, the individual researcher has lost a powerful tool to further our understanding of earthquake source properties. Within a few decades, the careless or simply uninformed investigator might be tempted to conclude that earthquakes violating scaling laws ceased to occur after 2014.

Incidentally, seismology is not alone in featuring a diversity of source properties acting to restrict the range of validity of scaling laws. For example, the size of hurricanes has been described through a Safir–Simpson "category" index, a single number quantifying several of their properties, such as geometrical extent, maximum wind velocity, and underpressure at the center of the eye; as such, it constitutes a measure conceptually similar to an earthquake magnitude. However, major hurricanes have shown that such an approach constitutes a drastic simplification, with Sandy in 2012 featuring a diameter of 1800 km for a minimum pressure of 940 mbar and winds of 185 km/h (Blake et al. 2013), while the 2015 Mexican hurricane Patricia registered a record low pressure of 872 mbar and winds of 345 km/h concentrated in a system not more than 800 km across (Kimberlain et al. 2016). Such limitations for hurricane scaling laws are not intrinsically different from the diversity expressed for seismic sources through variations in energy-to-moment ratios, and similarly prevent an accurate societal prediction of the level of natural hazard based on a single quantifier (magnitude or category) of the event. Another example would be the Volcanic Explosivity Index (Newhall and Self 1982), widely used to rank large volcanic eruptions, but ignoring specific properties significantly affecting their dynamics as well as their climatic and societal effects (Miles et al. 2004).

In this context, it behooves the seismological community to observe, respect, and fully document the occasional diversity in source properties of earthquakes, notably by restoring the now abandoned practice of systematic reporting of radiated seismic energy. This is a necessary step towards the furtherance of our understanding of seismic source properties in the context of their societal impact, in keeping with the pioneering work of Beno Gutenberg and Charles Richter.

Acknowledgements

I am grateful to Alexandr Rabinovich, Alexei Ivashchenko, and Igor Medvedev for providing copies of critical Russian references, and to the staff of the interlibrary loan desk at Northwestern University for their exceptional professionalism over the years. The paper was significantly improved by the comments of three anonymous reviewers. Discussions with Johannes Schweitzer are also acknowledged.

Appendix: Critical Details of Gutenberg and Richter's Derivations

In this Appendix, we present some critical details of the derivations of the slopes $Q = 1.8$ (Gutenberg and Richter 1942) and $Q = 1.5$ (Gutenberg and Richter 1956b), and in particular examine their robustness which eventually controls that of the inferred parameters Q.

A.1. Derivation of $Q = 1.8$ (Gutenberg and Richter 1942), Eqs. (21–30) of Main Text

Note that any increase in the slope σ_2 in (28) will lead to a decrease in Q. In the context of modern strong-motion seismology, it has become clear that an equation of the form (22) may not apply, and certainly not universally, between local magnitude and the logarithm of peak acceleration, with most modern models featuring nonlinear relationships (e.g., Abrahamson and Silva 1997). In practice, and as illustrated, e.g., by Bolt and Abrahamson (2003; their Fig. 3), a value of $\sigma_2 \approx 4$ may be legitimate around $M_L = 6$, but would grow as high as $\sigma_2 \approx 8$ at $M_L = 7$, illustrating the well-known effect of saturation of maximum acceleration with moment (Anderson and Lie 1994). In this context, it is difficult to justify on theoretical grounds any form of Eq. (22), let alone the value $\sigma_2 = 1.8$, which appears at any rate as a lower bound of values that could be derived in narrow ranges of magnitudes from modern datasets. In hindsight, σ_2 probably constitutes the least well constrained among the slopes σ_i in Gutenberg and Richter's (1942) approach.

As for the slope σ_3, it expresses the variation with M of the duration parameter t_0 defined by Gutenberg and Richter (1942) as the time over which "equal sinusoidal waves" are recorded at a near-field receiver. From a modern theoretical standpoint, t_0 would be expected to scale with the total duration of rupture on the fault, and hence as $M_0^{1/3}$. Assuming that the magnitude M is taken in its range of initial saturation (featuring a slope of 2/3 with $\log_{10} M_0$), σ_3 should then equal 1/2 (Trifunac and Brady 1975). By contrast, Gutenberg and Richter (1942) obtained empirically $\sigma_3 = 1/4$. Gutenberg and Richter (1956a) later revised this estimate to 0.32, based on a dataset of 29 measurements (their Fig. 1, p. 110), for which a modern regression yields a slope of 0.30 ± 0.02, or 0.29 ± 0.03 when excluding a lone measurement for a barely detectable shock (Richter and Nordquist 1948); at any rate, we have found that the quality of fit improves by only 5 % between slopes of 0.25 and 0.30.

Gutenberg and Richter's (1942) parameter t_0 may be comparable to the duration s defined in strong-motion seismology; in that framework, we note indeed a relationship of the form (23) between $\log_{10} s$ and M, with $\sigma_3 = 0.32$, quoted by several authors (e.g., Esteva and Rosenblueth 1964). However, further work has shown a poor dependence of $\log_{10} s$ on magnitude, and suggested a better fit using a linear relation between s and M (Housner 1965; Vanmarcke and Lai 1980). Various authors, including Housner (1965) and Trifunac and Brady (1975), have reported a *decrease* of duration with Mercalli intensity, especially when the former is measured using acceleration spectra, which for large earthquakes are expected to be strongly red-shifted; this led Housner (1965) to define a limit of "maximum duration" growing, perhaps linearly, albeit rather erratically, with magnitude. The bottom line of this discussion is that an equation of the form (23) is poorly fit by modern strong-motion data, and hence that the parameter σ_3, if at all justifiable, is prone to substantial variations, which in turn will impact the slope Q.

Note finally that the duration of strong motion, t_0, should not be confused with the coda duration upon which so-called duration magnitudes have been proposed, which incidentally also feature a broadly variable range of slope parameters σ (Lee and Stewart 1981).

In addition to these uncertainties in the values of σ_2 and σ_3 in (28), we stress, once again, that a fundamental difficulty with Gutenberg and Richter's (1942) proposed derivation remains that it assumes a harmonic oscillation in order to relate the displacement A_0 used in the definition of magnitude to the velocity V_0 inherent in the calculation of the energy flux. In particular, the fact that the period T_0 required to relate them disappears from the final Eq. (28) should not obscure that its implicit existence underlies the derivation.

An additional, and significant, problem stems from the assumption of a spherical wave front at the epicenter above a point source, for the calculation of the energy flux. There are two issues there which are bound to fail for large earthquakes whose fault dimension, L, becomes larger than the source depth. First, in the near field, ground motion is essentially proportional to the slip on the fault, and as such grows only like $M_0^{1/3}$, a result well known to practitioners of deformation codes such as Mansinha and Smylie's (1971) or Okada's (1985). Second, Gutenberg and Richter (1942) interpret the energy computed at an epicentral station as an energy flux, which they then integrate over a focal sphere whose radius is the hypocentral depth, thus clearly assuming a point source. By contrast, modern calculations of energy flux are carried out in the far field, many fault lengths away from the source, and as such are immune to that problem (e.g., Boatwright and Choy 1986).

More generally, we note that Gutenberg and Richter (1942) had appropriately recognized the time finiteness of the source by introducing its duration t_0 at the epicenter, and even its dominant period T_0, which inherently requires a finite source time, but they never addressed the problem of the *spatial* finiteness of the source, and their models always considered a point source in space. Only in the last section of their last paper (Gutenberg and Richter 1956a; p.142) would they recognize a future need to consider this problem, which was to be fully studied only after Gutenberg's death (Ben-Menahem 1961).

Gutenberg and Richter (1942; Eq. (41) p. 188) also sought to directly relate the acceleration at the epicenter, a_0, to the energy E of the earthquake. Eliminating M between (22) and (27) leads to

$$\log_{10} E = \sigma_8 \log_{10} a_0 + c_8 \qquad (A.1)$$

with

$$\sigma_8 = \sigma_5 - \sigma_1 + \sigma_2(\sigma_1 + \sigma_3) = 1 + \sigma_2(1 + \sigma_3)$$

$$\approx 3.25; \quad \frac{1}{\sigma_8} = 0.31.$$

$$(A.2)$$

This inverse slope being remarkably close to 1/3, the authors rounded it, and proposed in their Eq. (42) a cubic root dependence of acceleration a_0 on energy E. The modern reader should resist the temptation to associate this result to the growth of epicentral displacement A_0, itself controlled by seismic slip Δu (at least for sources shallower than their spatial extent), as $M_0^{1/3}$: for point source double-couples, Newton's (1687) law does predict a_0 directly proportional to M_0, but for finite sources, scaling laws would actually predict a corner frequency scaling as $M_0^{-1/3}$, leading to a *decrease* of peak acceleration as $M_0^{-1/3}$, a result generally in line with the empirically reported saturation of peak acceleration with earthquake size (Anderson and Lie 1994). We believe that the value $\sigma_8 \approx 3$ is a fortuitous result of the particular choice of slopes σ_2 and σ_3. For example, the "second" combination of $\sigma_2 = 4$ [as suggested by Bolt and Abrahamson (2003) for moderate earthquakes] and $\sigma_3 = 0.32$ [as used later by Gutenberg and Richter (1956a)] does lead to $Q = 1.57$, close to the values of 1.6 suggested by Gutenberg and Richter (1954) and 1.5 as given by (1), but this second combination nearly doubles σ_8, to a value of 6.28.

We next consider the slopes σ_6 and σ_7, defined in (29) and (30). As mentioned above, these slopes are difficult to reconcile with modern scaling laws, which would predict $\sigma_8 = 1/3$ under the assumption that T_0 is directly related to the inverse of a source corner frequency. We note, again, that σ_7 will vary strongly with the coefficients σ_2 and σ_3 determined empirically by Gutenberg and Richter (1942, 1956a). For example, for $(\sigma_2 = 4; \sigma_3 = 0.32)$, σ_6 rises to 0.38 and σ_7 doubles, to 0.24.

More systematically, we explore in Fig. 6 the variation of the slopes Q, σ_8 (through its physically more meaningful inverse, $1/\sigma_8$), and σ_7 as a function of the empirical slopes σ_2 and σ_3. The former is allowed to vary between 0 and 6, covering the range suggested by strong-motion investigations (Bolt and Abrahamson 2003), and the latter between 0.1 and 0.6, to include the theoretical value (0.5) expected

from scaling laws under the assumption that t_0 represents the inverse of a corner frequency of the source spectrum. The resulting values of Q, $1/\sigma_8$, and σ_7 are simply computed from (27), (29), (30), and (A.2) under the constraints $\sigma_1 = 1$, $\sigma_4 = 2$, and $\sigma_5 = 2$, and their contours are color-coded using individual palettes. The bull's eyes indicate the "first" combination ($\sigma_2 = 1.8; \sigma_3 = 0.25$) used by Gutenberg and Richter (1942), and the triangles the second combination [larger values of $\sigma_2 = 4$ suggested by strong-motion data; and $\sigma_3 = 0.32$ later adopted by Gutenberg and Richter (1956a)]. Figure 6 provides a quantitative illustration of the absence of precision of the slopes computed by Gutenberg and Richter, most notably of their energy-to-magnitude parameter Q. The last column of Table 1 summarizes their lack of robustness.

A.2. The Final Regression: $Q = 1.5$ (Gutenberg and Richter 1956b), Eqs. (31–40) of Main Text

First, we note that this final study reaches an extreme complexity in the definition of various magnitudes, considering no fewer than seven such scales. They are (1) Gutenberg's (1935) original one for locally recorded California shocks, appropriately renamed a local magnitude M_L; (2) the surface-wave magnitude M_s^G introduced by Gutenberg and Richter (1936) and formalized by Gutenberg (1945a); (3) the teleseismic body-wave magnitude m_B defined by Gutenberg (1945b, c); (4), its value corrected to M_B to make it more compatible with M_s^G; (5) conversely, an m_s applying the opposite correction to M_s^G in order to make it compatible with m_B; (6) a unified magnitude M (without subscript), consisting of a "weighted mean" between M_B and M_s^G (but with no details provided about the weighting process); and (7) a corresponding, "final" weighted mean m (without subscript) between m_B and m_s. We have used the temporary notation M_s^G to emphasize that this magnitude (2) is not a priori equivalent to the Prague M_s later defined by Vaněk et al. (1962), since the former measures an amplitude A and the latter the ratio A/T of amplitude to period; however, because of the prominence of 20-s waves in standard long-period teleseismic records, the two approaches are largely compatible; in addition, we note the similarity in

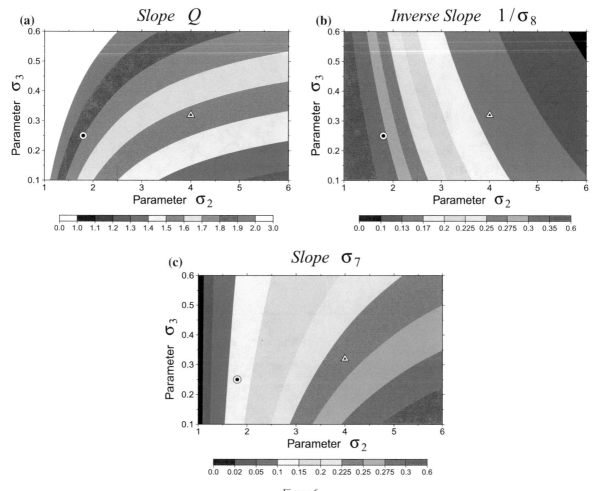

Figure 6

Variation of **a** slope Q defined in (1), **b** inverse slope $1/\sigma_8$ defined in (A.1), and **c** slope σ_7 defined in (30), as a function of the empirical slope parameters σ_2 (in abscissa) and σ_3 (in ordinate). Each frame uses a different palette, reproduced at its bottom. The bull's eye symbols refer to the first combination ($\sigma_2 = 1.8$; $\sigma_3 = 0.25$), used by Gutenberg and Richter (1942), the centered triangles to the second combination [$\sigma_2 = 4$ as suggested by modern strong-motion studies; $\sigma_3 = 0.32$ as used by Gutenberg and Richter (1956a)]

distance corrections between M_s^G [$1.656\log_{10}\Delta$ as opposed to $2\log_{10}\Delta$ in Gutenberg and Richter (1936)] and the Prague M_s ($1.66\log_{10}\Delta$), even though the Prague group did consider many previously proposed values of the slope (G. Purcaru, pers. comm., 1988), ranging from 1.31 (Nagamune and Seki 1958) to 1.92 (Bonelli and Esteban 1954). As a result, the two scales may indeed be essentially equivalent, allowing us to henceforth drop the superscript "G."

Among the critical issues regarding the derivation of $Q = 1.5$ by Gutenberg and Richter (1956b), we first address the effect of using (33) on the variable σ_6

defined in Sect. 4.1, and characterizing the variation of the dominant period T_0 with magnitude. By comparison with (29), and assuming μ_4'' can be neglected, Eq. (33) leads to

$$\sigma_6 = \frac{1 - \mu_4}{\sigma_1} = 0.2, \qquad (\text{A.3})$$

in good agreement with $\sigma_6 = 0.22$ as computed in Sect. 4.1 with the first combination ($\sigma_2 = 1.8$; $\sigma_3 = 0.25$) of parameters used by Gutenberg and Richter (1942). However, Eq. (33) regresses linearly with a slope $\mu_4 = 0.7$ between magnitudes of 1 and 9, but 0.66 between 5.5 and 8.5. The

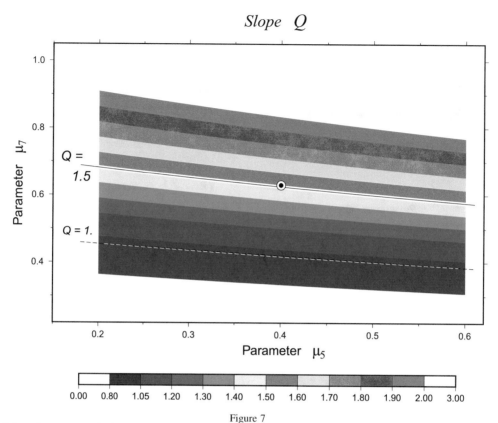

Figure 7
Variation of slope Q as computed in Gutenberg and Richter (1956b), as a function of the slopes μ_5 and μ_7 defined in (36) and (38). The bull's eye symbol identifies the parameters used by the authors, and the solid line the locus of other combinations yielding $Q = 1.5$. For reference, the dashed line similarly identifies the locus of $Q = 1$

corresponding values for σ_6 would be significantly altered, to 0.30 and 0.34, respectively.

We then consider the parameter μ_5 defined in (36), which takes the value 0.4 for $\mu_4 = 0.8$ and $\sigma_3 = 0.32$, as proposed by Gutenberg and Richter (1956a). We find that μ_5 can range from 0.31 under the first combination to as much as 0.48 for $(\sigma_3 = 0.32; \mu_4 = 0.66)$. Finally, excluding the lone datum for the barely detectable shock (Richter and Nordquist 1948) reduces the regressed slope μ_5 to the intermediate value 0.36 ± 0.09.

Next, we focus on Eq. (38), defining $\mu_6 = 2.4$, perhaps the most crucial step on the way to $Q = 1.5$. With μ_3 identically equal to 2, μ_6 is controlled by μ_5, itself the ratio of σ_3 to μ_4 (37). While Fig. 1 of Gutenberg and Richter (1956a) suggests an acceptable fit of (2) to a dataset of 21 points, their formal regression yields a higher slope ($\mu_6 = 2.57 \pm 0.19$), whose error bar includes the value 2.4, but which is

strongly controlled by Richter and Nordquist's (1948) minimal datum mentioned above, and by a nuclear test at Bikini, which clearly does not belong in a dataset examining the properties of earthquake sources. A regression of the remaining 19 points results in a greater, poorly constrained slope ($\mu_6 = 3.00 \pm 0.33$).

As summarized in Table 2, we conclude that, as in the case of their earlier calculation ($Q = 1.8$), the slope $Q = 1.5$ obtained by Gutenberg and Richter (1956b) is not robust when considering the various assumptions underlying it in the context of modern seismological theory.

A.3. Discussion and Conclusion

Given the empirical nature of the slopes σ_2 (relating magnitude to acceleration) and σ_3 (magnitude and duration), it would be in principle possible to read Fig. 6 backwards, i.e., to investigate which values of σ_2 and σ_3 would be necessary to obtain the parameter $Q = 1.5$ proposed by Gutenberg and Richter (1956b) (we recall that the slopes σ_1, σ_4, and σ_5 must have their fixed values of 1, 2, and 2, respectively). We are motivated by the fact that the slope σ_2 for example, relating acceleration to magnitude, is documented from modern studies to be very poorly constrained and indeed could be very high; the "duration" slope σ_3 is also poorly constrained by strong-motion studies.

In this context, some constraints could, at least in principle, come from the parameters σ_8 and σ_7, which may be expected to take predictable values under seismic laws. But we have seen that the definition of σ_8 runs into the problem of saturation of acceleration with earthquake size (σ_8 could even be negative); only the slope σ_7 relating the (dominant) period T_0 to energy bears some legitimate hope of obeying scaling laws (with $\sigma_7 = 1/3$), since it relates two quantities of a fundamentally physical nature, which would be expected to scale predictably with earthquake size. Figure 6c shows that $\sigma_7 = 1/3$ will require extreme values of both σ_2 (greater than 5) and σ_3 (less than 0.2), clearly very different from those used by Gutenberg and Richter (1956a), but perhaps not impossible; incidentally, they will result in $Q \leq 1.4$.

However, the above discussion has shown that the various quantities measured by Gutenberg and Richter are all of a high-frequency nature (if for no other reason, because of the instruments they used in the near field), and thus sensitive to effects such as the fine structure of the source [e.g., large heterogeneities of slip on the fault plane especially for the largest sources (Lay et al. 2011)]; as a result, the corresponding slopes, such as σ_7, may depart significantly from their theoretical values. In conclusion, the derivation of $Q = 1.8$ proposed by Gutenberg and Richter (1942), and corrected to 1.6 by Gutenberg and Richter (1956a), stems from a particular choice of critical slopes (σ_2, σ_3), which may represent best fits obtained empirically from then available datasets, but which are neither robust, given now available strong-motion datasets, nor justifiable theoretically under the canons of modern source theory.

A similar approach could be attempted with the slopes μ_5 and μ_7, in search of constraints on $Q = 1.5$

as derived by Gutenberg and Richter (1956b). Figure 7 examines systematically the dependence of Q on those empirical parameters. Our discussion in Sect. 4.2 has shown that $\mu_5 = 0.4$ could be either under- or overestimated, with values between 0.5 and 0.3 being plausible. As for μ_7, while the value 0.63, reported by Gutenberg and Richter (1956b), is in excellent agreement with the slope of 2/3 expected under the first stage of saturation, where m_B (but not yet M_s) would have started to feel the effects of source finiteness, Fig. 8 predicts that the slope 2/3 should apply only at low magnitudes ($M_s < 5$), with significantly lower values expected for larger earthquakes. A combination of $\mu_5 = 0.3$ and $\mu_7 = 0.5$ would yield $Q = 1.15$. Also, Gutenberg and Richter (1956a) note that a particular operational algorithm applies to measurements of the 1952 Kern County aftershocks; when those are excluded from the dataset in their Fig. 9, the slope μ_7 falls to 0.57 and Q to 1.37 (keeping $\mu_5 = 0.4$).

Finally, we note that, even though the authors had introduced nonlinear terms [e.g., in (35)], the

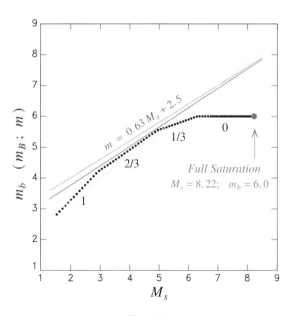

Figure 8
Variation of m_b as a function of M_s as predicted from a combination of (19) and (20) (solid dots). The theoretical slopes are indicated in black for each regime of partial or total saturation. Superimposed in red is the best fit (38) obtained for m versus M_s by Gutenberg and Richter (1956b), in blue the earlier regression by Gutenberg and Richter (1956a), and in green their fit to the 1952 Kern County aftershocks

derivation of $Q = 1.5$ ignores them. In this context, it is interesting to note that Gutenberg and Richter (1956a; p. 133) report that their colleague V.H. Benioff had suggested interpreting the curvature expressed in many of their graphs (best illustrated by their Fig. 3) as consisting of "two straight lines intersecting near magnitudes 5 and 6 [sic]," a visionary remark in view of the later description of staggered steps in magnitude saturation (Brune and Engen 1969; Geller 1976). Rather, Gutenberg and Richter elected to keep a quadratic form for (33), and then ignored the nonlinear terms in their derivation of $Q = 1.5$ (Gutenberg and Richter 1956b).

REFERENCES

Abrahamson, N. A., & Silva, W. J. (1997). Empirical response spectral attenuation relations for shallow crustal earthquakes. *Seismological Research Letters, 68*, 94–127.

Aki, K. (1966). Generation and propagation of *G* waves from the Niigata earthquake of June 16, 1964 - Part 2. Estimation of moment, released energy, and stress-strain drop from the *G*-wave spectrum. *Bull. Earthq. Res. Inst. Tokyo Univ., 44*, 73–88.

Aki, K. (1967). Scaling law of seismic spectrum. *Journal of Geophysical Research, 72*, 1217–1231.

Ambraseys, N. N., & Adams, R. (1998). The Rhodes earthquake of 26 June 1926. *Journal of Seismology, 2*, 267–292.

Ambraseys, N. N., & Bilham, R. (2012). The Sarez-Pamir earthquake and landslide of 18 February 1911. *Seismological Research Letters, 83*, 294–314.

Anderson, D. L., & Archambeau, C. B. (1964). The anelasticity of the Earth. *Journal of Geophysical Research, 69*, 2071–2084.

Anderson, J. G., & Lei, Y. (1994). Nonparametric description of peak acceleration as a function of magnitude, distance and site in Guerrero, Mexico. *Bulletin of the Seismological Society of America, 84*, 1003–1017.

Bakum, W. H., & Hopper, M. G. (2004). Magnitudes and locations of the 1811–1812 New Madrid, Missouri and of the 1886 Charleston, South Carolina, earthquakes. *Bulletin of the Seismological Society of America, 94*, 64–75.

Båth, M. (1955). The relation between magnitude and energy of earthquakes. *Proc. Am. Geophys. Union, 36*, 861–865.

Båth, M. (1967). Earthquake energy and magnitude. *Physics and Chemistry of the Earth, 7*, 115–165.

Ben-Menahem, A. (1961). Radiation of seismic surface waves from finite moving sources. *Bulletin of the Seismological Society of America, 51*, 401–435.

Ben-Menahem, A., & Toksöz, M. N. (1963). Source mechanism from the spectra of long-period seismic surface waves. 2. The Kamchatka earthquake of November 4, 1952. *Journal of Geophysical Research, 68*, 5207–5222.

Benioff, V. H. (1932). A new vertical seismograph. *Bulletin of the Seismological Society of America, 22*, 155–169.

Blake, E. S., Kimberlain, T. B., Berg, B. J., Cangialosi, P. J., & Beven II, J. L. (2013). *Tropical cyclone report: Hurricane Sandy (AL182012)*, 157 pp. Miami: Natl. Ocean. Atmos. Admin.

Boatwright, J., & Choy, G. L. (1986). Teleseismic estimates of the energy radiated by shallow earthquakes. *Journal of Geophysical Research, 91*, 2095–2112.

Bolt, B. A., & Abrahamson, N. A. (2003). Estimation of strong motion ground motions. In W. H. K. Lee, H. Kanamori, P. C. Jennings, & C. Kisslinger (Eds.), *International handbook of earthquake and engineering seismology* (pp. 983–1001). New York: Academic Press.

Bonelli R., J. M., & Esteban C., L. (1954). *La magnitud de los sismos en Toledo*, 14 pp. Madrid: Instituto Geografico y Catastral.

Bormann, P., Fujita, K., Mackey, K. G., & Gusev, A. (2012). The Russian *K*-class system, its relationships to magnitudes and its potential for future development and application. In P. Bormann (Ed.), *New manual of seismological observatory practice 2* (pp. 1–27). Potsdam: Deutsches GeoForschungsZentrum.

Brune, J. N., & Engen, G. R. (1969). Excitation of mantle Love waves and definition of mantle wave magnitude. *Bulletin of the Seismological Society of America, 59*, 923–933.

Brune, J. N., & King, C.-Y. (1967). Excitation of mantle Rayleigh waves of period 100 s as a function of magnitude. *Bulletin of the Seismological Society of America, 57*, 1355–1365.

Bullen, K. E. (1953). On strain energy and strength in the Earth's upper mantle. *Proc. Am. Geophys. Union, 34*, 107–109.

Bune, V. I. (1955). O klassifikatsii zemletryasenĭ po energii uprugikh voln, izluchaemykh iz ochaga. *Doklady Akad. Nauk Tajik. SSR, 14*, 31–34. (**in Russian**).

Bune, V. I. (1956). Ob ispol'zovanii metoda Golitsyna dlya priblizhennoĭ otsenki energii blizkikh zemletryasenĭ. *Trudy Akad. Nauk Tajik. SSR, 54*, 3–27. (**in Russian**).

Choy, G. L., & Boatwright, J. L. (1995). Global patterns of radiated seismic energy and apparent stress. *Journal of Geophysical Research, 100*, 18205–18228.

Choy, G. L., McGarr, A., Kirby, S. H., & Boatwright, J. L. (2006). An overview of the global variability in radiated energy and apparent stress. *Amer. Geophys. Union Geophys. Monog., 170*, 43–57.

Convers, J. A., & Newman, A. V. (2013). Rapid earthquake rupture duration estimates from teleseismic energy rates, with application to real-time warning. *Geophysical Research Letters, 40*, 1–5.

Cooley, J. W., & Tukey, J. W. (1965). An algorithm for the machine calculation of complex Fourier series. *Mathematics of Computation, 19*, 297–301.

Dahlen, F. A. (1977). The balance of energy in earthquake faulting. *Geophysical Journal of the Royal astronomical Society, 48*, 239–261.

Di Filippo, D., & Marcelli, L. (1950). Magnitudo ed energia dei terremoti in Italia. *Annali Geofis., 3*, 337–348.

Ekström, G., & Dziewoński, A. M. (1988). Evidence of bias in the estimation of earthquake size. *Nature, 332*, 319–323.

Elliott, A.J., Parsons, B., Elliott, J.R., & Hollingsworth, J. (2017). 3-D displacements in the 7 December 2015 *M* = 7.2 Murghob, Tajikistan earthquake from optical imagery, stereo topography and InSAR, and constraints on the 1911 Sarez event, *Eos Transactions American Geophysical Union, 98* (53), T22A-03 (**abstract**).

Esteva, L., & Rosenblueth, E. (1964). Espectros de temblores moderadas y grandes. *Boletin Sociedad Mexicana de Ingenieria Sísmica, 2*, 1–18.

Ewing, W. M., & Press, F. (1954). An investigation of mantle Rayleigh waves. *Bulletin of the Seismological Society of America, 44*, 127–147.

Fedotov, S. A. (1963). The absorption of transverse seismic waves in the upper mantle and energy classification of near earthquakes of intermediate depth, *Izv. Akad. Nauk SSSR Geofiz. Ser., 7*, 509–520. (**in Russian**).

Fry, B., Benites, R., & Kaiser, A. (2011). The character of acceleration in the $M_w = 6.2$ Christchurch earthquake. *Seismological Research Letters, 82*, 846–852.

Geller, R. J. (1976). Scaling relations for earthquake source parameters and magnitudes. *Bulletin of the Seismological Society of America, 66*, 1501–1523.

Gibowicz, S. J., Young, R. P., Talebi, S., & Rawlence, D. J. (1991). Source parameters of seismic events at the underground research laboratory in Manitoba, Canada: scaling relations for events with moment magnitude smaller than 2. *Bulletin of the Seismological Society of America, 81*, 1157–1182.

Golitsyn, B. B. (1915). Sur le tremblement de terre du 18 février 1911. *Comptes Rendus Acad. Sci. Paris, 160*, 810–814.

Gutenberg, B. (1945a). Amplitudes of surface waves and magnitudes of shallow earthquakes. *Bulletin of the Seismological Society of America, 35*, 3–12.

Gutenberg, B. (1945b). Amplitudes of *P*, *PP*, and *S* and magnitude of shallow earthquakes. *Bulletin of the Seismological Society of America, 35*, 57–69.

Gutenberg, B. (1945c). Magnitude determination for deep-focus earthquakes. *Bulletin of the Seismological Society of America, 35*, 117–130.

Gutenberg, B. (1956). On the energy of earthquakes, *Q. J. Geol. Soc. (London), 112*, 1–14.

Gutenberg, B. (1958). Wave velocities in the earth's core. *Bulletin of the Seismological Society of America, 48*, 301–314.

Gutenberg, B., & Richter, C. F. (1936). On seismic waves (Third paper), *Gerlands Beitr. z. Geophys., 47*, 73–131.

Gutenberg, B., & Richter, C.F. (1941). Seismicity of the Earth, *Geological Society of America Special Paper, 34*, 125 pp.

Gutenberg, B., & Richter, C. F. (1942). Earthquake magnitude, intensity, energy and acceleration. *Bulletin of the Seismological Society of America, 32*, 163–191.

Gutenberg, B., & Richter, C. F. (1949). *Seismicity of the Earth and associated phenomena*, 273 pp. Princeton University Press.

Gutenberg, B., & Richter, C. F. (1954). *Seismicity of the Earth and associated phenomena*, 310 pp. Princeton University Press.

Gutenberg, B., & Richter, C. F. (1956a). Earthquake magnitude, intensity, energy and acceleration (second paper). *Bulletin of the Seismological Society of America, 46*, 105–145.

Gutenberg, B., & Richter, C. F. (1956b). Magnitude and energy of earthquakes. *Annali di Geofisica, 9*, 1–15.

Hamburger, M., Kopnichev, Y., Levshin, A., Martynov, V., Mikhailova, N., Molnar, P., et al. (2007). In memoriam: Vitaly Ivanovich Khalturin (1927–2007). *Seismol. Res. Lett., 78*, 577–578.

Hanks, T. C., & Kanamori, H. (1979). A moment magnitude scale. *Journal of Geophysical Research, 84*, 2348–2350.

Haskell, N. A. (1964). Total energy and energy spectral density of elastic wave radiation from propagating faults. *Bulletin of the Seismological Society of America, 54*, 1811–1841.

Haskell, N. A. (1966). Total energy and energy spectral density of elastic wave radiation from propagating faults, Part II: a statistical source model. *Bulletin of the Seismological Society of America, 56*, 125–140.

Herrin, E. (1968). 1968 seismological tables for *P* phases. *Bulletin of the Seismological Society of America, 58*, 1193–1239.

Holden, C. (2011). Kinematic source model of the 22 February 2011 $M_w = 6.2$ Christchurch earthquake using strong motion data. *Seismological Research Letters, 82*, 783–788.

Housner, G. W. (1965). Intensity of earthquake ground shaking near the causative fault. In *Proc. 3rd World Conf. Earthq. Eng.*, Auckland (pp. 94–115).

Hutt, C. R., Bolton, H. F., & Holcomb, L. G. (2002). US contribution to digital global seismograph network. In W. H. K. Lee, H. Kanamori, P. Jennings, & C. Kisslinger (Eds.), *International handbook of earthquake and engineering seismology* (pp. 319–332). New York: Academic.

Ide, S., & Beroza, G. C. (2001). Does apparent stress vary with earthquake size? *Geophysical Research Letters, 28*, 3349–3352.

Ishii, M., Shearer, P. M., Houston, H., & Vidale, J. E. (2007). Teleseismic *P*-wave imaging of the 26 December 2004 Sumatra-Andaman and 28 March 2005 Sumatra earthquake ruptures using the Hi-net array. *Journal of Geophysical Research, 112*(B11), B11307, 16 pp.

Jeffreys, H. (1923). The pamir earthquake of 1911 February 18, in relation to the depths of earthquake foci. *Monthly Notices of the Royal astronomical Society, Geophysical Supplement, 1*, 22–31.

Jeffreys, H. (1929). *The Earth, its origin, history and physical constitution* (2nd ed.), 346 pp. Cambridge: Cambridge University Press.

Johnston, A. C. (1996). Seismic moment assessment of earthquakes in stable continental regions—III. New Madrid 1811–1812, Charleston, 1886, and Lisbon 1755. *Geophysical Journal International, 126*, 314–344.

Julian, B. R., & Anderson, D. L. (1968). Travel times, apparent velocities and amplitudes of body waves. *Bulletin of the Seismological Society of America, 58*, 339–366.

Kanamori, H. (1972). Mechanism of tsunami earthquakes. *Physics of the Earth and Planetary Interiors, 6*, 346–359.

Kanamori, H. (1977). The energy release in great earthquakes. *Journal of Geophysical Research, 82*, 2981–2987.

Kanamori, H. (1983). Magnitude scale and quantification of earthquakes. *Tectonophysics, 93*, 185–199.

Kanamori, H., & Anderson, D. L. (1975). Theoretical basis of some empirical relations in seismology. *Bulletin of the Seismological Society of America, 65*, 1073–1095.

Kárník, V., Kondorskaya, N. V., Riznitchenko, Ju V, Savarensky, E. F., Soloviev, S. L., Shebalin, N. V., et al. (1962). Standardization of the earthquake magnitude scale. *Studia Geophysica et Geodetica, 6*, 41–48.

Kim, W.-Y., Sykes, L. R., Armitage, J. H., Xie, J. K., Jacob, K. H., Richards, P. G., et al. (2001). Seismic waves generated by aircraft impacts and building collapses at World Trade Center, New York City. *Eos, Transactions of the American Geophysical Union, 82*(47), 565, 570–571.

Kimberlain, T. B., Blake, E. S., & Cangialosi, J. P. (2016). *National Hurricane Center Tropical Cyclone Report, Hurricane Patricia (EP202015)*, 10 pp., Miami: Natl. Ocean. Atmos. Admin.

Klotz, O. (1915). The earthquake of February 18, 1911: a discussion of this earthquake by Prince Galitzin, with comments thereon. *Journal of the Royal Astronomical Society of Canada, 9*, 428–437.

Knopoff, L., & Gilbert, J. F. (1959). Radiation from a strike-slip earthquake. *Bulletin of the Seismological Society of America, 49*, 163–178.

Lamb, H. (1916). On waves due to a travelling disturbance, *Philosophical Magazine, 13*, 386–399 and 539–548.

Lay, T., Ammon, C. J., Kanamori, H., Xue, L., & Kim, M. (2011). Possible large near-trench slip during the 2011 M_w = 9.0 off the Pacific coast of Tohoku earthquake. *Earth, Planets and Space, 63*, 687–692.

Lee, W. H. K., & Stewart, S. W. (1981). *Principles and applications of microearthquakes networks* (pp. 155–157). New York: Academic.

Lim, V. V., Akdodov, U., & Vinnichenko, S. (1997). *Sarez Lake is a threatening dragon of central Asia*, 54 pp., Geol. Manag. Gov. Rep., Dushanbe, Tajikistan

López, A. M., & Okal, E. A. (2006). A seismological reassessment of the source of the 1946 Aleutian "tsunami" earthquake. *Geophysical Journal International, 165*, 835–849.

Macelwane, J. B. (1926). Are important earthquakes ever caused by impact? *Bulletin of the Seismological Society of America, 16*, 15–18.

Mansinha, L., & Smylie, D. E. (1971). The displacement fields of inclined faults. *Bulletin of the Seismological Society of America, 61*, 1433–1440.

Mayeda, K., Gök, R., Walter, W.R. & Hofstetter, A. (2005). Evidence for non-constant energy/moment scaling from coda-derived source spectra, *Geophysical Research Letters, 32*(10), L10306, 4 pp.

McGarr, A. (1999). On relating apparent stress to the stress causing earthquake fault slip. *Journal of Geophysical Research, 104*, 3003–3011.

Mendenhall., T.C., (1888). *On the intensity of earthquakes, with approximate calculations of the energy involved*, 8 pp., Am. Assoc. Adv. Sci.: Proc.

Miles, M. G., Grainger, R. G., & Highwood, E. J. (2004). Volcanic aerosols: the significance of volcanic eruption strength and frequency for climate. *Journal of the Royal Meteorological Society, 130*, 2361–2376.

Milne, J. (1898). *Seismology*, 320 pp., London: K. Paul Trench, Trübner & Co.

Musson, R. (2007). British earthquakes. *Proceedings of the Geologists' Association, 118*, 305–337.

Nagamune, T., & Seki, A. (1958). Determination of earthquake magnitude from surface waves for Matsushiro seismological observatory and the relation between magnitude and energy. *Geophys. Mag., 28*, 303–308.

Negmatullayev, S.Kh., Ulubiyeva, T.R., & Djuraev, R.U. (2018). Sarezskoye earthquake of December 7, 2015, *Proc. Xth Intl. Conf. Monitoring Nuclear Tests and Their Consequences*, Almaty, Kazakhstan [abstract].

Newhall, C. G., & Self, S. (1982). The volcanic explosivity index (VEI). *Journal of Geophysical Research, 87*, 1231–1238.

Newman, A. V., & Okal, E. A. (1998). Teleseismic estimates of radiated seismic energy: The E/M_0 discriminant for tsunami earthquakes. *Journal of Geophysical Research: 103*, 26885–26898.

Newton, I. S. (1687). *Philosophiæ naturalis principia mathematica*. London: Pepys.

Okada, Y. (1985). Surface deformation due to shear and tensile faults in a half-space. *Bulletin of the Seismological Society of America, 75*, 1134–1154.

Okal, E. A. (1989). A theoretical discussion of time-domain magnitudes: the Prague formula for M_s and the mantle magnitude M_m. *Journal of Geophysical Research, 94*, 4194–4204.

Okal, E. A. (1992). Use of the mantle magnitude M_m for the reassessment of the seismic moment of historical earthquakes. I: Shallow events. *Pure Appl. Geophys., 139*, 17–57.

Okal, E. A. (2003). Normal modes energetics for far-field tsunamis generated by dislocations and landslides. *Pure and Applied Geophysics, 160*, 2189–2221.

Okal, E. A. (2013). From 3-Hz P waves to $_0S_2$: No evidence of a slow component to the source of the 2011 Tohoku earthquake. *Pure and Applied Geophysics, 170*, 963–973.

Okal, E. A., & Kirby, S. H. (2002). Energy-to-moment ratios for damaging intraslab earthquakes: Preliminary results on a few case studies. USGS Open File Rep., 02-328 (pp. 127–131).

Okal, E. A., & Newman, A. V. (2001). Tsunami earthquakes: the quest for a regional signal. *Physics of the Earth and Planetary Interiors, 124*, 45–70.

Okal, E. A., & Talandier, J. (1989). M_m: A variable period mantle magnitude. *Journal of Geophysical Research, 94*, 4169–4193.

Okal, E. A., & Talandier, J. (1990). M_m: Extension to Love waves of the concept of a variable-period mantle magnitude. *Pure and Applied Geophysics, 134*, 355–384.

Okal, E. A., Kirby, S. H., & Kalligeris, N. (2016). The Showa Sanriku earthquake of 1933 March 2: a global seismological reassessment. *Geophysical Journal International, 206*, 1492–1514.

Oldham, R. D. (1923). The Pamir earthquake of 18th February 1911. *Quarterly Journal of the Geological Society London, 79*, 237–245.

Prieto, G.A., Shearer, P.M., Vernon, F.L. & Kilb, D. (2004). Earthquake source scaling and self-similarity estimation from stacking P and S spectra, *Journal of Geophysical Research, 109*(B08), B08310, 13 pp.

Polet, J., & Kanamori, H. (2000). Shallow subduction zone earthquakes and their tsunamigenic potential. *Geophysical Journal International, 142*, 684–702.

Radulian, M., & Popa, M. (1996). Scaling of source parameters for Vrancea (Romania) intermediate-depth earthquakes. *Tectonophysics, 261*, 67–81.

Rautian, T. G. (1960). The energy of earthquakes. In Y. V. Riznichenko (Ed.), *Methods for the detailed study of seismicity* (pp. 75–114). Moscow: Izd. Akad. Nauk SSSR.

Rautian, T. G., Khalturin, M. I., Fujita, K., Mackey, K. G., & Kendall, A. D. (2007). Origins and methodology of the Russian K-class system and its relationship to magnitude scales. *Seismological Research Letters, 78*, 579–590.

Reid, H. F. (1910). *The California earthquake of April 18, 1906*, Rept. State Earthq. Invest. Comm (Vol. II), 192 pp., Washington: Carnegie Inst. Wash.

Reid, H. F. (1912). The energy of earthquakes. In R. de Kövesligethy (Ed.), *C.R. Séances 4ème Conf. Commis. Perm. et 2ème Assemb. Gén. Assoc. Intern. Sismologie, Manchester, 18-21 juil. 1911* (pp. 268–272). Budapest: Hornyánszky.

Richter, C. F. (1935). An instrumental earthquake magnitude scale. *Bulletin of the Seismological Society of America, 25*, 1–32.

Richter, C. F. (1958). *Elementary seismology*, 768 pp. San Francisco: W.H. Freeman.

Richter, C. F. (1962). Memorial to Beno Gutenberg (1889–1960), *Geol. Soc. Am. Proc. Vol. Annu. Rep.*, *1960*, 93–104.

Richter, C. F., & Nordquist, J. M. (1948). Minimal recorded earthquakes. *Bulletin of the Seismological Society of America*, *38*, 257–261.

Sagisaka, K. (1954). On the energy of earthquakes. *Geophys. Mag.*, *26*, 53–82.

Satô, Y. (1958). Attenuation, dispersion, and the wave guide of the *G* wave. *Bulletin of the Seismological Society of America*, *48*, 231–251.

Shpil'ko, G. A. (1914). Zemletryasenie 1991 goda na Pamirakh i evo posliedstviya. *Izv. Imperat. Russ. Geograf. Obshchestva*, *50*, 69–94. (**in Russian**).

Shpil'ko, G.A. (1915). Noviya sviedieniya ob Usoyskom zavalye i Sarezskom ozerye, *Izv. Turkestans. Otdiel'a Imperat. Russ. Geograf. Obshchestva, 11,* Tashkent, 2 pp. (**in Russian**).

Schuster, R. L., & Alford, D. (2004). Usoi landslide dam and Lake Sarez, Pamir Mountains, Tajikistan. *Environmental & Engineering Geoscience*, *10*, 151–168.

Solov'ev, S. L., & Solov'eva, O. N. (1967). The relationship between the energy class and magnitude of Kuril earthquakes. *Bull. Acad. Sci. USSR Earth Sci. Ser.*, *3*, 79–84.

Sonley, E., & Abercrombie, R. E. (2006). Effects of methods of attenuation correction on source parameter determination. *Geophysical Monograph-American Geophysical Union*, *170*, 91–97.

Tanioka, Y., Ruff, L., & Satake, K. (1997). What controls the lateral variation of large earthquake occurrence along the Japan Trench? *Island Arc*, *6*, 261–266.

Thomson, W. (1855). Note on the possible density of the luminiferous medium, and the mechanical value of a cubic mile of sunlight. *Phil. Mag.*, *9*, 36–40.

Trifunac, M. D., & Brady, G. (1975). A study on the duration of strong earthquake ground motion. *Bulletin of the Seismological Society of America*, *65*, 581–626.

Vaněk, J., Zátopek, A., Kárník, V., Kondorskaya, N. V., Riznichenko, Yu V, Savarenskiĭ, E. F., et al. (1962). Standardizatsya shkaly magnitud. *Izv. Akad. Nauk SSSR, Ser. Geofiz*, *2*, 153–158. (**in Russian; English translation. Bull. USSR Acad. Sci., 2, 108–111**).

Vanmarcke, E. H., & Lai, S.-S. P. (1980). Strong-motion duration and RMS amplitude of earthquake records. *Bulletin of the Seismological Society of America*, *70*, 1293–1307.

Vassiliou, M. S., & Kanamori, H. (1982). The energy release in earthquakes. *Bulletin of the Seismological Society of America*, *72*, 371–387.

Veith, K. F., & Clawson, G. E. (1972). Magnitude from short-period *P*-wave data. *Bulletin of the Seismological Society of America*, *62*, 435–462.

Venkataraman, A., & Kanamori, H. (2004). Effect of directivity on estimates of radiated seismic energy, *Journal of Geophysical Research, 109* (B4), B04301, 12 pp.

Vvedenskaya, A. V. (1956). Opredelenie poleĭ smeshcheniĭ pri zemletryaseniyakh s pomoshchyu teoriĭ dislokatsiĭ, Izv. Akad. Nauk SSSR. Ser. Geofiz., *6*, 277–284. (**in Russian**).

Wald, D. J., Kanamori, H., Helmberger, D. V., & Heaton, T. H. (1993). Source study of the 1906 San Francisco earthquake. *Bulletin of the Seismological Society of America*, *83*, 981–1019.

Walter, W. R., Mayeda, K., Gök, R., & Hofstetter, A. (2006). The scaling of seismic energy with moment: Simple models compared with observations. *American Geophysical Union Geophysical Monograph, 170*, 25–41.

Weinstein, S. A., & Okal, E. A. (2005). The mantle wave magnitude M_m and the slowness parameter Θ: Five years of real-time use in the context of tsunami warning. *Bull. Seismol. Soc. Am.*, *95*, 779–799.

Wrinch, D., & Jeffreys, H. (1923). On the seismic waves from the Oppau explosion of 1921 September 21. *Month. Not. R. Astron. Soc. Geophys. Suppl.*, *1*, 16–22.

Wu, F.T. (1966). *Lower limit of the total energy of earthquakes and partitioning of energy among seismic waves,* Ph.D. Dissertation, Part I, Pasadena: California Institute of Technology, 161 pp.

Ye, L., Lay, T., Kanamori, H., & Rivera, L. (2016). Rupture characteristics of major and great ($M_w \geq 7.0$) megathrust earthquakes from 1990 to 2015: 1. Source parameter scaling relationships. *Journal of Geophysical Research, 121*, 826–844.

Zöppritz, K., Geiger, L., & Gutenberg, B. (1912). Über Erdbenbenwellen. V, *Nachr. k. Gess. d. Wiss. z. Göttingen, math.-phys. Kl.,* 121-206.

(Received March 31, 2018, revised August 27, 2018, accepted September 8, 2018, Published online October 17, 2018)

Pure Appl. Geophys. 176 (2019), 3851–3865
© 2019 Springer Nature Switzerland AG
https://doi.org/10.1007/s00024-019-02207-2

Pure and Applied Geophysics

A Damped Dynamic Finite Difference Approach for Modeling Static Stress–Strain Fields

RONGRONG LIN,[1,2] (iD) XINDING FANG,[1,2] YUANDI GAN,[1] and YINGCAI ZHENG[1]

Abstract—Modeling dynamic and static responses of an elastic medium often employs different numerical schemes. By introducing damping into the system, we show how the widely used time-marching staggered finite difference (FD) approach in solving elastodynamic wave equation can be used to model time-independent elastostatic problems. The damped FD method can compute elastostatic stress and strain fields of a model subject to the influence of an external field via prescribed boundary conditions. We also show how to obtain an optimized value for the damping factor. We verified the damped FD approach by comparing results against the analytical solutions for a borehole model and a laminated model. We also validated our approach numerically for an inclusion model by comparing the results computed by a finite element method. The damped FD showed excellent agreement with both the analytical results and the finite element results.

Key words: Finite difference, static stress–strain fields, numerical modeling, damped wave equation, effective medium.

1. Introduction

Time marching finite difference methods are widely used in modeling dynamic seismic wave propagation and scattering (Kelly et al. 1976; Virieux 1984, 1986; der Levan 1988; Schuster 2017). However, in many cases, we need to understand the static properties of an elastic medium which can be a heterogeneous or composite medium with several different materials.

For example, when the heterogeneity scale in a rock sample is much smaller than the wavelength, the sample can be viewed as a homogeneous effective medium but anisotropic. In this case, the effective elastic properties of the sample are to be sought and we need to know the bulk strain/deformation given applied stresses. We then can compute the effective moduli which relate bulk average of the strains to the bulk average of the stresses.

Many methods were proposed to compute the effective moduli. For instance, Saenger et al. (2000) calculated the effective elastic moduli by propagating a dynamic pulse through a heterogeneous medium using a finite difference method and used the travel-times of the P and S waves to obtain average elastic properties. This finite difference dynamic pulse propagation method is frequency dependent for a heterogeneous model. Therefore, results obtained using this method are not strictly static solutions.

On the other hand, finite element (FE) methods are commonly used in engineering problems for static stress–strain analyses owing to its geometrical flexibility and numerical accuracy. It can also solve dynamic problems such as fault rupture (Meng 2017; Meng and Wang 2018). The finite element method partitions a model into small meshing elements with nodes. Interpolation functions across nodes are used to convert the partial differential equations into a matrix equation system (e.g., Reddy 2006). However, for static problems with large degree of freedoms, meshing the model into elements and solving the matrix equation with implicit FE can be computationally challenging.

Although our goal is to solve the elastostatic equation for the static stress/strain fields, we will show that by solving a correspondent damped elastodynamic equation, the obtained dynamic solution can converge to the same static fields of interest. In this case, we can take advantage of the FD schemes in solving the damped elastodynamic equation and obtain the static stress/strain field of model without the effort of inverting a large matrix otherwise

---

[1] Department of Earth and Atmospheric Sciences, University of Houston, Houston, TX, USA. E-mail: rlin5@uh.edu; fangxd@sustc.edu.cn

[2] School of Earth and Space Sciences, Southern University of Science and Technology China, Shenzhen, China.

needed in an implicit finite element method for static problems.

2. Methods

In the following, we will show that the time-dependent dynamic solution converges to the static solution for the elastic problems. We will first use a 1D mass-spring system to illustrate this idea and then move to 2D and 3D problems.

2.1. One-Dimensional Example of a Spring-Mass System

2.1.1 Problem Statement

Given a massless spring of a natural length L_0 and elastic constant k with one end fixed at the ceiling and the other end free, if a force, in this case, the gravity of a ball with mass m, is applied at the lower end of the spring (Fig. 1a), what is the axial strain under such stress boundary condition?

2.1.2 Elastostatic Solution

We can do a force balance analysis to figure out the axial strain. We assume the elongation of the spring is ΔL and use the Hooke's law: $k\Delta L = mg$ to get the axial strain, $\varepsilon = \frac{\Delta L}{L_0} = \frac{mg}{kL_0}$.

2.1.3 Modified Problem

If we put the spring system in a viscous fluid (Fig. 1b) and we ignore the fluid buoyance, the amount of static spring elongation is the same as in the previous case (Fig. 1a). This is because the viscous force matters only when the mass has a relative velocity with respect to the fluid. In the static situation, the force balance is the same as the previous problem (Fig. 1a) and the axial strain of the spring is exactly the same as in the case of no viscous fluid applied.

In the elastostatic case, we can see that both systems (viscous or not) give the same strain. The interesting case is the correspondent dynamic problem. If we suddenly hook the mass to the lower end of the spring at time zero, the mass will oscillate in the viscous fluid up and down around an equilibrium position. As time increases, the oscillation amplitude decays because of the damping effect of the viscous fluid. The viscous force will eventually be zero and the system will reach the static equilibrium that we seek for.

2.2. Dynamic Solution in Viscous Case

If we immerse the mass-spring system in a viscous fluid (Fig. 1b), the equation of motion of the mass is:

$$m\frac{dx^2}{dt^2} + \eta\frac{dx}{dt} + kx = mg, \qquad (1)$$

where $x(t)$ is time (t) dependent displacement and g is the gravity accelaration. The solution of the equation is (Udias and Buforn 2018)

$$x(t) = A exp\left(-\frac{\omega_0}{2Q}t\right)\sin(\omega t + \phi) + \frac{mg}{k}, \qquad (2)$$

with $\omega = \omega_0\sqrt{1 - \frac{1}{4Q^2}}, Q = \frac{m\omega_0}{\eta}$, and $\omega_0 = \sqrt{k/m}$ being the harmonic oscillation frequency corresponding to zero damping (i.e. $\eta = 0$) (See "Appendix" for details).

Equation 2 represents a damped oscillation. As time $t \rightarrow +\infty$, the mass will stop at the displacement $x(t \rightarrow +\infty) = \frac{mg}{k}$, which produces exactly the same axial strain as in the static case. This shows that at large times, the time-dependent dynamic solution

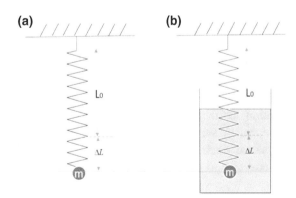

Figure 1
1D elastostatic problem. **a** Spring with a mass; **b** spring with a mass immersed in a viscous fluid

converges to the static solution. In the following, we will extend this concept to 2D and 3D cases.

2.2.1 Extension to 2D and 3D Cases

The reason that we can obtain the elastostatic solution by solving a damped dynamic equation is clear for the spring-mass system in a viscous fluid. At time zero, an external force (i.e., a step function) is suddenly imposed on the system and is present ever since. This sudden force will excite the spring to vibrate. However, the kinetic energy (due to vibration) will die out eventually because of the viscous damping and make the system stabilized at the same equilibrium state with the static problem.

Likewise, in a 2D or 3D case, we can view an elastic medium as a web of inter-connected mass points and springs. We can immerse this system in a viscous fluid and solve the dynamic problem. This viscous immersion can be realized by adding a damping term in a dynamic equation such as the second term in Eq. 1. This has an obvious advantage in numerical calculation where we can employ the time-marching finite difference method to solve for stress and strain associated with the equation. In 2D and 3D elastic cases where a large number of model grids are used, solving for elastodynamic equation in a time marching fashion can be easier and more efficient than solving the static equations.

3. Finite Difference Modeling of Damped Wave Equation

In static loading modeling, the governing equation is

$$\nabla \cdot \boldsymbol{\tau} = 0, \qquad (3)$$

where $\boldsymbol{\tau}(x, y, z)$ is the time-independent stress tensor field in the interior of the computational domain. Our goal is to solve this equation with the appropriate boundary conditions (e.g., stress applied on the boundaries).

Instead of solving the static Eq. 3, we solved the damped wave equation of the form (Morse and Feshback 1953)

$$\rho \frac{\partial \vec{v}}{\partial t} + \rho \, \vec{v} d = \nabla \cdot \boldsymbol{\tau}, \qquad (4)$$

where $\vec{v}(x, y, z, t)$ is the particle motion velocity field and $\boldsymbol{\tau}(x, y, z, t)$ is the stress tensor field and both fields depend on time t and spatial coordinates x, y, and z. $\rho(x, y, z)$ is density, and d is a constant positive damping factor. Equation 4 is a time evolution equation for the particle velocity and stress fields. As time elapses, the kinetic energy in the system is being progressively reduced. When the kinetic energy damps out, the left side of Eq. 4 vanishes and the displacement field will not change anymore. The resultant stress field $\boldsymbol{\tau}$ is the static field that satisfies both the elastostatic Eq. 3 and the boundary conditions.

3.1. Equation in Component Form

In the following theoretical analysis, for simplicity, we only discuss two-dimensional (x, z) modeling as an example, and 3D cases will be shown in numerical examples.

Assume the model material is isotropic, Eq. 4 can be expressed as the following first-order velocity-stress equations:

$$
\begin{cases}
\frac{\partial v_x}{\partial t} + d v_x = b \left(\frac{\partial \tau_{xx}}{\partial x} + \frac{\partial \tau_{xz}}{\partial z} \right) \\
\frac{\partial v_z}{\partial t} + d v_z = b \left(\frac{\partial \tau_{xz}}{\partial x} + \frac{\partial \tau_{zz}}{\partial z} \right) \\
\frac{\partial \tau_{xx}}{\partial t} = (\lambda + 2\mu) \frac{\partial v_x}{\partial x} + \lambda \frac{\partial v_z}{\partial z}, \\
\frac{\partial \tau_{zz}}{\partial t} = (\lambda + 2\mu) \frac{\partial v_z}{\partial z} + \lambda \frac{\partial v_x}{\partial x} \\
\frac{\partial \tau_{xz}}{\partial t} = \mu \left(\frac{\partial v_x}{\partial z} + \frac{\partial v_z}{\partial x} \right)
\end{cases}
\qquad (5)
$$

where b is the reciprocal of density; λ and μ are the Lamé constants; v_x and v_z represent velocity fields of the x and z components, respectively; τ_{xx} and τ_{zz} are normal stresses and τ_{xz} is the shear stress. The damping factor d is a constant that should be determined before formally running the modeling process.

3.2. Equation in Discretized Form

We propose to solve Eq. 5 using the staggered finite difference method (e.g., Virieux 1986; Fang et al. 2014). The finite difference formulas for velocities v_x and v_z with second-order accuracy in both space and time are expressed as:

$$U_{i,j}^{k+1/2} = \frac{1}{1+\frac{d\Delta t}{2}} \left[\left(1 - \frac{d\Delta t}{2}\right) U_{i,j}^{k-1/2} \right.$$
$$+ B_{i,j}\frac{\Delta t}{\Delta x}\left(\Sigma_{i+1/2,j}^{k} - \Sigma_{i-1/2,j}^{k}\right) \qquad (6)$$
$$\left. + B_{i,j}\frac{\Delta t}{\Delta z}\left(\Xi_{i,j+1/2}^{k} - \Xi_{i,j-1/2}^{k}\right) \right],$$

$$V_{i+1/2,j+1/2}^{k+1/2} = \frac{1}{1+\frac{d\Delta t}{2}} \left[\left(1 - \frac{d\Delta t}{2}\right) V_{i+1/2,j+1/2}^{k-1/2} \right.$$
$$+ B_{i+1/2,j+1/2}\frac{\Delta t}{\Delta x}\left(\Xi_{i+1,j+1/2}^{k} - \Xi_{i,j+1/2}^{k}\right)$$
$$\left. + B_{i+1/2,j+1/2}\frac{\Delta t}{\Delta z}\left(T_{i+1/2,j+1}^{k} - T_{i+1/2,j}^{k}\right) \right],$$
$$(7)$$

where Σ represents the normal stress τ_{xx}, Ξ represents the shear stress τ_{xz}, Ξ represents the normal stress τ_{zz}, U represents velocity v_x and V represents velocity v_z. Δt is the time grid step, Δx and Δz are grid step in x-axis and z-axis directions, respectively, and B is the buoyancy of the model (Virieux 1986). The superscript means time step and the subscripts are x and z grid numbers separated by a comma.

The static stress or strain boundary conditions can be imposed to the boundaries of the model at time zero. They produce elastic deformations that contain both static and dynamic components. The static and dynamic elastic energy can be respectively characterized by the model's strain energy and kinetic energy (Aki and Richards 1980) that evolve over time. In order to obtain the static response, we need to choose an appropriate value for the damping factor, d, to damp out the kinetic energy. Based on this reasoning, an optimal value for d may exist. Once the system reaches an equilibrium, the stress and strain fields can be used to derive the effective elastic moduli of the model.

4. Choice of the Damping Factor

It is heuristic to first study a 1D damped oscillation. For a damped harmonic oscillator as we have discussed in Eq. 1, there exists a critical damping coefficient c which will make the oscillation system reach equilibrium with minimal amount of time. Based on Eq. 1, the critical damping coefficient C_C can be solved as (see "Appendix")

$$C_C = 2m\sqrt{k/m} = 2m\omega_n, \qquad (8)$$

where ω_n is the natural frequency or resonance frequency of the oscillator corresponding to no damping. Since we can view the elastic medium as a web of inter-connected mass points and springs, each mass point can be seen as a damped harmonic oscillator, the relationship between the natural frequency of the system and critical damping coefficient should persist in the 2D model as described in Eq. 5.

The following 2D example shows the relationship between the natural frequency and the critical damping. We can set the damping parameter d in Eq. 4 to be zero and observe the motion of the nodes in the model under prescribed stresses as boundary conditions. We used a 200 m by 200 m homogeneous 2D model with P-wave velocity $V_P = 4382$ m/s, S-wave velocity $V_S = 2530$ m/s, and density $\rho = 3000$ kg/m^3. Spatial steps in x and z directions are $\Delta x = \Delta z = 2$ m, time step is $\Delta t = 0.15$ ms. The Δx, Δz and Δt values are chosen to satisfy the numerical stability condition for the explicit scheme that $V_P\frac{\Delta t}{\Delta x} < \frac{1}{\sqrt{2}}$ (Virieux 1986). The model is prescribed with normal stresses $\sigma_{xx} = 1$ MPa, $\sigma_{zz} = 1$ MPa on all its four boundaries. The motion plots of the three selected locations at coordinates (4 m, 4 m), (98 m, 98 m), (180 m, 180 m) (their positions in the model space can be derived from Fig. 2) are presented (Fig. 3).

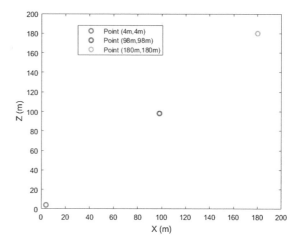

Figure 2
Illustration of the 2D model domain and the three picked field locations at coordinates (4 m, 4 m), (98 m, 98 m) and (180 m, 180 m)

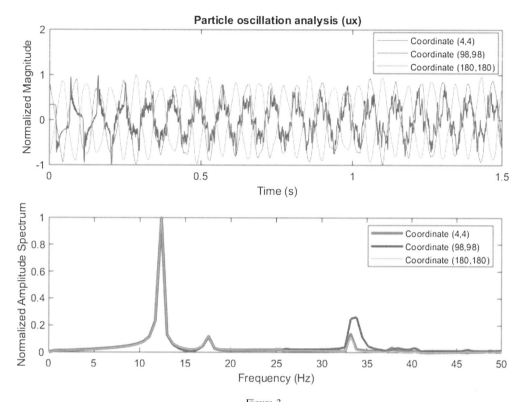

Figure 3
Particle displacement in the x direction (top) and the corresponding amplitude spectra (bottom) imply that a consistent harmonic oscillation frequency may be calculated to estimate the damping coefficient. The displacement along x direction (u_x) of nodes at three picked locations are plotted. The other nodes in the model are not plotted here but they obey the same pattern. The displacement along z direction (u_z) is not displayed but it behaves the same as u_x in amplitude spectra

We find that in this homogeneous isotropic model the nodes oscillate at the same frequency (Fig. 3) and in this particular model the primary frequency is 12.37 Hz which will be used as the natural frequency of the model with Eq. 8 to calculate the critical damping coefficient. From Eq. 8, the normalized damping parameter in Eq. 4 takes the form of

$$d = \frac{C_C}{m} = 2\omega_n = 2 \times 2\pi \times f_n. \qquad (9)$$

Since the damping parameter d is used in the form of $d\Delta t$ in Eqs. 6 and 7, it is convenient to denote $D = d\Delta t$. The parameter d has the unit of rad/s, while D is unitless. In the following, we use $D = d\Delta t$ instead of d.

A comparison of the evolution of the total strain energy and kinetic energy in the system for different damping parameters can be computed (Fig. 4). The model is said to reach the static equilibrium when the strain energy converges to a stable value and the kinetic energy drops below a certain small threshold. With a damping parameter $D = 0.0233$, the system reaches equilibrium with the least amount of time. As expected, the strain energy converges to the same value regardless of the damping parameter as long as it's positive. The model convergence under different damping parameters are also analyzed and shows the existence of an optimal damping value (Fig. 5).

Additionally, if we use bigger Δt, the time steps to reach equilibrium will be surely less, as long as it satisfies the numerical stability condition for the explicit scheme that $V_P \frac{\Delta t}{\Delta x} < \frac{1}{\sqrt{2}}$. For example, under the same damping parameter d, when we used $\Delta t = 0.3$ ms, the time steps to reach equilibrium are half of the case of $\Delta t = 0.15$ ms. Furthermore, if we use bigger spatial steps Δx and Δz, bigger Δt can be tolerated, even less time steps are required.

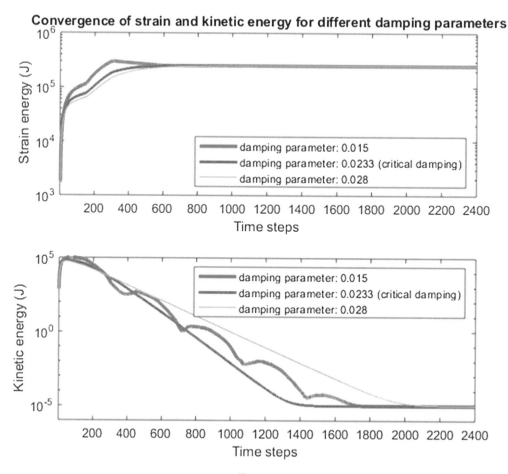

Figure 4
Strain energy (top) and kinetic energy (bottom) (in log scale) in the entire model versus time steps in the finite difference implementation. Here we set the equilibrium threshold for the kinetic energy as 10^{-5} J. Curves in different colors show the energy evolution with different damping parameters, D. The model reaches equilibrium after 1400, 1780, and 2100 time steps for damping parameters D set to 0.0233, 0.015, and 0.028, respectively

One thing to notice is that the natural frequency is independent of the Δx, Δz and Δt. The natural frequency is related to the material properties.

5. Natural Frequency and Standing Waves

The forces applied on the model boundaries generate seismic waves which bounce back and forth within the model to produce reflections. Under this resonance condition, the reflections interfere with the direct waves and result in standing waves. For the condition with four boundaries of the 2D square model fixed, the fundamental harmonics frequency is

$$f = \frac{v}{2L}, \qquad (10)$$

where v is velocity, L is the length of the side of square model (e.g. Blackstock 2000). Since in this case only normal stresses are applied, P-wave velocity is the primary influence, we can get the fundamental frequency approximated as $f = \frac{4382}{2 \times 200} = 10.96$ Hz, which is close to the 12.37 Hz obtained from particle oscillation analysis.

This gives us a more practical and much faster way to estimate the critical damping parameter without the effort to run the particle oscillation analysis. Although the fundamental natural frequency we derived from Eq. 10 is not the most accurate, the damping performance is still good. For this model,

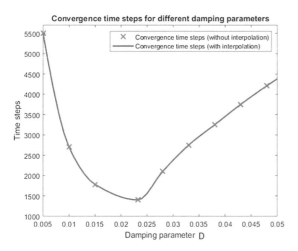

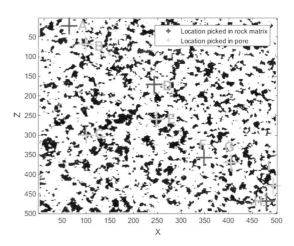

Figure 5

Convergence time steps for different damping parameters for a 2D model. The optimal damping occurs at around 0.023

Figure 6

Illustration of the 2D porous sandstone model and its nine picked nodes (crosses, A to I) at different coordinates. The particle velocity waveforms for these 9 nodes are shown in Fig. 7

the damping parameter we get from fundamental frequency 10.96 Hz is $D = 0.0206$, and the time steps it takes to reach equilibrium is still approximately 1400 (Fig. 5).

5.1. Inhomogeneous Material

When the model is inhomogeneous, we can choose several field locations and use their average natural frequency as the optimal damping coefficient. We note that there is a flexibility to choose the damping parameter. Using the optimal D will allow the dynamic model to reach its static equilibrium in the least amount of time. However, using a sub-optimal value D will also allow us to obtain the same static equilibrium state but may not in a quickest way. A 500 by 500 grids (with both x and z direction spatial steps 2 m) porous sandstone rock model with porosity 0.24 is used as an example. The rock matrix of such model is set to quartz and the pores are saturated with water (Fig. 6). The same boundary conditions are applied as in the homogeneous example. We plotted the displacement at several locations (Fig. 7). From the spectral analysis, the average resonance frequency of the model is approximately 2 Hz.

Again, we can estimate the natural frequency of the inhomogeneous media simply using Eq. 10. However, the velocity term needs to be the averaged P-wave velocity over the compositions of the model.

Raymer et al. (1980) suggested an empirical porosity-velocity relationship for low porosity sandstone

$$V = (1 - \emptyset)^2 V_0 + \emptyset V_{fl}, \tag{11}$$

where V, V_0 and V_{fl} are the velocities of the rock, the mineral matrix and the pore fluid, respectively and $\emptyset$ is the pore porosity.

In this case, $\emptyset = 0.24$, $V_0 = 6018.8$ m/s, which is the P-wave velocity of quartz, and $V_{fl} = 1500$ m/s, which is the P-wave velocity of water. Thus we get $V = 3836.5$ m/s. Using Eq. 10, we have the fundamental frequency as 1.92 Hz, which is very close to the 2 Hz obtained as in Fig. 7.

6. Verification Using Analytical Models

6.1. Circular Borehole Model

The stresses on the cross-sectional plane of a circular borehole under the compression of far field stresses and internal fluid pressure can be expressed as (Fjar 2008):

$$\sigma_r = \frac{\sigma_H + \sigma_h}{2}\left(1 - \frac{R_w^2}{r^2}\right)$$
$$+ \frac{\sigma_H - \sigma_h}{2}\left(1 + 3\frac{R_w^4}{r^4} - 4\frac{R_w^2}{r^2}\right)cos2\theta + P_w\frac{R_w^2}{r^2}, \tag{12}$$

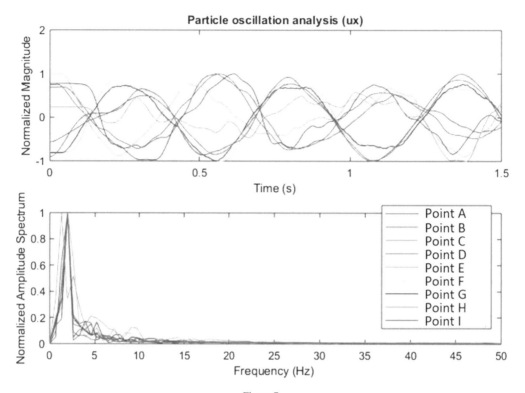

Figure 7
Displacements in the x direction (top) at the chosen field locations (Fig. 6) and the corresponding amplitude spectra (bottom)

$$\sigma_\theta = \frac{\sigma_H + \sigma_h}{2}\left(1 + \frac{R_w^2}{r^2}\right)$$
$$- \frac{\sigma_H - \sigma_h}{2}\left(1 + 3\frac{R_w^4}{r^4}\right)cos2\theta - P_w\frac{R_w^2}{r^2}, \quad (13)$$

$$\tau_{r\theta} = -\frac{\sigma_H - \sigma_h}{2}\left(1 - 3\frac{R_w^4}{r^4} + 2\frac{R_w^2}{r^2}\right)sin2\theta, \quad (14)$$

where σ_r and σ_θ are radial and tangential normal stresses, respectively; $\tau_{r\theta}$ is the shear stress; σ_H and σ_h respectively represent the far field maximum and minimum horizontal stresses; R_w is borehole radius; P_w is the borehole fluid pressure; r represents the distance measured from the borehole center; θ is the azimuth angle measured counter-clockwise from the x-axis. Following equations are used to convert cylindrical polar coordinates stresses σ_r, σ_θ and $\tau_{r\theta}$ calculated by Eqs. 12–14 to the Cartesian coordinates (x, y)

$$\sigma_r = \frac{1}{2}(\sigma_x + \sigma_y) + \frac{1}{2}(\sigma_x - \sigma_y)cos2\theta + \tau_{xy}sin2\theta,$$
$$(15)$$

$$\sigma_\theta = \frac{1}{2}(\sigma_x + \sigma_y) - \frac{1}{2}(\sigma_x - \sigma_y)cos2\theta - \tau_{xy}sin2\theta,$$
$$(16)$$

$$\tau_{r\theta} = \frac{1}{2}(\sigma_y - \sigma_x)sin2\theta + \tau_{xy}cos2\theta. \quad (17)$$

For a 10 m by 10 m model with a wellbore of radius 10 cm in the center of the model, suppose the maximum horizontal stress and minimal horizontal stress in the region are 40 MPa and 30 MPa respectively and the borehole fluid pressure is 20 MPa, the normal and shear stress fields calculated by the analytical solution (Eqs. 12–17) and the finite difference method are compared against each other (Figs. 8, 9, 10).

To apply the damped finite difference simulation, we firstly assigned fixed constant value $\sigma_x = P_w$ and $\sigma_y = P_w$ in the wellbore area of the 10 m by 10 m model. Then we assumed the boundaries are at infinite far distance and prescribed the four model boundaries with $\sigma_x = \sigma_H$ and $\sigma_y = \sigma_h$. Finally the simulation program was ran with modeling

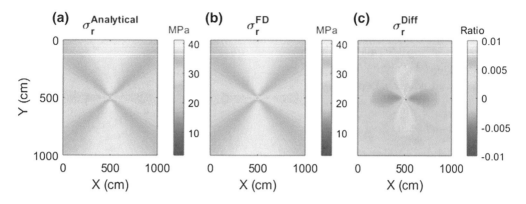

Figure 8

a σ_r calculated from the analytical solution and **b** the finite difference method as well as (**c**) the numerical error

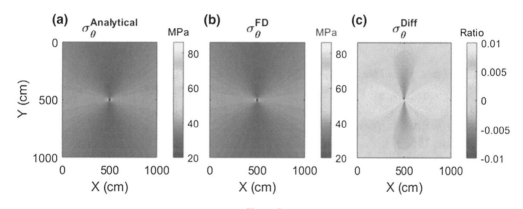

Figure 9

a σ_r calculated from the analytical solution and **b** the finite difference method as well as (**c**) the numerical error

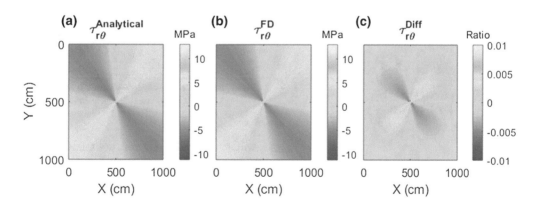

Figure 10

a $\tau_{r\theta}$ calculated from the analytical solution and **b** the finite difference method as well as (**c**) the numerical error

parameters $\Delta x = \Delta z = 0.4$ cm, and $\Delta t = 6 \times 10^{-7}$ s. The model was assigned with properties of P-wave velocity $V_P = 4382$ m/s, S-wave velocity $V_S = 2530$ m/s and density $\rho = 3000$ kg/m^3. Damping coefficient ($D = 0.0017$) was obtained using Eq. 10

neglecting the effect on model velocity from water in the wellbore. One thing to clarify is that we don't need to add any pinpoint to prevent rigid body motion, since we solve for the velocity/stress field instead of the displacement field.

From the results we can see that the damped finite difference simulation results on stress calculation match well with the analytical results with difference (calculated as (Analytical-FD results)/Analytical) less than 1%, which we call numerical error.

6.2. A Laminated Model in 3D Space

A laminated model that consists of a thin-layered sequence of dolomite and shale was used to verify the applicability of the proposed damped finite difference method for static stress modeling purpose. The materials' properties of P-wave velocity, S-wave velocity and density are as follows (Mavko et al. 2009):

Dolomite: $V_P = 5200$ m/s, $V_S = 2700$ m/s, $\rho = 2450$ kg/m^3;

Shale: $V_P = 2900$ m/s, $V_S = 1400$ m/s, $\rho = 2340$ kg/m^3.

The laminated structure results in transverse isotropy with the effective stiffness tensor given analytically as (Backus 1962):

$$\begin{bmatrix} A & B & F & 0 & 0 & 0 \\ B & A & F & 0 & 0 & 0 \\ F & F & C & 0 & 0 & 0 \\ 0 & 0 & 0 & E & 0 & 0 \\ 0 & 0 & 0 & 0 & E & 0 \\ 0 & 0 & 0 & 0 & 0 & M \end{bmatrix} \quad (18)$$

with

$$A = \left\langle 4\rho V_s^2\left(1 - \frac{V_S^2}{V_p^2}\right)\right\rangle + \left\langle 1 - 2\frac{V_S^2}{V_p^2}\right\rangle^2 \left\langle \left(\rho V_p^2\right)^{-1}\right\rangle^{-1},$$

$$B = \left\langle 2\rho V_s^2\left(1 - \frac{2V_S^2}{V_p^2}\right)\right\rangle + \left\langle 1 - 2\frac{V_S^2}{V_p^2}\right\rangle^2 \left\langle \left(\rho V_p^2\right)^{-1}\right\rangle^{-1},$$

$$C = \left\langle \left(\rho V_p^2\right)^{-1}\right\rangle^{-1},$$

$$E = \left\langle (\rho V_s^2)^{-1}\right\rangle^{-1},$$

$$F = \left\langle 1 - 2\frac{V_S^2}{V_p^2}\left(\rho V_p^2\right)^{-1}\right\rangle^{-1},$$

$$M = \langle \rho V_s^2 \rangle,$$

$$B = A - 2M,$$

$$(19)$$

where the brackets $\langle \rangle$ represent volume averaging of the properties weighted by their volume fractions,

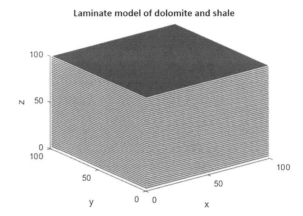

Figure 11
Laminate model (3D geometry) for a thin-layered sequence of 50% dolomite and 50% shale. The blue color indicates dolomite and the yellow indicates shale

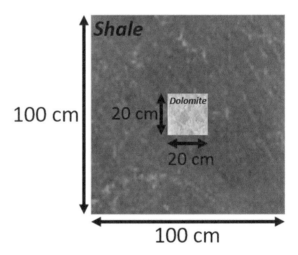

Figure 12
Illustration of the 100 cm by 100 cm 2D model of shale with a 20 cm by 20 cm cubic inclusion of dolomite

which are both 0.5 for dolomite and shale. The laminated model (Fig. 11) used in the modeling contains 50 horizontal layers of dolomite-shale interbeds. The model has 100 grids in x, y, and z directions.

In our damped FD method, we can compute the stress (σ)-strain (ε) tensors at each grid in the model which is a six-faced cube. From the Hooke's law, the following relationship can be used to find the elastic moduli (A, M, C, E, and F):

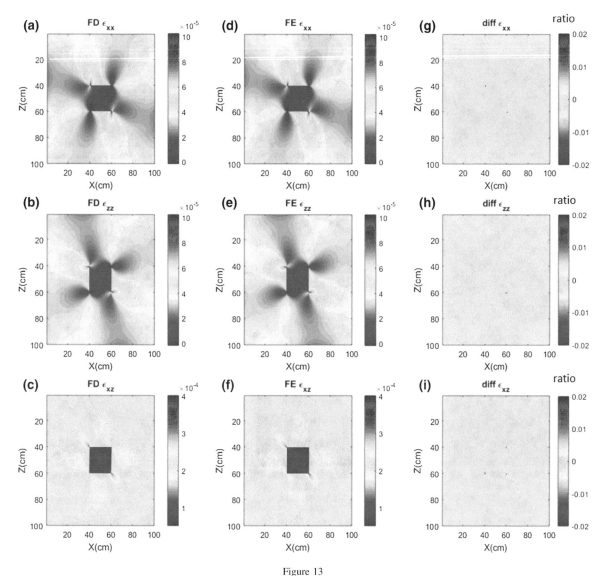

Figure 13
Comparison of the damped finite difference and finite element results. Under the same stress boundary conditions, the strain fields calculated by FD method (**a**, **c**, **e**) are similar to the ones calculated by the FE method (**b**, **d**, **f**). The ratios of difference are mostly less than 1%, with relatively bigger differences on the edge of the inclusioin boundaries (**g**, **h**, **i**)

$$
\begin{bmatrix} \sigma_{xx} \\ \sigma_{yy} \\ \sigma_{zz} \\ \sigma_{yz} \\ \sigma_{xz} \\ \sigma_{xy} \end{bmatrix} = \begin{bmatrix} A & A-2M & F & 0 & 0 & 0 \\ A-2M & A & F & 0 & 0 & 0 \\ F & F & C & 0 & 0 & 0 \\ 0 & 0 & 0 & E & 0 & 0 \\ 0 & 0 & 0 & 0 & E & 0 \\ 0 & 0 & 0 & 0 & 0 & M \end{bmatrix} \begin{bmatrix} \varepsilon_{xx} \\ \varepsilon_{yy} \\ \varepsilon_{zz} \\ \varepsilon_{yz} \\ \varepsilon_{xz} \\ \varepsilon_{xy} \end{bmatrix}.
\tag{20}
$$

To solve for the moduli using the damped finite difference method, we run two calculations:

(1) with the boundary conditions {$\sigma_{xx} = \sigma_{yy} = \sigma_{zz} = \sigma_{xy} = \sigma_{xz} = \sigma_{yz} = 1$ MPa}, the volume-averaged strains we obtained from the damped finite difference method are {$\varepsilon_{xx} = 1.1603 \times 10^{-5}$, $\varepsilon_{zz} = 2.150962 \times 10^{-5}$, $\varepsilon_{yy} = 1.1603 \times 10^{-5}$, $\varepsilon_{yz} = 1.37022 \times 10^{-4}$, $\varepsilon_{xy} = 8.9524 \times 10^{-5}$, $\varepsilon_{xz} = 1.37022 \times 10^{-4}$};

(2) with the boundary conditions {$\sigma_{xx} = \sigma_{xy} = \sigma_{xz} = \sigma_{yz} = 1$ MPa, $\sigma_{zz} = 3$ MPa, $\sigma_{yy} = 2$ MPa}, the volume-averaged strains we got from the damped finite difference method is {$\varepsilon_{xx} = 2.2313 \times 10^{-5}$,

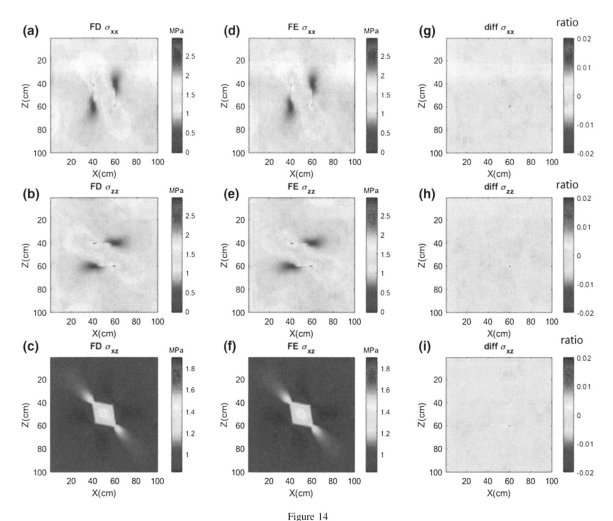

Figure 14
Comparison of the damped finite difference and finite element results. Under the same stress boundary conditions, the stress fields calculated by the FD method (**a–c**) are similar to the ones calculated by the FE method (**d–f**). The ratios of difference are mostly less than 1%, with relatively bigger differences on the edge of the inclusioin boundaries (**g–i**)

$\varepsilon_{zz} = 9.9309985 \times 10^{-5}$, $\varepsilon_{yy} = 2.2568542 \times 10^{-5}$, $\varepsilon_{yz} = 1.3702295 \times 10^{-4}$, $\varepsilon_{xy} = 8.9524 \times 10^{-5}$, $\varepsilon_{xz} = 1.3702295 \times 10^{-4}$}.

We applied the same damping parameter for both cases with $D = 0.0191$ based on Eq. 10 and a volume averaged velocity of 4050 m/s.

Using these stresses and strains, we solved for Eq. 20 and obtained $A = 40.32$ GPa, $M = 11.164$ GPa, $F = 15.032$ GPa, $C = 30.1744$ GPa and $E = 7.29845$ GPa. Comparing our numerical results with the Backus average (Backus 1962) results: $A = 40.63$ GPa, $M = 11.22345$ GPa, $F = 15.0917$ GPa, $C = $ 30.345 GPa and $E = 7.2986$ GPa, we find that the relative errors are between 0.001% to 0.76%.

6.3. Comparison With Finite Element Method

We used an open-source finite element code which has been widely used in digital rock studies (Garboczi 1998) to compare with our damped finite difference method. The finite element code used a conjugate gradient relaxation for iterative solutions.

We tested a 2D 100 cm by 100 cm model of shale ($V_P = 2900$ m/s, $V_S = 1400$ m/s, $\rho = 2340$ kg/m$^{3)}$ with a 20 cm by 20 cm cubic inclusion of dolomite

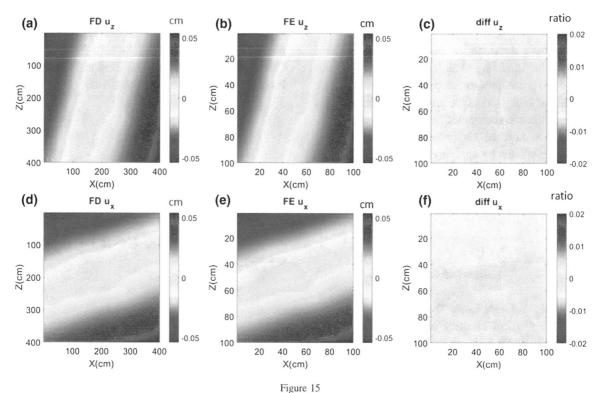

Figure 15
Comparison of the damped finite difference and finite element results. Under the same stress boundary conditions, the displacement fields calculated by the FD method (**a, d**) are similar to the ones calculated by the FE method (**b, e**). The ratios of difference are less than 1% (**c, f**)

($V_P = 5200$ m/s, $V_S = 2700$ m/s, $\rho = 2450$ kg/m^3) in the middle of the model (Fig. 12).

The strain and stress fields were calculated using finite difference and finite element methods with the same boundary conditions of $\{\sigma_{xx} = \sigma_{zz} = \sigma_{xz} = 1$ MPa$\}$. The damped finite difference and finite element results are close (Figs. 13, 14) and their relative differences are below 1%. The differences are attributed to the different numerical schemes. The strain fields from damped finite difference method are computed from displacement field (time integral of particle velocity field) using first order forward finite difference, while the strain field from finite element method is the averaged strain of each pixel from integral of pixel nodes' strains. The displacement fields from both methods are displayed as well (Fig. 15). As a result, these two different schemes may result in minor differences of stress and strain field around the inclusion boundaries.

As for the computational efficiency, it is relatively difficult to compare the numerical cost for two different methods when they employ different numerical schemes. To benchmark two methods, we used 400 by 400 grids model with $\Delta x = \Delta z = 0.25$ cm. For the damped finite difference code, we used $\Delta t = 3.4 \times 10^{-7}$ s and damping prameter $D = 0.0064$. It took 19.5 s CPU time and 14 MB memory. The finite element code took 20.1 s and 20 MB memory use when running on the same workstation. Since the finite element code is optimized with conjugate gradient method, the computing time advantage of damped FD method was not obvious on such a small model. However, the memory usage is 30% less for damped FD than FE, even though the FE code uses the conjugate gradient method which is known for its memory efficiency among other FE methods. Additionally, there is much room to optimize the FD method significantly in particular for computing large 3D models in a parallel computing architect.

7. Conclusions

We have developed a damped finite difference scheme for modeling static stress–strain and introduced the procedures to compute the effective elastic moduli of rock models. A great advantage is that a finite difference code modeling seismic wave propagation can now be easily adapted to solve static loading problems by applying damping. The numerical results accurately match with existing analytical expressions, such as the stress field around a wellbore, and the Backus average for a thin-layered model.

Acknowledgements

We thank Professor Leon Thomsen for discussions about the physical meaning of the damping parameter. We thank Dr. Nishank Saxena for pointing to the finite element code. We appreciate the financial support from UH and SUSTC. We appreciate the UH Xfrac group for providing clusters for computations. We also appreciate the comments from Dr. Hao Hu and Dr. Chen Qi. R. Lin and Y. Zheng are partially supported by NSF EAR-1833058. X. Fang is supported by National Natural Science Foundation of China (grant 41704112) and National Key R&D Program of China (2018YFC0310105).

Appendix

Equation 1 is a homogeneous second order differential equation with the solution in the form of:

$$x(t) = e^{\lambda t}. \tag{21}$$

Substituting $x(t)$ in Eq. 1 using Eq. 21 results a quadratic equation:

$$m\lambda^2 + c\lambda + k = 0, \tag{22}$$

and the root of the equation is:

$$\lambda = \frac{-c \pm \sqrt{c^2 - 4mk}}{2m}. \tag{23}$$

When $c > 2\sqrt{mk}$, λ is always a negative real number and the system is overdamped. When $c < 2\sqrt{mk}$, λ is a complex number with negative real

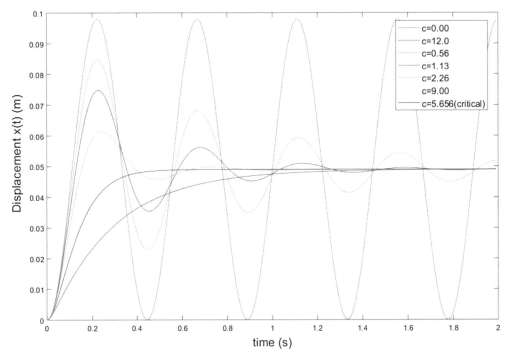

Figure 16
Spring displacement for different viscous damping parameters

part and the system is underdamped. The critical damping occurs when $c = 2\sqrt{mk}$ and the particle will be returning to its equilibrium status in the shortest time under such condition. (Gregory 2006)

For example, in the case of Eq. 1, if spring constant $k = 40$ N/m and natural length $L_0 = 0.08$ m, at time zero, a mass $m = 0.2$ kg is hooked to the spring, the plots of spring displacement for different viscous damping parameter c is shown (Fig. 16). When $c = 2\sqrt{k \times m} = 2\sqrt{40 \times 0.2} = 5.656$ kg/s, the system reaches the equilibrium of displacement $x = 0.049$ m with the least amount of time around 0.6 s among all other damping parameters.

Publisher's Note   Springer Nature remains neutral with regard to jurisdictional claims in published maps and institutional affiliations.

REFERENCES

Aki, K., & Richards, P. (1980). *Quantitative seismology: Theory and methods*. Sausalito: University Science Books.

Backus, G. E. (1962). Long-Wave elastic anisotropy produced by horizontal layering. *Journal of Geophysical Research, 67*(11), 4427–4440.

Blackstock, D. (2000). *Fundamentals of physical acoustics*. Oxford: Wiley.

der Levan, A. (1988). Fourth-order finite difference P-SV seismograms: *eophysics, 53*, 1425–1436.

Fang, X., Fehler, M. C., & Cheng, A. (2014). Simulation of the effect of stress-induced anisotropy on borehole compressional wave propagation. *Geophysics, 79*(4), D205–D216.

Fjar, E. (2008). *Petroleum related rock mechanics*. Amsterdam: Elsevier.

Garboczi, E. J. (1998). *Finite element and finite difference programs for computing the linear electric and elastic properties of digital image of random materials* (p. 6269). Rep: National Institute of Standards and Technology.

Gregory, R. (2006). *Douglas*. Classical Mechanics: Cambridge University Press.

Kelly, K., Ward, S., Treitel, S., & Alford, M. (1976). Synthetic seismograms: A finite difference approach. *Geophysics, 41*, 2–27.

Mavko, G. T., Mukerji, T., & Dvorkin, J. (2009). *The rock physics handbook* (2nd ed.). Cambridge: Cambridge University Press.

Meng, C. (2017). Benchmarking Defmod, an open source FEM code for modeling episodic fault rupture. *Computers & Geosciences, 100*, 10–26.

Meng, C., & Wang, H. (2018). A finite element and finite difference mixed approach for modeling fault rupture and ground motion. *Computers & Geosciences, 113*, 54–69.

Morse, P., & Feshback, W. (1953). *Methods of theoretical physics* (Vol. 2). New York: McGraw-Hill Book Co.

Raymer, L. L., Hunt, E. R., & Gardner, J. S. (1980). An improved sonic transit time to porosity transform. *21st Annual Logging Symposium, Trans. Soc. Prof. Well Log Analysts*, 8–11 July, 1–12.

Reddy, J. N. (2006). *An introduction to the finite element method*. New York: McGraw-Hill.

Saenger, E. H., Gold, N., & Shapiro, S. A. (2000). Modeling the propagation of elastic waves using a modified finite difference grid. *Wave Motion, 31*(1), 77–92.

Schuster, G. (2017). *Seismic inversion*. Tulsa: Society of Exploration Geophysicists. https://doi.org/10.1190/1.9781560803423.ch8

Udias, A., & Buforn, E. (2018). *Principles of seismology*. Cambridge: Cambridge University Press.

Virieus, J. (1986). P-SV wave propagation in heterogeneous media: Velocity-stress finite difference method. *Geophysics, 51*, 889–901.

Virieus, J. (1984). SH-wave propagation in heterogeneous media: Velocity-stress finite difference method. *Geophysics, 49*, 1933–1942.

(Received January 22, 2019, revised April 23, 2019, accepted April 25, 2019, Published online May 9, 2019)

Pure Appl. Geophys. 176 (2019), 3867–3891
© 2018 The Author(s)
https://doi.org/10.1007/s00024-018-1991-x

Interseismic Coupling and Slow Slip Events on the Cascadia Megathrust

SYLVAIN MICHEL,[1,2] (iD) ADRIANO GUALANDI,[2,3] and JEAN-PHILIPPE AVOUAC[2]

Abstract—In this study, we model geodetic strain accumulation along the Cascadia subduction zone between 2007.0 and 2017.632 using position time series from 352 continuous GPS stations. First, we use the secular linear motion to determine interseismic locking along the megathrust. We determine two end member models, assuming that the megathrust is either a priori locked or creeping, which differ essentially along the trench where the inversion is poorly constrained by the data. In either case, significant locking of the megathrust updip of the coastline is needed. The downdip limit of the locked portion lies ∼ 20–80 km updip from the coast assuming a locked a priori, but very close to the coast for a creeping a priori. Second, we use a variational Bayesian Independent Component Analysis (vbICA) decomposition to model geodetic strain time variations, an approach which is effective to separate the geodetic strain signal due to non-tectonic and tectonic sources. The Slow Slip Events (SSEs) kinematics is retrieved by linearly inverting for slip on the megathrust the Independent Components related to these transient phenomena. The procedure allows the detection and modelling of 64 SSEs which spatially and temporally match with the tremors activity. SEEs and tremors occur well inland from the coastline and follow closely the estimated location of the mantle wedge corner. The transition zone, between the locked portion of the megathrust and the zone of tremors, is creeping rather steadily at the long-term slip rate and probably buffers the effect of SSEs on the megathrust seismogenic portion.

Key words: Cascadia megathrust, slow slip events, interseismic coupling, variational bayesian independent component analysis.

Electronic supplementary material The online version of this article (https://doi.org/10.1007/s00024-018-1991-x) contains supplementary material, which is available to authorized users.

[1] Bullard Laboratories, Department of Earth Sciences, University of Cambridge, Madingley Road, Cambridge, Cambridgeshire CB3 0EZ, UK. E-mail: sylvain_michel@live.fr

[2] Department of Geology and Planetary Sciences, California Institute of Technology, 1200 E California Blvd, Pasadena, CA 91125, USA. E-mail: adriano.geolandi@gmail.com; avouac@gps.caltech.edu

[3] Jet Propulsion Laboratory, California Institute of Technology, 4800 Oak Grove Dr, Pasadena, CA 91109, USA.

1. Introduction

The Juan de Fuca (JdF) plate subducts beneath the North American plate along the Cascadia megathrust off the coast of southwestern Canada and northwestern United States (Fig. 1). This subduction system is a typical warm case example, due to the subducting plates young age (< 10 Ma) and moderate convergence rates (27–45 mm/year) (e.g. Hyndman et al. 1997; DeMets and Dixon 1999).

The Cascadia megathrust hosted an $M \sim 9$ earthquake in 1700, producing a tsunami that was reported in Japan (Satake et al. 2003). Additional $M \sim 9$ in the last 10 000 years have been documented from turbidite sequences (Goldfinger et al. 2017), tsunami deposits and coastal geological evidence (Atwater 1987; Clague and Bobrowsky 1994; Atwater and Hemphill-Haley 1997; Kelsey et al. 2005; Wang et al. 2013; Kemp et al. 2018). However, since 1700, the megathrust has remained relatively silent. At present, the background seismicity is very quiet except at its Northern and Southern ends (Wang and Trehu 2016).

The interseismic loading retrieved from surface geodetic measurements allowed to estimate the degree of locking of the megathrust and create maps of interseismic coupling (defined as the ratio of slip deficit rate along the megathrust, determined from the geodetic interseismic data, and the long-term slip rate). Interseismic coupling maps are useful for seismic hazard assessment as locked portions of the megathrust are indeed expected to potentially slip in large interplate earthquakes as has been observed in several seismically active regions (e.g. Chlieh et al. 2008; Moreno et al. 2010; Loveless and Meade 2011). Previous inversions of geodetic strain rate measured onshore indicate that the megathrust is locked to some degree in the interseismic period

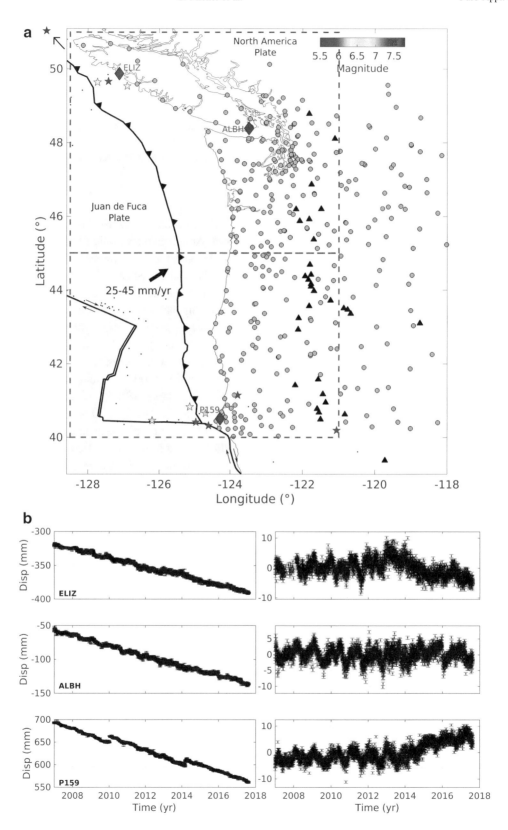

◀Figure 1
a Distribution of GPS stations (yellow dots) along the Cascadia subduction zone. The black lines are the plate boundaries and the black triangles indicate active volcanoes locations (USGS). The black dots indicate M5.5 + seismicity from 1984 to 2017 (ANSS catalogue). The stars indicate the 11 earthquakes that affected the GPS stations, the colour indicating their magnitude. The red diamonds indicate, from North to South, the location of the stations ELIZ, ALBH and P159. The two dashed red rectangles delimit the North and South sections as indicated in Sect. 3.1.2. **b** Left panels indicate the East component raw time series for the stations ELIZ, ALBH and P159. The right panels show the detrended, time series of those 3 stations corrected for offsets and regional tectonics using the block model of Schmaltze et al. (2014)

(Hyndman et al. 1997; Wang et al. 2003). We use the secular velocity field from 352 continuous GPS (cGPS) stations corrected for post-glacial rebound using the model of Peltier et al. (2015), to determine an interseismic coupling map for the 2007–2017 period. We also determine temporal variations of slip along the megathrust over that time period to clarify the relative position of the portions of the megathrust that are locked in the interseismic period and those that are producing transient slip events. The geodetic position time series from Cascadia show strong temporal variations which indicate episodic aseismic slip events, commonly called Slow Slip Events (SSEs) (Dragert et al. 2001; Miller et al. 2002). These slip events are mostly aseismic but are correlated in time and space with tremors (Rogers and Dragert 2003; Wech and Bartlow 2014), which presumably occur on the plate interface (Wech and Creager 2007). It is, therefore, likely that SSEs are related to episodic slip along the megathrust. SSEs are, however, not the only source of variation of the geodetic time series from their secular trend. These variations can be related to various causes. The larger earthquakes in the area may have caused some offsets and transient post-seismic deformation. The geodetic time series also show variations that are seasonal and likely due to surface load variations as has been observed in various other settings (Blewitt et al. 2001; Bettinelli et al. 2008; Fu et al. 2012). Temporal variations in geodetic series can also be spurious due to the ITRF realization generally used to express position in a global reference frame (Dong et al. 2002).

In this study, we analyse the geodetic position time series from Cascadia with the objective of separating the terms due to the various temporally varying factors and to interseismic coupling. To that effect, we apply to the detrended position time series a variational Bayesian Independent Component Analysis (vbICA) (Choudrey and Roberts 2003), a signal processing technique which has proven to be effective in extracting different sources of deformation (afterslip, SSEs, and seasonal signals) in geodetic time series (Gualandi et al. 2016, 2017a, b, c). We next perform an inversion of the Independent Components (ICs) related to SSEs, following the general approach of the principal component analysis-based inversion method (Kositsky and Avouac 2010). With this vbICA-based Inversion Method, (vbICAIM), we

Table 1

Earthquakes affecting GPS time series from the North and South section (USGS)

Section	Magnitude	Year	Month	Day	Latitude	Longitude	Depth
North	6.4	2011	9	9	49.535	126.893	22
	7.8	2012	10	28	52.788	132.101	14
	5.5	2013	8	4	49.661	127.429	10
	6.5	2014	4	24	49.639	127.732	10
South	6.5	2010	1	10	40.652	124.693	28.7
	5.9	2010	2	4	40.412	124.961	23
	5.6	2012	2	13	41.143	123.79	27.4
	5.7	2013	5	24	40.192	121.06	8
	6.8	2014	3	10	40.829	125.134	16.4
	5.7	2015	1	28	40.318	124.607	16.9
	6.6	2016	12	8	40.454	126.194	8.5

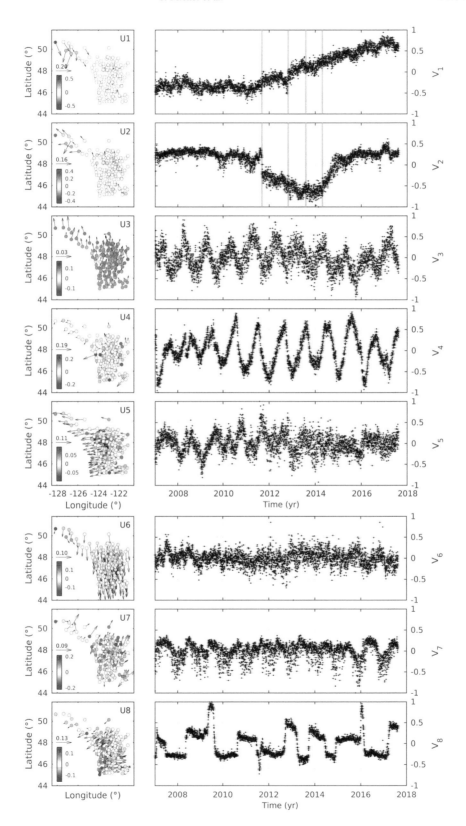

◄Figure 2

Spatial and temporal functions of the independent components obtained from applying a vbICA decomposition to the data from the Northern area (see location in Fig. 1) for 8 components. The left panels show the spatial functions (matrix U). The arrows and the coloured dots indicate horizontal and vertical motion, respectively. The right shows the temporal functions (matrix V). The green lines indicate the timing of the earthquakes of the northern section as indicated in Table 1. We consider components 1 and 2 to be related to post-seismic displacements following earthquakes in the North (Fig. 1 and Table 1)

produce a kinematic model of the spatio-temporal variation of slip on the Megathrust from 2007 to 2017. We also determine the pattern of interseismic coupling along the megathrust from inverting the secular geodetic signal due to interseismic strain.

The remaining of the manuscript is organized as follows. We first introduce the data considered in this study in Sect. 2. We then describe how we extract the signals related to the different sources of interest. In Sects. 4 and 5, we present the results of the inversion to retrieve interseismic coupling and the SSEs kinematics, respectively. Final remarks and further discussions conclude the manuscript.

2. Data

We use daily sampled position time series in the IGS08 reference frame from the Pacific Geodetic Array (PANGA) and the Plate Boundary Observatory (PBO) maintained by UNAVCO and processed by the Nevada Geodetic Laboratory (http://geodesy.unr. edu, last access August 2017). Most of the available continuous GPS (cGPS) stations were deployed in 2007, and we consider the time range that goes from 2007.0 to 2017.632. We use only time series with at most 40% of missing data and we exclude all the stations in the proximity (< 15 km) of volcanoes to avoid contamination by volcanic signals. We also discard station BLYN because of spurious large displacements of unknown origin that were clearly not observed at nearby stations. The final selection includes $N_{GPS} = 352$ cGPS stations (Fig. 1a). We then refer all the stations to the North America reference frame using the regional block model of Schmaltze et al. (2014). The position time series are then organized in a $M \times T$ matrix X_{obs}, where

$M = 3 \times N_{GPS}$ is the total number of time series (East, North, and Vertical direction per each station), and $T = 3883$ is the total number of observed epochs.

Two tremor catalogues are used in this study, one from Ide (2012) between 2007 and 2009.595, and one from the Pacific Northwest Seismic Network (PNSN) catalogue from 2009.595 to 2017.632 (https://pnsn. org/tremor).

3. Signal Extraction

Our goal is to retrieve: (1) the secular strain rate due to the average pattern of locking of Cascadia's megathrust, and (2) the SSEs history on the megathrust in the considered time span.

To attain our first goal, we use the long-term linear trend estimated from a 'trajectory' model (Equation S1) in which each position time series is modelled using a combination of predetermined functions. We use a linear combination of a linear trend, annual and semi-annual terms, co-seismic step functions and exponential post-seismic functions. This approach provides a reasonable estimate of the linear trend with limited bias introduced by the non-linear terms. We do not model at this step the SSEs, but they do have a limited influence on the secular motion estimate since their typical returning period is much smaller than the overall observed time span, as it will be verified a posteriori. The secular geodetic velocities estimated from this approach are used to determine interseismic locking ratio on the Cascadia megathrust (Sect. 4). We use the long-term linear trend and the offsets that are simultaneously estimated by the trajectory model (Equation S1 and Supplementary Material) to correct the position time series. These time series are then the input for the variational Bayesian Independent Component Analysis (vbICA) algorithm. A brief description of the vbICA algorithm is offered in the next Sect. 3.1.1. We then describe its application to the extraction of post-seismic relaxation (Sect. 3.1.2), seasonal deformation (Sect. 3.1.3), and SSEs displacement (Sect. 3.1.4).

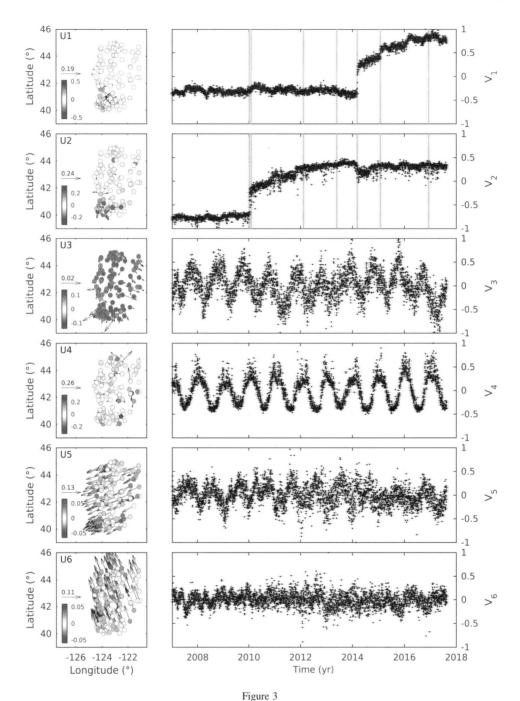

Figure 3
Spatial and temporal functions of the independent components obtained from applying a vbICA decomposition to the data from the Southern area (see location in Fig. 1) for 6 components. The left panels show the spatial functions (matrix U). The arrows and the coloured dots indicate horizontal and vertical motion, respectively. The right panels indicate the components temporal evolution (matrix V). The green lines indicate the timing of the earthquakes of the southern section as indicated in Table 1. We consider components 1 and 2 to be related to post-seismic displacements following earthquakes in the South (Fig. 1 and Table 1)

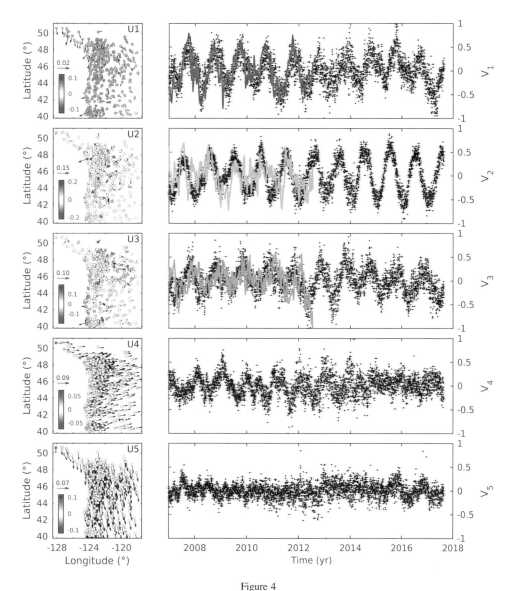

Figure 4
Spatial pattern and temporal evolution of each component of an 11-component decomposition (Sect. 3.1.3.). The left panels show the spatial pattern (matrix U). The arrows and the coloured dots indicate horizontal and vertical motions, respectively. The right panels show the components temporal functions (matrix V). The red, blue and green lines in the temporal evolution of components 1, 2 and 3 correspond to the temporal evolution of the components 1, 3 and 2, obtained from the decomposition of the theoretical geodetic time series due to surface load variations derived from GRACE. For clarity, they are plotted (right panels) separately and together with the temporal function of their associated components derived from the GPS time series (black dots). We consider components 1–3 to be seasonal deformation, 4 and 5 to be common mode, 6 to be a local effect, 7 to be seasonal scattered noise and 8–11 to be related to SSEs

3.1. Sse Surface Deformation Extraction

3.1.1 vbICA Algorithm

The vbICA algorithm allows the centered position time series (X) to be reconstructed by a linear combination of a limited number (R) of Independent Components (ICs). The centring step is performed by removing from a time series its mean value. Each IC is fully characterized by a spatial distribution (U_r), a temporal function (V_r), and a relative weight (S_r). None of these quantities is a priori determined, allowing the data to reveal their inner structure. These

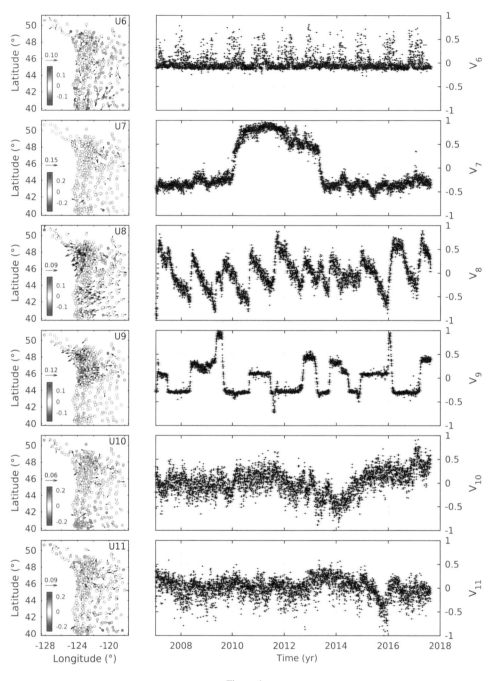

Figure 4
continued

quantities are organized in matrices, such that the following approximation holds (Gualandi et al. 2016):

$$X_{M \times T} \approx U_{M \times R} S_{R \times R} V_{R \times T}^{T} + N_{M \times T} \qquad (1)$$

where U and V have unit norm columns, S is a diagonal matrix, and N characterizes the noise, assumed to be Gaussian. Recall $M = 3 \times N_{GPS}$ is the total number of time series (East, North, and Vertical direction per each station) and $T = 3883$ is the total number of observed epochs. This notation is similar to that of Principal Component Analysis (PCA), but here the constraint of orthogonality between the columns of U is relaxed, as well as between the columns of V. Moreover, any ICA algorithm is not aiming at the diagonalization of the variance–covariance matrix of the data and, thus, the diagonal values in S do not represent a percentage of the dataset variance explained. Both U and V are non-dimensional, while S and N are in the same unit as X (mm). In contrast to standard ICA algorithms (e.g. FastICA, Hyvarinnen and Oja, 1997), vbICA allows for more flexibility in the description of the temporal sources V since it can model multimodal time function distributions. More details can be found in Gualandi et al. (2016) and references therein.

3.1.2 Post-seismic Relaxation

The largest deformation signal in the corrected position time series is due to post-seismic relaxation following the earthquakes listed in Table 1.

Due to the complexity of the tectonically active region in the proximity of northern Vancouver island in the North and of the triple junction point in the South (Table 1), it is challenging to obtain a clean source separation using the data from the entire network of stations. A first attempt to apply the vbICA to the whole dataset did not bring a clean separation between post-seismic events occurring at the northern and southern edges of the subducting plate. We, therefore, decided to first divide the study area into two sectors, split by the latitude 45° and limited to the east by the longitude − 121°, as shown in Fig. 1a.

We run a first vbICA on each section to extract the post-seismic deformation, and we retrieve 8 and 6

ICs for the northern and southern section, respectively (Figs. 2 and 3). The amount of variance explained by the decomposition is over 68.27%, considered here as a threshold for the selection of the number of components to be retained. For the southern section, we interpret IC1 and IC2 as related to post-seismic deformation in the region. The temporal functions show clear non-linear patterns starting at the time of occurrence of the large earthquakes (Fig. 3). Moreover, their spatial functions (U) show a signal localized in the southwestern section, in the proximity of the earthquake epicentres. The same argument is valid for the analysis of the northern section, where IC1 and IC2 are interpreted as post-seismic sources (Fig. 2). We admit that we cannot isolate the contribution of every single post-seismic deformation episode, and certainly there is cross-talk between the post-seismic ICs. Nonetheless, due to the spatially localized responses, we are confident that the information thus extracted is referring to the post-seismic deformation.

In the South, six M5.5 + events (Table 1) have visibly induced detectable geodetic signals (stars in Fig. 1a and green lines in Fig. 3) and all are strike-slip events (USGS). In the North, similarly, three strike-slip M5.5 + earthquakes have occurred and affected GPS stations (Table 1, Figs. 1a, 2). There is one $M7.8$ with a reverse mechanism that occurred in 2012 (Table 1, star out of frame in Figs. 1a and 2), which might have induced post-seismic deformation on the megathrust or on a crustal fault of the North American plate (Hayes et al. 2017). The mainshock occurred 363 km from the closest GPS station (HOLB). Its rupture extended ∼ 100–150 km along strike. This event seems to have affected the northernmost stations. In either case, we make a distinction between afterslip and SSEs and thus correct our dataset X for the displacement associated with all the retrieved post-seismic ICs. The stations corrected for post-seismic deformation are listed in Table S1.

3.1.3 Seasonal Deformation

After the removal of post-seismic deformation, the largest source of deformation is due to seasonal effects. Several mechanisms can be at the origin of seasonal deformation, but the strongest is related to

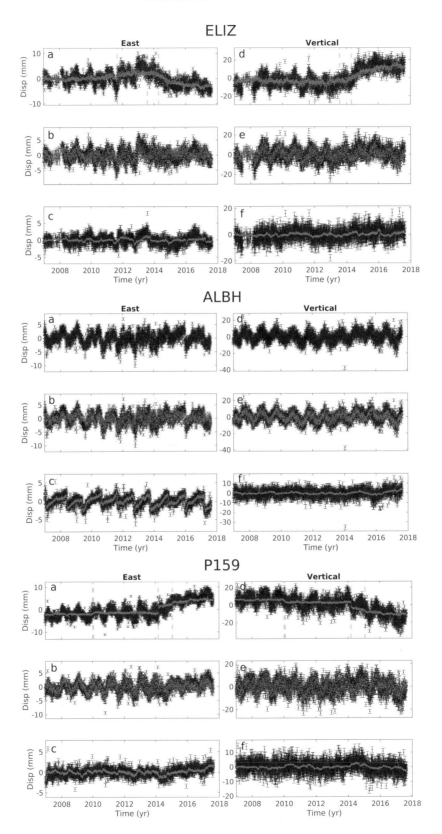

◀Figure 5

Comparison between the corrected GPS time series used as input for the vbICA decomposition and the modelled signal for station ELIZ, ALBH and P159, at each correction step (Sects. 3.1.2., 3.1.3. and 3.1.4.). The left and right panels indicate the East and vertical directions of the GPS time series, respectively. For each station, **a** and **d** show the detrended GPS time series (black dots and associated errorbars) and the displacement modelled from the ICs related to post-seismic deformation (red line). **b** and **e** show the detrended GPS time series corrected from post-seismic displacement (black dots and associated errorbars) and the displacement modelled from the ICs related to noise, and seasonal and local effects deformations (red line). **c** and **f** show the detrended GPS time series corrected from noise and post-seismic, seasonal and local effects deformation (black dots and associated errorbars) and the displacement modelled from the ICs related to SSEs (red line). ALBH has no post-seismic ICA model since we did not correct any post-seismic deformation at this station

the load induced by surface mass variations (e.g. Blewitt et al. 2001; Dong et al. 2002; Bettinelli et al. 2008). Thanks to satellite gravity records from the Gravity Recovery and Climate Experiment (GRACE) (Tapley et al. 2004), it is possible to estimate the load in equivalent water thickness (EWT). From this dataset, and assuming a certain composition of the

Earth, it is possible to model the predicted seasonal deformation due to this source (e.g. Chanard et al. 2014, 2018). We perform a vbICA on both the GPS position time series and the deformation predicted from the GRACE data at the same locations (Fig. 4). From the GPS decomposition, we identify components 1, 2 and 3 as seasonal sources of deformation. A decomposition with 3 ICs on the GRACE-derived time series explains 100% of the variance. A comparison of the temporal evolution of the GPS-derived and GRACE-derived ICs shows a good correlation (> 60%) between the first GPS-derived and GRACE-derived components (Fig. 4) (Table S2). We conclude that the first GPS-derived component is most likely associated with hydrological loading. The other two seasonal GPS-derived components have poor correlations with any GRACE-derived component and thus cannot be interpreted as related to surface hydrology. Their origin is unknown and might be due to other seasonal non-tectonic effects here not taken into account [e.g. thermal strain (Xu et al. 2017)]. Given the uniform spatial pattern of ICs

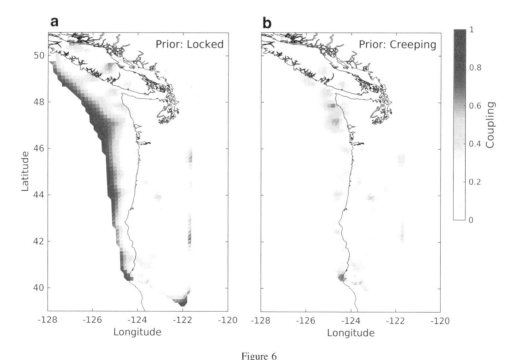

Figure 6

Interseismic coupling maps obtained with different a priori hypothesis on fault coupling. Interseismic coupling is defined as the ratio of the slip rate deficit derived from the modelling of interseismic geodetic strain over the long-term slip rate on the megathrust predicted by the block model. **a** The megathrust is assumed fully locked a priori. **b** The megathrust is a priori assumed to be creeping at the long-term slip rate. The black line corresponds to the coastline

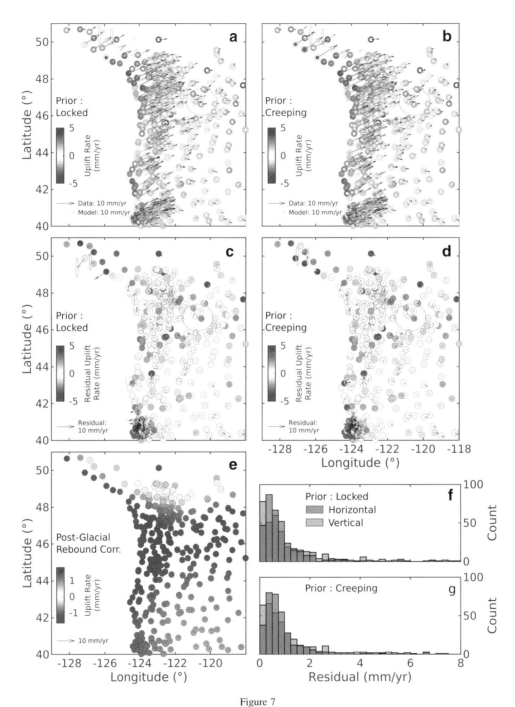

Figure 7
a and **b** show the comparison between the observed and modelled secular rates for the models assuming a priori that the megathrust is locked and creeping, respectively. **c** and **d** show the corresponding residuals assuming locking or creep a priori, respectively. **e** shows the post-glacial rebound correction based on Peltier et al. (2015). The arrows and the coloured dots indicate horizontal and vertical motion, respectively. The inner and outer dots in (**a**) and (**b**) represent the data and model vertical secular rates, respectively. The green ellipses indicate 1-sigma uncertainties. **f** and **g** show histograms of the residuals assuming locking or creep a priori, respectively

4 and 5 (Fig. 4), we make the hypothesis that they are related to a network Common Mode Error (CME). IC6 (Fig. 4) can be described as a seasonal scatter, potentially due to the effect of snow and consequent multipath, and is considered as noise. The spatial pattern of IC7 (Fig. 4) is very localized and we assume that it is a local effect. The deformation related to the ICs from 1–7 is thus removed from the time series, which have at this point been corrected from inter-block motion, long-term linear tectonic motion, co-seismic and instrumental offsets, post-seismic relaxation, seasonal signals, network errors and local effect deformations.

3.1.4 SSEs Displacement

Finally, we apply a vbICA on the residual displacement time series. The SSEs signal is buried in the noise and the percentage of variance explained grows very slowly as the number of components increases. The number of components is chosen based on the Negative Free Energy (NFE) of the decomposition, which balances the fit to the data and the complexity of the model (Choudrey and Roberts 2003; Gualandi et al. 2016). The NFE indicates how close the approximating probability density function (pdf) of the hidden variables of the model is to the true posterior pdf. When the NFE reaches its maximum value, the Kullback–Leibler divergence between the aforementioned pdfs is minimum. When passing from 32 to 33 ICs the NFE decreases, and we thus select 32 components for our final decomposition (Fig. S1). Among these 32 ICs, we ascribe 15 of them to the kinematics of SSEs (components 1–15 in Fig. S2). The remaining 17 sources are considered as either noise (components 16–25) or local effects (components 26–32) (Fig. S2).

The comparison between the GPS position time series and the vbICA reconstruction at each extraction step (i.e. Sects. 3.1.2., 3.1.3. and 3.1.4.) is shown in Fig. 5 for the 3 stations indicated in Fig. 1.

4. Interseismic Coupling

We perform a static inversion of the velocity field estimated from the trajectory model. We correct the long-term linear trends from post-glacial rebound displacements using estimations from Peltier et al. (2015) (Fig. 7e). We use a downsampled version of the fault geometry determined by McCrory et al. (2012). The total number of sub-faults is $P = 3339$, and they have triangular shape, with a characteristic length of ~ 15 km. The Green's functions are calculated using Okada (1992) dislocation model. We use the same regularization scheme of Radiguet et al. (2011), which allows to give an interseismic back-slip a priori model as input. Two a priori interseismic back-slip models are tested, one with a fully locked fault (coupling equal to 1), and the other with a fully creeping fault (coupling equal to 0). We expect the poorly resolved areas, specifically near the trench where no data are available, to stay near the a priori model. Those two a priori models define thus the 2 extreme cases where the fault trench is either fully locked or creeping. A rake direction constraint is imposed using the plate rate direction from Schmalzle et al. (2014) block model. We adopt as smoothing parameter the a priori uncertainty on the model parameters σ_0, while fixing the correlation distance λ to the average patch size. Taking the locked a priori model, varying σ_0 produces the L-curve shown in Fig. S3. We select $\sigma_0 = 10^{-0.1}$ for our preferred inversion that yields the best trade-off between misfit and smoothness of the slip distribution. We use the same parametrization for the creeping a priori model. We calculate the interseismic coupling map dividing each sub-fault's back-slip rate, given by the inversion, by its associated block model plate rate. The coupling maps for the two cases are given in Fig. 6. The fit to the data and the residuals are shown in Fig. 7. The resolution and restitution maps are shown in Fig. S4. The sensitivity to the regularisation parametrisation is shown in Fig. S5.

The two interseismic coupling maps (Fig. 6) show a partially locked shallow portion of the fault. The position of the locked zones, however, differs from one model to the other. Interestingly, the a priori fully creeping fault has locked areas that are closer to the coast than the a priori fully locked fault.

5. Slow Slip Events Kinematics

5.1. SSES Inversion

Once the surface deformations associated with the SSEs are isolated, we perform a static inversion of the spatial distribution related to the 15 extracted temporal functions. This approach follows the one adopted in Gualandi et al. (2017a, b, c). The principle is the same as the one described in Kositsky and Avouac (2010), but a vbICA is replacing the PCA decomposition of the data. We invert the selected ICs on the same geometry used for the interseismic velocities, and we adopt the same regularization scheme. A null a priori model is imposed, which implies that we expect a null a posteriori solution where the data do not require for slip on the fault. No constraints on the rake direction are imposed. Varying σ_0 produces the L-curve shown in Fig. S6. We select $\sigma_0 = 10^{-1.5}$ for our preferred inversion that yields the best trade-off between misfit and smoothness of the slip distribution. We invert the 15 ICs using the same parametrization for each IC and recombine them linearly.

We thus obtain the spatio-temporal evolution of slip deficit related to SSEs ($\delta_{\text{deficit}}(p, t)$) with respect to the long-term slip deficit given by the locking ratio (see Sect. 4). In order to get a coherent slip deficit rate ($\dot{\delta}_{deficit}(p,t)$) time evolution ($t = 1,...,$ T) on each sub-fault ($p = 1, ..., P$), we take the time derivative of the low-pass filtered slip deficit. In particular, we apply an equiripple low-pass filter with passband frequency of $1/21$ days^{-1}, stopband of $1/35$ days^{-1}, passband ripple of 1 dB with 60 dB of stopband attenuation. If the slip deficit increases on a given patch (i.e. positive slip deficit rate), then that patch is accumulating strain, i.e. it is loading. When the slip deficit decreases (i.e. negative slip deficit rate), then the patch undergoes slip. SSEs thus correspond to time periods of negative slip deficit rate. A movie of the SSEs kinematics is available in the supplement (Movie S1). The sensitivity of the kinematic model to the regularisation parametrisation is also shown in the supplement (Movie S2, S3 and S4).

Figure 8 ▶
Catalogue of 64 detected SSEs which occurred between 2007 and 2017. The bottom panels indicate the normalized slip distribution of each SSEs based on the initial low-pass filtering of δ_{deficit} with a cutoff at 21 days^{-1}. The top panels with the red curve indicate their moment rate functions based on the δ_{deficit} 8-days window smoothing. The pink shading represents the moment rate function uncertainty based on the standard deviation of the mean on the rough version of δ_{deficit} using a 16-days moving window centered in the epoch of interest

5.2. SSEs Spatial and Temporal Determination

To isolate each SSE in space and time, we proceed as follows. We consider only periods where slip deficit rate is negative, thus, associated with the SSEs unloading. To enable the detection of SSEs over the noise level of our kinematic model, we make the hypothesis that at a given epoch t a given sub-fault p is experiencing an SSE if $\dot{\delta}_{\text{deficit}}(p, t) < V_{\text{thresh}} = -40$ mm/year. For the specific epoch under exam, we identify all the sub-faults with slip rates below this threshold by applying a contour at V_{thresh}. The number of contours defines the number of SSEs occurring at time t. We do the same at time $t + 1$ and verify which contours at time $t + 1$ overlap with the contours at time t. If there is an overlap, we assume that they are the same SSE. For every SSE, we thus automatically select a starting and ending time (t_k^{in} and t_k^{fin}, with $k = 1, ..., N_{\text{SSEs}}$). Special cases can arise where an SSE splits itself or two SSEs merge together. In both scenarios, the SSEs at time t and $t + 1$ are considered to be the same. With this procedure, we identify 119 potential SSEs between 2007 and 2017. From those 119, some involve only 1 or 2 sub-faults and are very short in time. We consider them as noise. We also remove any candidate under 60 km depth. Our final selection of SSEs gathers $N_{\text{SSEs}} = 64$ events shown in Fig. 8.

With this procedure, we are able to estimate the temporal and spatial evolution of each SSE. To get their moment rate function, we proceed as follows. Since SSEs can migrate, we estimate the full area involved in the k-th event as the union of all the sub-faults participating to it, denoted by the set $\{s_k\} = \left\{ p | \dot{\delta}_{\text{deficit},k}(p, t) < V_{\text{thresh}} \text{ for } t_k^{\text{in}} \le t \le t_k^{\text{fin}} \right\}$. However, the V_{thresh} parameter might truncate

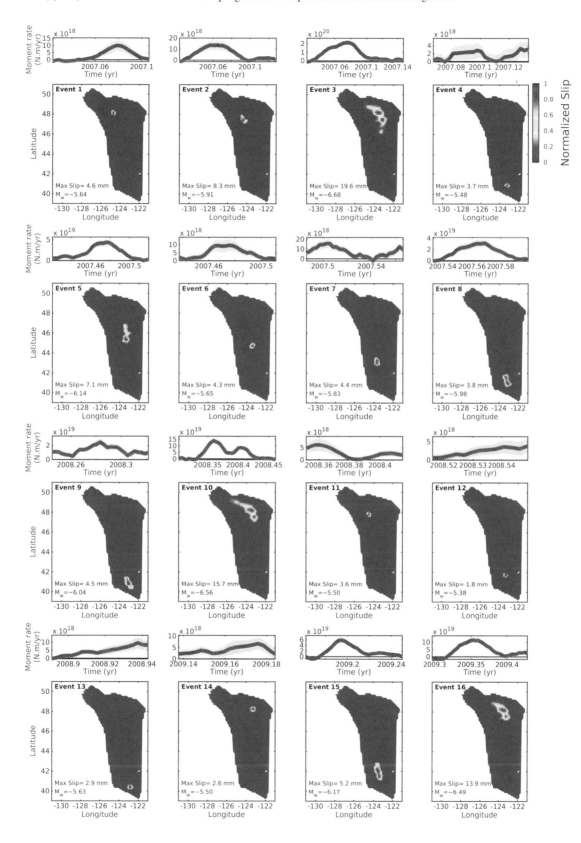

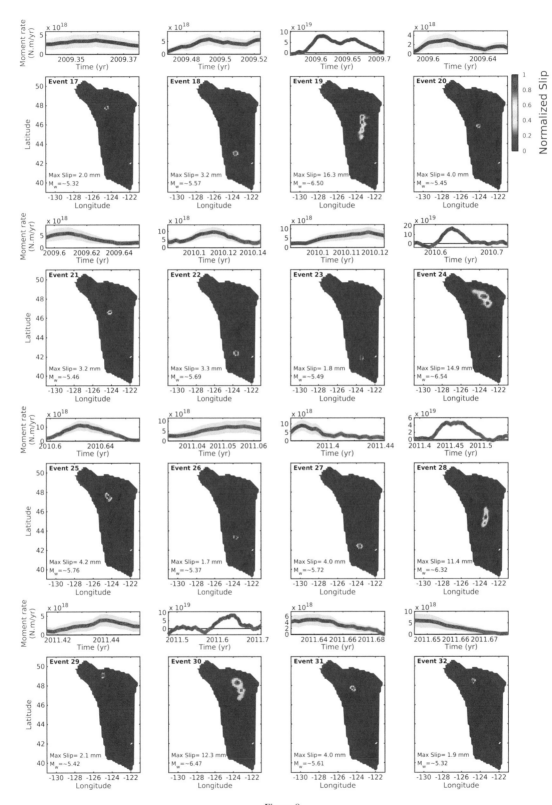

Figure 8
continued

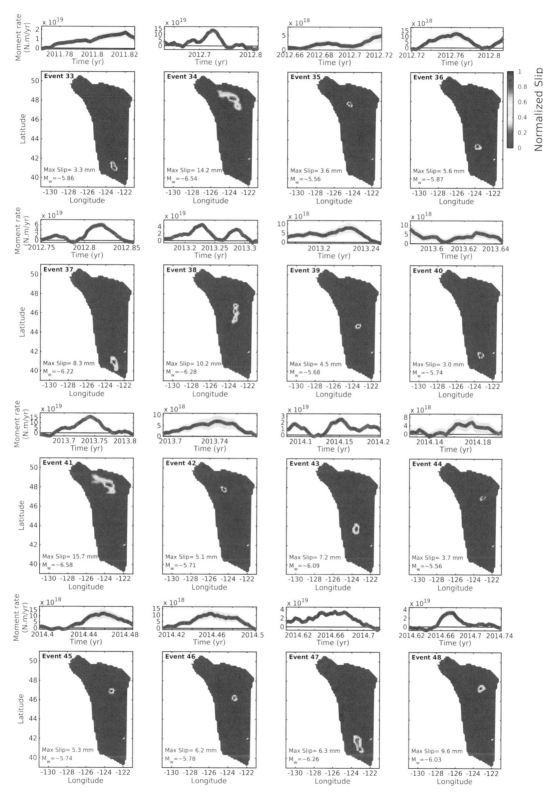

Figure 8
continued

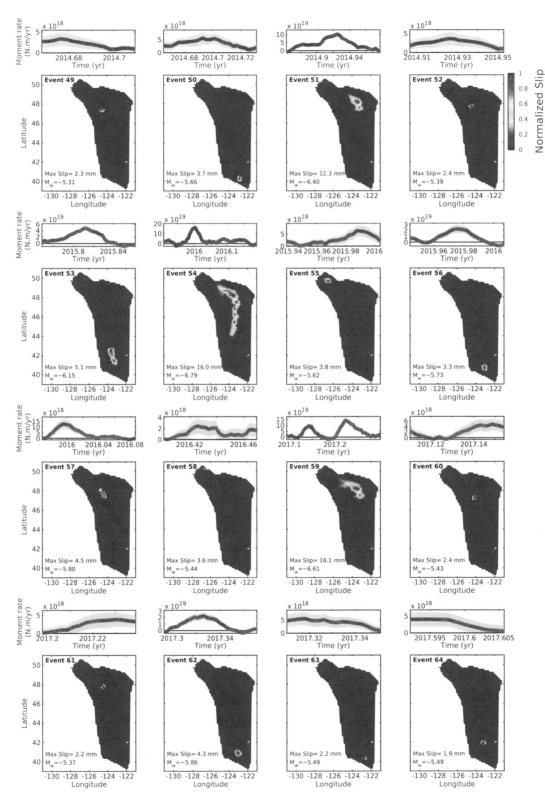

Figure 8
continued

spatially and temporally our estimated SSEs. To relax slightly this constrain, we extend the spatial influence of each SSE to their neighbouring sub-faults and extend their time delimitation by 1 day before and after. Note that during and within the area of influence of each SSE, the slip deficit rate of sub-faults can be negative or positive depending on the SSE's history of propagation. For one event, within its area of influence and time period, we calculate the moment rate function at time t, $\dot{M}_0(t)$, using equation:

$$\dot{M}_{0,k}(t) = -\sum_{s_k=1}^{S_k} \mu A_{s_k} \dot{\delta}_{\text{deficit},k}(s_k, t), \quad \text{for} \quad (2)$$
$$t_k^{\text{in}} \leq t \leq t_k^{\text{fin}} \quad \text{and} \quad \text{with} \quad k = 1, \ldots, N_{\text{SSEs}},$$

where μ is the shear modulus (here fixed to 30 GPa), S_k is the total number of sub-faults belonging to the k-th SSE's area of interest, A_{s_k} is the area of sub-fault s_k. The negative sign in front of the sum is added in order to have positive moment rates during SSEs, reference that we will keep for the rest of this study.

Applying this moment rate calculation methodology on the filtered δ_{deficit} described in Sect. 5.1 (passband frequency of 1/21 days^{-1}, stopband of 1/35 days^{-1}) results, however, in very smoothed moment rate functions. To retrieve more detailed moment rate functions, we perform instead a zero-phase digital filtering on the rough δ_{deficit} using a 8-days window, but focus only on the area and time period of each SSE as estimated before from the very smoothed data. The moment rate of each SSE acquired from the rougher version of δ_{deficit} is shown in Fig. 8. Those moment rates do not take into account interseismic loading during SSEs. To estimate the uncertainty on the SSEs moment rate function, we assume that it is represented by their short-term variability before filtering. The uncertainties represented in Fig. 8 are calculated based on the standard deviation of the mean on the rough version of δ_{deficit} using a 16-day moving window centered in the epoch of interest.

To evaluate SSEs propagation speed from our kinematic model, we estimate SSE propagation front locations as follows. A representative line of the average along-strike location of SSEs is first chosen (red line in Fig. 9a, Table S3). The intersection between this line and the contour of the cumulative slip area of SSEs at each time step ($\dot{\delta}_{\text{deficit}} < V_{\text{thresh}}$) defines the position of the SSE propagation fronts. The distance, along the SSEs location representative line, between the propagation fronts and the onset location of each SSE is then calculated at each time step. The onset location of SSEs is assumed to be the projected barycentre of their first slip area contour on the SSE location representative line. Figure 9 shows the position of 14 SSEs front as a function of time.

5.3. SSEs Kinematics

Our SSE catalogue (Fig. 8) contains events with moment magnitude from 5.3 to 6.8 and duration between 14 and 106 days. Their propagation speed is of the order of ∼ 2000–4000 km/year (respectively, ∼ 5.5–11 km/day) (Fig. 9).

The spatio-temporal evolution of the tremors and SSEs is systematically correlated (Movie S1) as shown for example for event 34 in Fig. 10, as had already been shown in some previous studies (Wech and Bartlow 2014). The SSEs located slightly offshore, along the coast between 47° and 49° of latitude (e.g. events 1, 2), are most likely inversion artefacts, since a limited number of stations are present in the offshore portion of the megathrust.

From our catalogue (Figs. 8 and 11c), we observe that SSEs often rupture the northern (i.e. event 10, 16, 24, 34, 41, 51, and 59), centre (i.e. events 5, 19, 28 and 38) or southern (i.e. events 8, 9, 15, 33, 37, 47 and 53) segments of Cascadia independently, but sometime multiple segments are ruptured during a single event (i.e. event 54). Segments are thus not completely independent from one another (Schmidt and Gao 2010; Wech and Bartlow 2014). We also observe that SSEs can rupture segments from either North to South or South to North, showing that the direction of propagation is variable (Schmidt and Gao 2010). Some events propagate also bilaterally (Schmidt and Gao 2010; Dragert and Wang 2011; Wech and Bartlow 2014) as is observed for the event 24 in our inversion (Movie S1).

The combined moment rate of all the SSEs from our catalogue between 2007 and 2017 is shown in Fig. 11a. Two endmembers are shown, one assuming that interseismic loading is negligible during SSEs and the other assuming that the fault is loaded at the

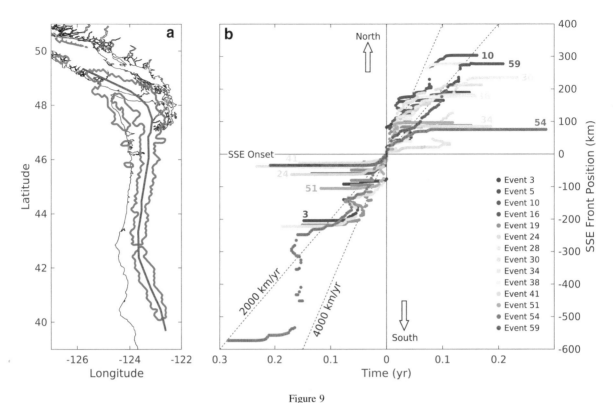

Figure 9

SSE front position evolution of 14 events. **a** Map showing the representative line of the average along-strike location of SSEs (red line). The intersection between this line and the SSEs cumulative slip contour at each time step indicates location of the SSEs fronts. The blue line indicates the cumulative slip area contour of all SSEs. The black lines indicate the coast. **b** SSEs front positions as a function of time in reference to the SSEs onset locations. Positive position is to the North, negative to the South. Time axes are symmetric (positive on both sides), recording the time after the SSE onset. The two green dashed lines are reference propagation rates of 2000 (∼ 5.5 km/day) and 4000 km/year (∼ 11 km/day)

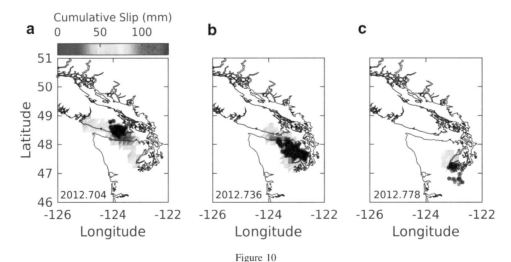

Figure 10

Three snapshot of the slip cumulated over 1 day during the propagation of SSE 34. The black dots correspond to tremors for the same days. The black lines indicate the coast. The bottom left number corresponds to the date. The tremors closely track the propagation of the SSE

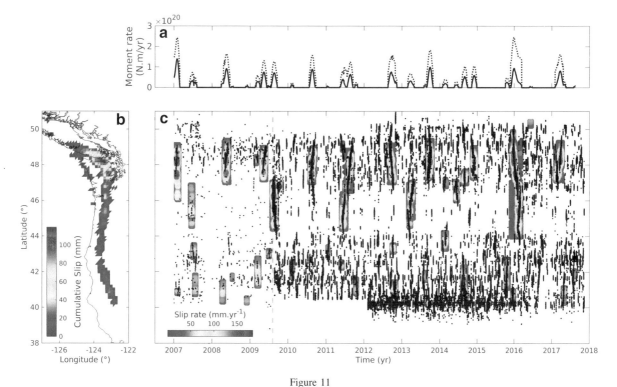

Figure 11

a Combined moment rate functions of all the detected SSEs from our catalogue. The continuous and dashed black lines correspond to the moment rate taking and without taking into account interseismic loading during SSEs, respectively. To place an upper bound on the moment release during SSEs, the dashed lines are calculated by comparison with the moment deficit that would have occurred during each SSE had the fault remained fully locked. Those moment rate functions are based on the low-pass filtered δ_{deficit} with the passband at 21 days^{-1}. **b** SSEs cumulative slip. **c** Occurrence of SSEs (colour shading) as a function of time. The black dots indicate tremors. The catalogue from Ide (2012) is used until 2009.595, the catalogue from PNSN is used thereafter

long-term plate rate during this period. They represent the two possible extremes of moment rate released by the SSEs from our catalogue. Bias in segment determination is introduced from the selection of V_{thresh}. For example, certain SSEs might not be detected or a same SSE could have been cut into pieces (e.g. events 28 and 30 in 2011). Increasing V_{thresh} to -35 mm/year, we retrieve instead 81 SSEs, several events are merged (events 1,2 and 3 in 2007 or 28 and 30 in 2011), smaller events appear, but the global dynamics remains unchanged (Fig. S7). Additionally, increasing V_{thresh} also increases the risk of introducing noise.

5.4. Potential Biases and Limitations

The combination of the uncertainty on GPS measurements, the low-pass filtering, the value of V_{thresh} and the interpreted fault geometry, specifically in the North, hinders our possibility to retrieve small

SSEs that remains within the noise level. In the most Northern part of the fault, our kinematic model is difficult to interpret due to the proximity of the fault's border which is prone to noise from the inversion. The southern part of Cascadia seems more prone to having small SSEs, which are more difficult to detect. This can be due to the slower convergence rate of the region (27 mm/year) compared to the northern segments (45 mm/year). The spatial distribution of GPS stations plays also a role in that matter.

Two opposite effects may affect the duration estimation of the SSEs. The temporal slip deficit filtering tends to augment SSEs duration. On the other hand, the selection of small enough (i.e. large enough in absolute value, in order to limit the effects of noise) V_{thresh} cuts the onset (t_k^{in}) and end (t_k^{fin}) of SSEs. Moreover, it additionally impacts SSEs' spatial extent, disregarding areas that do not slide fast enough. The value of the estimated peak slip during an SSE (Fig. 8) depends on the regularization of the inversion. The fact that the inversions are ill-posed and require regularization thus probably explains why different inversions can yield significantly different peak slip values. For example, the SSEs peak slip reported in our study ($\sim$ 1.5 cm) are smaller than the peak slip indicated in Dragert and Wang (2011) ($\sim$ 3–5 cm) for the May 2008 event.

There is also a possibility that the number of selected ICs used for the SSE model (Sect. 3.1.4) biases the kinematic description. In our case, 15 components were selected to describe the kinematics. However, the selection is initially based on a pool of 32 components with 17 components seemingly noise. Increasing or decreasing the total number of components produced by the vbICA changes slightly the number of ICs related to SSEs, but does not change qualitatively the dynamics observed in our model. On the contrary, by increasing the total number of ICs, the vbICA extracts further components estimated as local effects or noise. We tested up to 44 ICs.

The uncertainty on the position of the SSEs propagation front also depends on the factors mentioned above (i.e. GPS measurements, low-pass filtering, value of V_{thresh}, interpreted fault geometry and number of ICs). Additionally, the choice of the representative line of the average along-strike location of SSEs also plays a role, even though minor. Note that the initial phase of the SSEs front positions might represent more a shadow of the slip deficit rate of SSEs decreasing under V_{thresh} rather than the actual SSEs front propagation. The values of SSE front propagation rate estimated from our kinematic model are anyways similar to the ones estimated from tremor propagation since SSEs and tremors are most often correlated.

6. Implications and Conclusion

We determined the secular model of interseismic coupling on the megathrust (the time-average pattern of locking ratio) and the kinematics of SSEs from 2007 to 2017 in a consistent and coherent way. The vbICA has shown to be very effective at separating the SSEs from the various other sources of temporal variations in the position time series, such as those due to surface hydrology, post-seismic signals, local effects or common mode motion.

With our vbICA-based Inversion Method (vbI-CAIM), we were therefore able to describe the kinematics of fault slip on the Cascadia megathrust between 2007 and 2017 using 352 cGPS position time series. Our kinematic model provides the most comprehensive view of the SSEs along Cascadia. We were able to produce a catalogue of 64 events over the 2007–2017 time period. As already documented in previous studies of SSEs (e.g. Bartlow et al. 2011; Wech and Bartlow 2014), we find a remarkable systematic correlation in space and time between SSEs and tremors.

Our analysis demonstrates a clear along-dip segmentation of the mode of slip along the megathrust (Fig. 12). Some locked zone is clearly needed

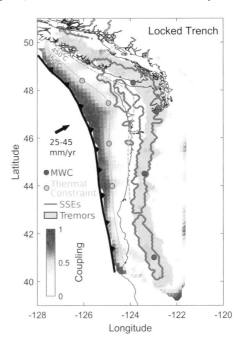

Figure 12
Interseismic coupling (white to red shading) and tremor (grey dots) distribution. The thin black curve indicates the coast. The blue contour indicates the area influenced by SSEs as measured in this study. The red dots indicate the position of the intersection between the forearc Moho and the megathrust, so the tip of the mantle wedge corner (MWC), determined from geophysical profiles along which dV_s/V_s and V_s was determined (POLARIS at the southern tip of the Vancouver Island (Nicholson et al. 2005), CASC93 in central Oregon (Nabelek et al. 1996; Rondenay et al. 2001), CAFE at Puget Sound (McGary et al. 2011; Abers et al. 2009), and FAME, BDSN and USArray/TA in the south of Cascadia). The solid and dashed green lines indicate the location of the 350 and 450 °C isotherms on the megathrust (Hyndman et al. 2015). The green dots represent the location of thermal constraints used to estimate the isotherms

between the trench and the coastline to account for the secular compression and the secular pattern of subsidence and uplift (Fig. 7). The resolution near the trench is poor and we cannot determine if the locked zone extends all the way to the trench or not. The models obtained assuming either a priori creep or locking of the megathrust fit the data equally well. They yield somewhat different position of the downdip limit of the locked zone (Fig. 6). It lies about $\sim$ 20–80 km updip of the coastline at a depth of about $\sim$ 5–15 km when the megathrust is assumed locked a priori (Fig. 6a). It lies much closer to the coastline, corresponding to a depth of about $\sim$ 10–30 km, if the megathrust is assumed to be creeping a priori (Fig. 6b). These observations are consistent with most previous studies of interseismic coupling (e.g. Wang et al. 2001; McCaffrey et al. 2000, 2007; Bruhat and Segall 2016). The portion of the megathrust that ruptures during large interplate earthquakes must lie mostly updip of such transition between locked and creeping sections. The lack of seismicity over the transition zone hints for a fully locked shallow portion of the fault (Wang and Trehu 2016). We would also expect earthquakes to nucleate along the zone of stress accumulation, at the transition between the locked and creeping region, as is observed for example in the context of other megathrusts (e.g. Cattin and Avouac 2000; Bollinger et al. 2004). The lack of seismicity in this zone where stress is accumulating the fastest is intriguing. It suggests that the interseismic stress buildup has not yet compensated the last stress drop event to trigger earthquakes there. It could also suggest that large interplate earthquakes penetrate deeper than the lower edge of the locked fault zone. Jiang and Lapusta (2016) demonstrated via numerical simulations of the seismic cycle under a rate-and-state friction framework that this mechanism would indeed produce a protracted period of seismic quiescence.

The zone of SSEs and tremors is relatively well resolved and, when compared with the interseismic coupling model (Fig. 12), clearly reflects a downdip segmentation of the mode of slip on the megathrust. This zone lies inland from the coastline and is clearly disconnected from the locked zone as already pointed out in some past studies (e.g. Gao and Wang 2017). In this transition zone, which spans between $\sim$ 100

km and $\sim$ 150 km away from the trench, fault creep is remarkably stationary resulting in a slip rate close to the long-term slip rate along the megathrust. This zone thus seems to act as a buffer isolating the seismogenic zone of the megathrust from the zone of SSEs. This buffering zone probably reduces the risk that an SSE triggers a large megathrust rupture (Segall and Bradley 2012).

The downdip segmentation of the mode of slip along megathrust has long been noticed and considered to reflect the influence of both temperature and lithology (Hyndman et al. 1997; Scholz 1998; Oleskevitch et al. 1999). These two factors could also explain the existence of two separate zones of unstable frictional sliding (Gao and Wang 2017). The one closer to the trench would correspond to a zone of rate-weakening frictional sliding controlled by the frictional properties of continental rocks. For quartzo-feldspathic rocks, friction indeed transitions from rate-weakening to rate-strengthening as temperature exceeds ~ 350 °C (Blanpied et al. 1995). In Cascadia, this temperature is reached at a depth shallower than the intersection with the forearc Moho, and therefore probably determines the downdip limit of the shallower zone of unstable sliding (Hyndman et al. 1997; Scholz 1998; Oleskevitch et al. 1999) (Fig. 12). The second region is instead located around the mantle wedge corner, where high-pore fluid pressure might result in low permeability of the serpentinized mantle (Wada et al. 2008). This hypothesis is consistent with the correlation between the zone of SSEs and tremors with the forearc Moho (red dots in Fig. 12) as proposed by Gao and Wang (2017).

Acknowledgements

We thank Kelin Wang for his constructive review of this manuscript. This study was partially supported by NSF award EAR-1821853 to JPA. We thank also Zachary E. Ross for his help. We thank Kristel Chanard for discussions and providing the times series predicted based on the surface mass variations derived from GRACE.

REFERENCES

Abers, G. A., MacKenzie, L. S., Rondenay, S., Zhang, Z., Wech, A. G., & Creager, K. C. (2009). Imaging the source region of Cascadia tremor and intermediate-depth earthquakes. *Geology, 37*(12), 1119–1122.

Atwater, B. F. (1987). Evidence for great Holocene earthquakes along the outer coast of Washington State. *Science, 236*(4804), 942–944.

Atwater, B. F., & Hemphill-Haley, E. (1997) *Recurrence intervals for great earthquakes of the past 3,500 years at northeastern Willapa Bay, Washington* (No. 1576). Washington, D.C.: USGPO.

Bartlow, N. M., Miyazaki, S. I., Bradley, A. M., & Segall, P. (2011). Space-time correlation of slip and tremor during the 2009 Cascadia slow slip event. *Geophysical Research Letters, 38*(18), 1–6.

Bettinelli, P., Avouac, J. P., Flouzat, M., Bollinger, L., Ramillien, G., Rajaure, S., et al. (2008). Seasonal variations of seismicity and geodetic strain in the Himalaya induced by surface hydrology. *Earth and Planetary Science Letters, 266*(3–4), 332–344.

Blanpied, M. L., Lockner, D. A., & Byerlee, J. D. (1995). Frictional slip of granite at hydrothermal conditions. *Journal of Geophysical Research-Solid Earth, 100*(B7), 13045–13064.

Blewitt, G., Lavallée, D., Clarke, P., & Nurutdinov, K. (2001). A new global mode of earth deformation: Seasonal cycle detected. *Science, 294*(5550), 2342–2345.

Bollinger, L., Avouac, J. P., Cattin, R., & Pandey, M. R. (2004). Stress buildup in the Himalaya. *Journal of Geophysical Research: Solid Earth, 109*(B11), 1–8.

Bruhat, L., & Segall, P. (2016). Coupling on the northern Cascadia subduction zone from geodetic measurements and physics-based models. *Journal of Geophysical Research: Solid Earth, 121*(11), 8297–8314.

Cattin, R., & Avouac, J. P. (2000). Modeling mountain building and the seismic cycle in the Himalaya of Nepal. *Journal of Geophysical Research: Solid Earth, 105*(B6), 13389–13407.

Chanard, K., Avouac, J. P., Ramillien, G., & Genrich, J. (2014). Modeling deformation induced by seasonal variations of continental water in the Himalaya region: sensitivity to Earth elastic structure. *Journal of Geophysical Research: Solid Earth, 119*(6), 5097–5113.

Chanard, K., Fleitout, L., Calais, E., Rebischung, P., & Avouac, J. P. (2018). Toward a global horizontal and vertical elastic load deformation model derived from GRACE and GNSS station position time series. *Journal of Geophysical Research: Solid Earth, 123*(4), 3225–3237.

Chlieh, M., Avouac, J. P., Sieh, K., Natawidjaja, D. H., & Galetzka, J. (2008). Heterogeneous coupling of the Sumatran megathrust constrained by geodetic and paleogeodetic measurements. *Journal of Geophysical Research: Solid Earth, 113*(B5), 1–31.

Choudrey, R., & Roberts, S. (2003). *Variational Bayesian mixture of independent component analysers for finding self-similar areas in images.* Nara: ICA.

Clague, J. J., & Bobrowsky, P. T. (1994). Evidence for a large earthquake and tsunami 100–400 years ago on western Vancouver Island, British Columbia. *Quaternary Research, 41*(2), 176–184.

DeMets, C., & Dixon, T. H. (1999). New kinematic models for Pacific-North America motion from 3 Ma to present, I: Evidence for steady motion and biases in the NUVEL-1A model. *Geophysical Research Letters, 26*(13), 1921–1924.

Dong, D., Fang, P., Bock, Y., Cheng, M. K., & Miyazaki, S. I. (2002). Anatomy of apparent seasonal variations from GPS-derived site position time series. *Journal of Geophysical Research: Solid Earth, 107*(B4), 1–17.

Dragert, H., & Wang, K. (2011). Temporal evolution of an episodic tremor and slip event along the northern Cascadia margin. *Journal of Geophysical Research: Solid Earth, 116*(B12), 1–12.

Dragert, H., Wang, K., & James, T. S. (2001). A silent slip event on the deeper Cascadia subduction interface. *Science, 292*(5521), 1525–1528.

Fu, Y., Freymueller, J. T., & Jensen, T. (2012). Seasonal hydrological loading in southern Alaska observed by GPS and GRACE. *Geophysical Research Letters, 39*(15), 1–5.

Gao, X., & Wang, K. (2017). Rheological separation of the megathrust seismogenic zone and episodic tremor and slip. *Nature, 543*(7645), 416.

Goldfinger, C., Galer, S., Beeson, J., Hamilton, T., Black, B., Romsos, C., et al. (2017). The importance of site selection, sediment supply, and hydrodynamics: A case study of submarine paleoseismology on the Northern Cascadia margin, Washington USA. *Marine Geology, 384,* 4–46.

Gualandi, A., Nichele, C., Serpelloni, E., Chiaraluce, L., Anderlini, L., Latorre, D., et al. (2017a). Aseismic deformation associated with an earthquake swarm in the northern Apennines (Italy). *Geophysical Research Letters, 44*(15), 7706–7714.

Gualandi, A., Perfettini, H., Radiguet, M., Cotte, N., & Kostoglodov, V. (2017b). GPS deformation related to the Mw7. 3, 2014, Papanoa earthquake (Mexico) reveals the aseismic behavior of the Guerrero seismic gap. *Geophysical Research Letters, 44*(12), 6039–6047.

Gualandi, A., Avouac, J. P., Galetzka, J., Genrich, J. F., Blewitt, G., Adhikari, L. B., et al. (2017c). Pre-and post-seismic deformation related to the 2015, Mw7. 8 Gorkha earthquake, Nepal. *Tectonophysics, 714,* 90–106.

Gualandi, A., Serpelloni, E., & Belardinelli, M. E. (2016). Blind source separation problem in GPS time series. *Journal of Geodesy, 90*(4), 323–341.

Hayes, G. P., Meyers, E. K., Dewey, J. W., Briggs, R. W., Earle, P. S., Benz, H. M., & Furlong, K. P. (2017). *Tectonic summaries of magnitude 7 and greater earthquakes from 2000 to 2015* (No. 2016-1192). Reston: US Geological Survey.

Hyndman, R. D., McCrory, P. A., Wech, A., Kao, H., & Ague, J. (2015). Cascadia subducting plate fluids channelled to fore-arc mantle corner: ETS and silica deposition. *Journal of Geophysical Research: Solid Earth, 120*(6), 4344–4358.

Hyndman, R. D., Yamano, M., & Oleskevich, D. A. (1997). The seismogenic zone of subduction thrust faults. *Island Arc, 6*(3), 244–260.

Ide, S. (2012). Variety and spatial heterogeneity of tectonic tremor worldwide. *Journal of Geophysical Research: Solid Earth, 117*(B3), 1–18.

Jiang, J., & Lapusta, N. (2016). Deeper penetration of large earthquakes on seismically quiescent faults. *Science, 352*(6291), 1293–1297.

Kelsey, H. M., Nelson, A. R., Hemphill-Haley, E., & Witter, R. C. (2005). Tsunami history of an Oregon coastal lake reveals a 4600 yr record of great earthquakes on the Cascadia subduction zone. *Geological Society of America Bulletin, 117*(7–8), 1009–1032.

Kemp, A. C., Cahill, N., Engelhart, S. E., Hawkes, A. D., & Wang, K. (2018). Revising estimates of spatially variable subsidence during the AD 1700 Cascadia earthquake using a Bayesian foraminiferal transfer function. *Bulletin of the Seismological Society of America, 108*(2), 654–673.

Kositsky, A. P., & Avouac, J. P. (2010). Inverting geodetic time series with a principal component analysis-based inversion method. *Journal of Geophysical Research: Solid Earth, 115*(B3), 1–19.

Loveless, J. P., & Meade, B. J. (2011). Spatial correlation of interseismic coupling and coseismic rupture extent of the 2011 M-W = 9.0 Tohoku-oki earthquake. *Geophysical Research Letters, 38*, 1–5.

McCaffrey, R., Long, M. D., Goldfinger, C., Zwick, P. C., Nabelek, J. L., Johnson, C. K., et al. (2000). Rotation and plate locking at the southern Cascadia subduction zone. *Geophysical Research Letters, 27*(19), 3117–3120.

McCaffrey, R., Qamar, A. I., King, R. W., Wells, R., Khazaradze, G., Williams, C. A., et al. (2007). Fault locking, block rotation and crustal deformation in the Pacific Northwest. *Geophysical Journal International, 169*(3), 1315–1340.

McCrory, P. A., Blair, J. L., Waldhauser, F., & Oppenheimer, D. H. (2012). Juan de Fuca slab geometry and its relation to Wadati-Benioff zone seismicity. *Journal of Geophysical Research: Solid Earth, 117*(B9), 1–23.

McGary, R. S., Rondenay, S., Evans, R. L., Abers, G. A., & Wannamaker, P. E. (2011). A joint geophysical investigation of the Cascadia subduction system in central Washington using dense arrays of passive seismic and magnetotelluric station data. In *AGU Fall Meeting Abstracts.*

Miller, M. M., Melbourne, T., Johnson, D. J., & Sumner, W. Q. (2002). Periodic slow earthquakes from the Cascadia subduction zone. *Science, 295*(5564), 2423.

Moreno, M., Rosenau, M., & Oncken, O. (2010). 2010 Maule earthquake slip correlates with pre-seismic locking of Andean subduction zone. *Nature, 467*(7312), 198–U184.

Nabelek, J., Trehu, A., Li, X. Q., & Fabritius, A. (1996). *Lithospheric structure of the Cascadia Arc beneath Oregon.* New York: Transactions of the American Geophysical Union.

Nicholson, T., Bostock, M., & Cassidy, J. F. (2005). New constraints on subduction zone structure in northern Cascadia. *Geophysical Journal International, 161*(3), 849–859.

Okada, Y. (1992). Internal deformation due to shear and tensile faults in a half-space. *Bulletin of the Seismological Society of America, 82*(2), 1018–1040.

Oleskevich, D. A., Hyndman, R. D., & Wang, K. (1999). The updip and downdip limits to great subduction earthquakes: thermal and structural models of Cascadia, south Alaska, SW Japan, and Chile. *Journal of Geophysical Research-Solid Earth, 104*(B7), 14965–14991.

Peltier, W. R., Argus, D. F., & Drummond, R. (2015). Space geodesy constrains ice age terminal deglaciation: the global ICE-6G_C (VM5a) model. *Journal of Geophysical Research: Solid Earth, 120*(1), 450–487.

Radiguet, M., Cotton, F., Vergnolle, M., Campillo, M., Valette, B., Kostoglodov, V., et al. (2011). Spatial and temporal evolution of a long term slow slip event: the 2006 Guerrero Slow Slip Event. *Geophysical Journal International, 184*(2), 816–828.

Rogers, G., & Dragert, H. (2003). Episodic tremor and slip on the Cascadia subduction zone: the chatter of silent slip. *Science, 300*(5627), 1942–1943.

Rondenay, S., Bostock, M. G., & Shragge, J. (2001). Multiparameter two-dimensional inversion of scattered teleseismic body waves 3. Application to the Cascadia 1993 data set. *Journal of Geophysical Research: Solid Earth, 106*, 30795–30807.

Satake, K., Wang, K., & Atwater, B. F. (2003). Fault slip and seismic moment of the 1700 Cascadia earthquake inferred from Japanese tsunami descriptions. *Journal of Geophysical Research: Solid Earth, 108*(B11), 1–17.

Schmalze, G. M., McCaffrey, R., & Creager, K. C. (2014). Central Cascadia subduction zone creep. *Geochemistry, Geophysics, Geosystems, 15*(4), 1515–1532.

Schmidt, D. A., & Gao, H. (2010). Source parameters and time-dependent slip distributions of slow slip events on the Cascadia subduction zone from 1998 to 2008. *Journal of Geophysical Research: Solid Earth, 115*(B4), 1–13.

Scholz, C. H. (1998). Earthquakes and friction laws. *Nature, 391*(6662), 37.

Segall, P., & Bradley, A. M. (2012). Slow-slip evolves into megathrust earthquakes in 2D numerical simulations. *Geophysical Research Letters, 39*(18), 1–5.

Tapley, B. D., Bettadpur, S., Ries, J. C., Thompson, P. F., & Watkins, M. M. (2004). GRACE measurements of mass variability in the Earth system. *Science, 305*(5683), 503–505.

Wada, I., Wang, K., He, J., & Hyndman, R. D. (2008). Weakening of the subduction interface and its effects on surface heat flow, slab dehydration, and mantle wedge serpentinizatiofvn. *Journal of Geophysical Research: Solid Earth, 113*(B4), 1–15.

Wang, P. L., Engelhart, S. E., Wang, K., Hawkes, A. D., Horton, B. P., Nelson, A. R., et al. (2013). Heterogeneous rupture in the great Cascadia earthquake of 1700 inferred from coastal subsidence estimates. *Journal of Geophysical Research: Solid Earth, 118*(5), 2460–2473.

Wang, K., He, J., Dragert, H., & James, T. S. (2001). Three-dimensional viscoelastic interseismic deformation model for the Cascadia subduction zone. *Earth, Planets and Space, 53*(4), 295–306.

Wang, K., & Tréhu, A. M. (2016). Invited review paper: some outstanding issues in the study of great megathrust earthquakes—The Cascadia example. *Journal of Geodynamics, 98,* 1–18.

Wang, K., Wells, R., Mazzotti, S., Hyndman, R. D., & Sagiya, T. (2003). A revised dislocation model of interseismic deformation of the Cascadia subduction zone. *Journal of Geophysical Research: Solid Earth, 108*(B1), 1–13.

Wech, A. G., & Bartlow, N. M. (2014). Slip rate and tremor genesis in Cascadia. *Geophysical Research Letters, 41*(2), 392–398.

Wech, A. G., & Creager, K. C. (2007). Cascadia tremor polarization evidence for plate interface slip. *Geophysical Research Letters, 34*(22), 1–6.

Xu, X., Dong, D., Fang, M., Zhou, Y., Wei, N., & Zhou, F. (2017). Contributions of thermoelastic deformation to seasonal variations in GPS station position. *GPS Solutions, 21*(3), 1265–1274.

(Received June 10, 2018, revised August 27, 2018, accepted September 6, 2018, Published online September 24, 2018)

Pure Appl. Geophys. 176 (2019), 3893–3911
© 2019 Springer Nature Switzerland AG
https://doi.org/10.1007/s00024-019-02121-7

Pure and Applied Geophysics

Interseismic Coupling in the Central Nepalese Himalaya: Spatial Correlation with the 2015 Mw 7.9 Gorkha Earthquake

SHUIPING LI,[1] [ID] QI WANG,[1] GANG CHEN,[2] PING HE,[1] KAIHUA DING,[3] YUNGUO CHEN,[1] and RONG ZOU[1]

Abstract—Geodetic measurements conducted in the Himalaya over the last two decades have shown that the shallow portion of the main himalayan thrust (MHT) was entirely locked during the interseismic period. The induced elastic strain accumulated on the MHT beneath the Lesser Himalaya was not released until the 2015 Gorkha Mw 7.9 earthquake, which ruptured the north edge of the locked portion of the MHT. We utilized our own Global Positioning System (GPS) data from southern Tibet, combined with published geodetic velocities, to quantify the spatial variations of the coupling that prevailed before the Gorkha earthquake. The refined coupling model shows that the MHT was strongly locked (coupling > 0.5) in the uppermost 15 km of crust, corresponding to a downdip width of $\sim$ 100 km. This model suggests a sharp transition zone of strain accumulation, with a rapid decrease in the coupling coefficient from 1.0 to less than 0.2 along $\sim$ 50 km of the MHT, coinciding with the locations of microseismicity. We also determined slip models for the 2015 Gorkha earthquake and its Mw 7.3 aftershock, considering the ramp–flat–ramp–flat structure of the MHT. We found that $\sim$ 85% of the total moment released by the Gorkha earthquake was concentrated on the partially coupled transition portion of the MHT, indicating that the earthquake mainly ruptured the brittle/ductile transition zone. The coseismic Coulomb failure stress increased along the southern and western parts adjacent to the rupture zone, pushing these two regions closer to failure. The moment deficits that have accumulated in these regions could trigger Mw 8.0 and Mw 8.3 earthquakes, respectively.

Key words: GPS, convergence rate, interseismic coupling, 2015 Gorkha earthquake, brittle/ductile transition zone.

Electronic supplementary material The online version of this article (https://doi.org/10.1007/s00024-019-02121-7) contains supplementary material, which is available to authorized users.

[1] Hubei Subsurface Multi-scale Imaging Key Laboratory, Institute of Geophysics and Geomatics, China University of Geosciences, Wuhan 430074, China. E-mail: cug_lsp@foxmail.com

[2] College of Marine Science and Technology, China University of Geosciences, Wuhan 430074, China. E-mail: ddwhcg@cug.edu.cn

[3] Faculty of Information Engineering, China University of Geosciences, Wuhan 430074, China.

1. Introduction

The Himalayan orogenic belt is the most active intracontinental deformation region in the world, resulting from the continuing collision between the Indian and Eurasian plates since 50 Ma (Molnar and Tapponnier 1975). The Himalaya defines the southern margin of the Tibetan Plateau. To the south, the topography rises abruptly from an elevation of less than 1000 m in the foreland basin to more than 6000 m over a 200 km distance towards the plateau (Avouac 2003). The formation of the mountains is believed to be associated with the activation of three main thrust faults, namely the main central thrust (MCT), main boundary thrust (mbt), and main frontal thrust (MFT), which absorb a significant fraction of the shortening deformation in the Himalaya (Cattin and Avouac 2000; Lavé and Avouac 2000, 2001). Previous studies suggested that the three thrust faults imbricate within the upper crust and sole into the same mid-crust décollement, the main himalayan thrust (MHT), which dips gently to the north beneath the Higher Himalaya and southern Tibet, with depth of 30–40 km (Fig. 1) (Nábělek et al. 2009). Global positioning system (GPS) measurements show that the present-day crustal deformation across the Himalaya is characterized by a significant shortening rate of 20 mm/year, representing nearly one half of the total convergence rate between the Indian and Eurasian plates (Bilham et al. 1997; Wang et al. 2001). The majority of the shortening across the range has been accommodated by the slip along the MHT, resulting in force to trigger large earthquakes. The shallow portion of the MHT, south of the Higher Himalaya, has been demonstrated to be frictionally locked in terms of geodetic measurements and

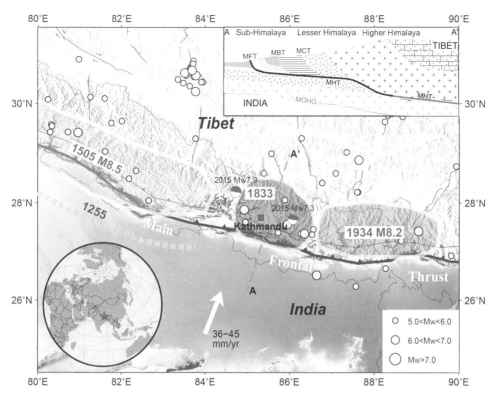

Figure 1

Seismotectonic setting and large historic earthquakes in central Nepalese Himalaya. The epicenters of the 2015 Gorkha Mw 7.9 earthquake and its Mw 7.3 aftershock are marked by two beach balls. From west to east, three earthquakes occurred sequentially in 1505 (green zone), 1833 (red zone), and 1934 (blue zone), respectively (Ambraseys and Douglas, 2004; Kumar et al. 2010; Sapkota et al. 2013). The blue dashed line indicate lateral extension of surface rupture of the 1255 earthquake. The light yellow circles indicate historical earthquakes with Mw > 5.0 in 1976–2016 from the CMT catalog. The location of Kathmandu is marked by a blue square. The thick white arrow shows Indian plate motion relative to Eurasia. The upper-right inset shows the topographical cross-section of transect A–A' modified from Lavé and Avouac (2001)

microseismic activities (Jouanne et al. 2004; Bettinelli et al. 2006), resulting in significant strain accumulation during the interseismic period. A portion of this strain is ultimately released during many large megathrust earthquakes, such as the 1505 M ∼ 8.5 earthquake and the 1934 M ∼ 8.2 earthquake in the central Himalaya (Fig. 1). More than eight Mw > 7.5 earthquakes have been recorded in the Himalaya over the past 500 years (Bilham and Ambraseys 2005). Although the coseismic ruptures of these larger events, in responding to interseismic strain accumulation, have partially released the elastic energy, a moment deficit remains to be balanced in the future, which could contribute to devastating earthquakes (Stevens and Avouac 2016; Xiong et al. 2017).

Interseismic coupling plays an important role in assessing earthquake potential and recurrence, and is commonly employed to model the strain accumulation state (Bollinger et al. 2004; Avouac et al. 2015). GPS measurements in several subduction zones, such as those of Chile, Sumatra, the Andes, and Japan, have exhibited heterogeneous coupling patterns on plate boundary faults, suggesting that the plate interface in the 0–40 km seismogenic depth range consists of interfingered patches that either remain locked or creep aseismically (Suwa et al. 2006; Prawirodirdjo et al. 2010; Chlieh et al. 2011). In the Himalaya, the coupling pattern on the MHT has been constrained by geodetic measurements. Ader et al. (2012) proposed a coupling model in the Nepal Himalaya, showing that the MHT is locked from the surface to a downdip width of $\sim$ 100 km. Stevens and Avouac (2015) estimated the strain accumulation along the whole Himalayan arc, suggesting that coupling on the MHT behaves homogeneously with a seismic moment accumulation rate of $15.1 \pm 1 \times 10^{19}$ Nm/year. These studies have described the first-order characteristics of the strain budget on the MHT. However, most previously established GPS stations are situated in the foothills of the Himalaya. GPS sites in southern Tibet are sparse. The coupling variation beneath the Higher Himalaya remains ambiguous. In addition, most previous studies have suggested that seismic slips during large earthquakes tend to occur in areas that remain locked during the interseismic period (Chlieh et al. 2011; Métois et al. 2012). Nonetheless, the correlation between coseismic slips and interseismic coupling in the Himalaya is still poorly understood, owing to the lack of large earthquakes documented in terms of seismic waveforms or geodetic observations.

The 2015 Mw 7.9 Gorkha earthquake occurred in central Nepal along the higher Himalaya. The hypocenter depth was 15 km, indicating that this event appears to have occurred at the interface of the MHT (Bai et al. 2016; McNamara et al. 2017; Arora et al. 2017). The Gorkha event is the first occurrence of a large continental thrust earthquake in Himalaya to be concurrently recorded in terms of geodetic and seismic measurements, providing a rare opportunity to explore the correlation between the interseismic coupling and coseismic rupture. It is well known that a large earthquake is likely to rupture the most strongly coupled segments. Analyzing the spatial correlation between the interseismic coupling and coseismic slip distribution is helpful for understanding the balance of the seismic moment accumulation and release (Morsut et al. 2017). In this study, we utilized GPS data surveyed in southern Tibet together with published geodetic velocities to derive a complete velocity field in the central Nepalese Himalaya. This new velocity field is then employed to refine the convergence rates across the central Nepalese Himalaya. Considering the ramp–flat–ramp–flat geometry of the MHT, we calculated the interseismic coupling using triangular dislocation. We also determined the coseismic slip distribution of the Gorkha earthquake to analyze the spatial correlation

between the interseismic coupling and coseismic rupture features.

2. GPS Data and Analysis

2.1. GPS Data Processing

The GPS velocities recorded in the central Nepalese Himalaya mainly consist of results from the Crustal Movement Observation Network of China (CMONOC) project (Wang et al. 2017) and published studies (Bettinelli et al. 2006; Bilham et al. 1997). In addition, we began recording GPS measurements in southern Tibet in 1999, and have obtained GPS data for approximately 40 sites with an occupation of more than 36 h for each survey. At least three periods of observation were conducted for each site. All the data were recorded prior to the occurrence of the 2015 Gorkha earthquake. The CMONOC data and our own data were processed using the GIPSY-OASIS-II software (Zumberge et al. 1997). The data analysis followed the procedures adopted by Fu and Freymueller (2012). We utilized the Jet Propulsion Laboratory's (JPL's) reanalyzed IGS08 orbit and clock products, and refined the absolute antenna phase center models for both the GPS receiver and satellite antennas. We utilized the global mapping function (GMF) tropospheric mapping function and the global pressure and temperature (GPT) model to reduce the tropospheric delay error, and the ocean tide model TPXO7.0 to correct for ocean tide loading. Finally, we obtained a set of GPS velocity fields under a unified reference frame by completing a seven-parameter Helmert transformation through some common sites that occur within the different datasets. The differences in the EW and NS components of the common stations between different datasets are generally smaller than 1.2 and 1.0 mm/year, respectively. Figure 2a shows the combined velocity field for $\sim$ 120 sites in the central Nepalese Himalaya relative to the stable Eurasian reference frame. The mean uncertainty of GPS velocities is $\sim$ 1.5 mm/year.

It should be noted that the majority of GPS velocities in southern Tibet are derived from campaign-model measurements. The use of vertical

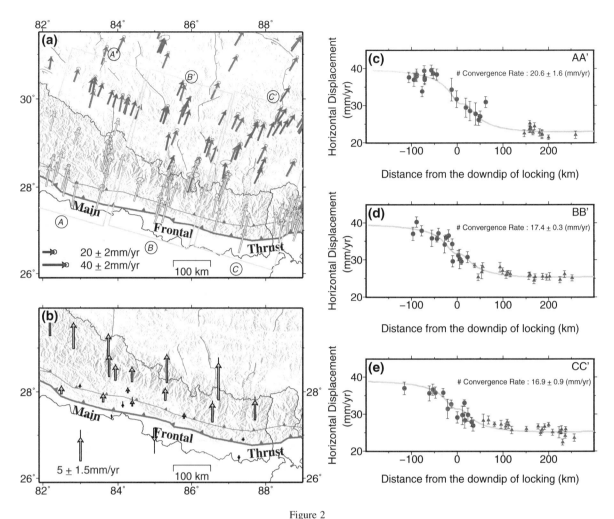

Figure 2

GPS velocity field in the central Nepalese Himalaya. **a** Horizontal GPS velocities relative to the Eurasia reference frame (with 95% confidence ellipses). The yellow arrows represent GPS velocities from published studies (Ader et al. 2012; Bettinelli et al. 2006). The red arrows show the velocities of the campaign and continuous sites from the CMONOC project. Our own ∼ 40 GPS velocities in southern Tibet are plotted as blue arrows. The three green boxes indicate the locations of the velocity profiles (AA′, BB′, CC′) perpendicular to the Himalayan arc. **b** GPS vertical velocities in Nepal relative to the India reference frame. **c–e** GPS horizontal velocity profiles and calculated convergence rates. The red circles and triangles with error bars represent the velocity components parallel to the azimuth of each profile with uncertainties of one standard deviation. The triangles denote the velocity components of our own data and the CMONOC data. The blue solid lines indicate the model predictions

velocities in southern Tibet remains challenging, owing to the limited observation timespan. Commonly, the vertical positioning precision of GPS measurements is typically 2–2.5 times lower than that of the horizontal. Furthermore, in southern Tibet the magnitude of the crustal vertical velocity is usually considerably smaller than that of the horizontal velocity (Liang et al. 2013). Therefore, the vertical velocities in southern Tibet are not adopted in the

modeling. In central Nepal, a network of continuous GPS (cGPS) stations have been installed since 1997 by Caltech in order to monitor and determine present-day velocities in the Himalaya (Ader et al. 2012). Such a long time span of observations makes it possible to acquire reliable vertical deformations in this region. Data from the time of installation of the cGPS stations to the day before the Gorkha earthquake were analyzed by Jouanne et al. (2017) using the Bernese software. These vertical velocities (relative to the India reference frame, as shown in Fig. 2b) enhance the spatial density of observations for the coupling model. These are utilized in the coupling inversion approach presented below, although we find that they do not add much constraint to the model.

2.2. Estimation of Convergence Rate

An important goal in Himalayan studies during recent decades has been to refine the Himalayan convergence rate, as this is responsible for the productivity of Himalayan earthquakes. In coupling inversion, the long-term convergence rate across the Himalaya is also an important reference value, as described in Sect. 3. Here, we incorporate the new GPS data from southern Tibet to further constrain the Himalayan convergence rate. Following most previous investigations (e.g., Bilham et al. 1997; Bettinelli et al. 2006), we modeled the interseismic strain resulting from slips along a creeping dislocation embedded in an elastic half-space. These dislocations represent the aseismic shear north of the locked portion of the Himalayan detachment system. Three velocity profiles were projected along the azimuth of convergence (N16°E). The length and width for each profile were $\sim$ 200 km and $\sim$ 110 km, respectively. We assumed that the thrusting slip on a detachment fault dominated in every profile across the Himalayan orogenic belt, ignoring lateral variations in the thrust slip in accordance with the two-dimensional (2D) edge dislocation model. The analytical solution for a pure dip slip fault can be written as follows (Freund and Barnett 1976):

$$v = \frac{s}{\pi} \left[\frac{h^2 \cos \alpha + xh \sin \alpha}{h^2 + x^2} + \sin \alpha \tan^{-1}(x/h) \right] \quad (1)$$

where v represents the site velocity normal to the Himalayan arc, s denotes the thrusting rate on a major detachment fault, and h and α are the locking depth and fault dip, respectively. Furthermore, x indicates the horizontal distance to the trace of the fault.

Moderate-sized earthquakes ($5 < Ms < 6$) and smaller magnitude events (mL < 4) in the central Himalaya are confined to 10–20 km in depth (Molnar and Chen 1984; Pandey et al. 1995). The INDEPTH

seismic reflection profile in eastern Nepal exhibits a discrete fault plane dip of $\sim 9^\circ$ (Zhao et al. 1993). Supposing that the aforementioned fault geometry can be applied elsewhere in the central Nepalese Himalaya, the locking depth of the MHT can be fixed at 20 km with a dip of $\sim 9^\circ$. The weighted least-squares method was adopted to estimate the convergence rate.

The 2D dislocation model provided an adequate fit to the observed GPS velocities (Fig. 2c–e). The root-mean-square (RMS) errors of the post-fit residuals for each velocity profile were always less than 3 mm/year, which is broadly compatible with the formal uncertainties of observed velocities, suggesting that few if any unmodeled biases remained. The inverted slip rates were not strongly affected (a change of less than 2 mm/year) by the parameters of the locking depth and fault dip in the range of 15 km < locking depth < 25 km and 4° < fault dip < 12°, indicating that the trade-off effect between the convergence rate and fault geometry was moderate.

In west-central Nepal (profile AA'), the estimated convergence rate is 20.6 ± 1.6 mm/year, which is comparable to the 20.2 ± 1.1 mm/year estimated by Stevens and Avouac (2015). In the central Nepalese Himalaya (profile BB'), GPS observations yield a convergence rate of 17.4 ± 0.3 mm/year, which is slightly lower than the 21.5 ± 2.5 mm/year determined by the Holocene river terraces (Lavé and Avouac 2000). This result implies that a small fraction of the active convergence of the Himalayan orogen might not be accommodated by slips along the MHT. In the eastern Nepalese Himalaya (profile CC'), GPS measurements suggest a convergence rate of 16.9 ± 0.9 mm/year, which is consistent with the slip rate of 16 mm/year determined by Banerjee et al. (2008). In general, the convergence rate decreases from the west-central to the eastern Nepalese Himalaya. The along-strike variations for convergence rates could reflect the postseismic viscoelastic relaxation effects caused by large earthquakes, because many GPS sites lie within or adjacent to the epicentral areas of large Himalayan earthquakes, and the GPS velocities are inevitably more or less biased by the residual postseismic deformations from large megathrust earthquakes. The postseismic

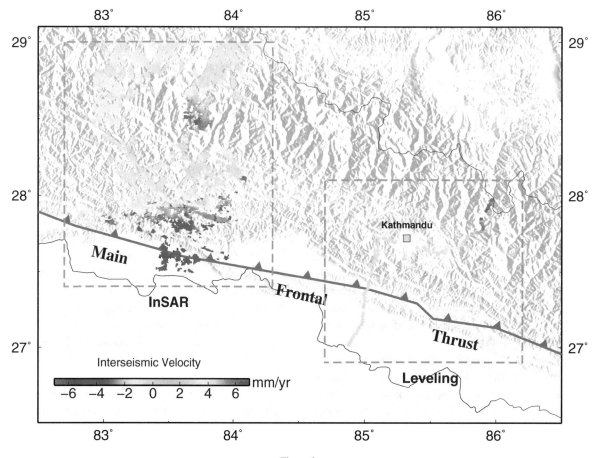

Figure 3
Observed interseismic InSAR and leveling velocities in central Nepalese Himalaya

effects caused by the 1934 Bihar–Nepal M $\sim$ 8.2 earthquake on the interseismic convergence rates are evaluated and discussed in Sect. 5.1. It is worth noting that because of our broader and denser distribution of GPS data in southern Tibet, the estimated convergence rates can provide more tightly constrained upper bounds.

3. Modeling the Interseismic Coupling in Central Nepal

The resolved coupling image can benefit from a refined spatial density of observations, including vertical deformation rates. In addition to the GPS vertical velocities in Nepal, we also collected all the available interseismic interferometric synthetic aperture radar (InSAR) and leveling data in the central

Himalaya in order to determine the coupling parameters, although they are only localized in two narrow profiles (Fig. 3) (Grandin et al. 2012; Jackson and Bilham 1994). Owing to the highly rugged and heavy vegetation terrains in the Himalaya, as of now only one stripe of SAR C-band images acquired by the ENVISAT satellite on track 119 (descending) between 2003 and 2010 has successfully been processed. For satellite SAR images, the small baseline subsets (SBAS) processing strategy was exploited to produce interferograms. In addition, a slope-adaptive spectral shift range filtering method was applied during interferogram formation to reduce the geometric decorrelation resulting from steep terrain slopes in central Nepal. After interferogram unwrapping and geocoding, the line-of-sight (LOS) displacement velocity maps and corresponding time-series were retrieved (Readers interested in the information of ENVISAT SAR processed with the SBAS method may refer to the work of Grandin et al. (2012)). The InSAR profile shows a gentle increase of the uplift rate toward the North, with a peak uplift rate of 7 mm/year (~ 100 km north of the MFT). To reduce the burden of calculation in inversion, a quad-tree algorithm was employed to down-sample the velocities (He et al. 2016). For the leveling data, we followed previous studies to remove the points in the Kathmandu Valley that are undergoing obvious sub-sidence (Grandin et al. 2012). Finally, 63 InSAR and 152 leveling velocities are included in the coupling inversion.

We adopted the back-slip modeling approach (Savage 1983) to solve for the slip rate deficit along the MHT. This model assumes that the hanging wall does not deform over the long term, and only applies to the case of a planar fault. However, it remains a valid approximation even if the megathrust is not strictly planar in reality (Vergne et al. 2001). In our modeling, the geometry of the MHT was character-ized by four connected fault portions dipping northward from the MFT (Elliott et al. 2016): first, a shallow ramp with a dip of 30° between the surface and a 5 km depth; second, a flat detachment with a dip of 7° extending from a depth of 5 km to 15 km; third, a mid-crust ramp that has a larger dip angle of 20° at 15 to 25 km; and finally a sub-horizontal

décollement with a dip of 6° beneath the Higher Himalaya. These four fault segments constitute the "ramp–flat–ramp–flat" structure of the MHT (Fig. 4). The strike of the interface is fixed at 285°. We extended the model fault downward with a width of 400 km from the MFT to avoid edge effects. Mean-while, such an expanded downdip width can better constrain the slip rate deficit at a large depth using our own GPS data in southern Tibet. The entire interface of the MHT is discretized into a 93 × 20 matrix of rectangular subfaults, each with dimensions of 20 km × 20 km. To achieve seamless gridding on the model plane, each subfault was approximated by two triangular dislocation elements (TDEs). As a result, a total of 1860 TDEs were employed, for which we calculated the Green's function using angular dislocation in an elastic half-space (Meade 2007).

We calculated the coupling coefficient (ϕ) for each patch based on the following equation (Cheloni et al. 2014):

$$\phi = \frac{V_d}{V_c} \qquad (2)$$

where V_d is the slip rate deficit on a fault patch and V_c represents the long-term block convergence rate over many earthquake cycles. The slip deficit rate is con-strained to be positive and no larger than the long-

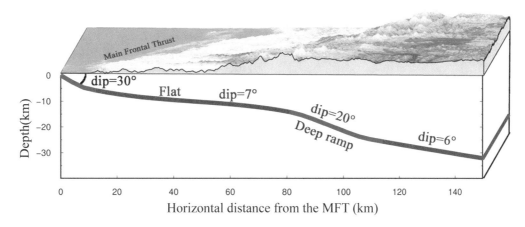

Figure 4
Cross section of the MHT interface used in the modeling. The two blue lines indicate the shallow and mid-crust ramps. The red lines denote the flat structures of the MHT

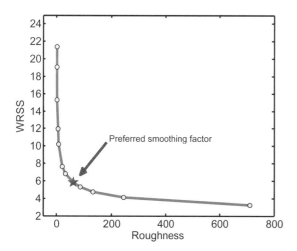

Figure 5
Trade-off curve between model roughness and WRSS during the inversion of the coupling distribution. The red pentagram represents the optimal smoothing factor

observations. In the joint inversion, the datasets are weighted based on the trade-off curve between solution misfit and the optimum weight ratio between GPS and InSAR-leveling data (Xu et al. 2009; Yi et al. 2017). The non-negative least-squares algorithm was applied to solve the slip on each patch (Lawson and Hanson 1974). We utilized a scale-dependent umbrella operator to regularize the fault-slip distribution (Maerten 2005). The operator is defined as follows:

$$\nabla^2 s_i = \frac{2}{L_i} \sum_{j=1}^{3} \frac{s_j - s_i}{h_{ij}}, \qquad (5)$$

where h_{ij} denotes the distance from the center of the element i to the adjacent jth element, and L_i and s_i are the sum of the element center distances and the slip vector of the ith element, respectively.

4. Results

4.1. Coupling Characteristics

The best-fitting coupling model along the MHT, derived from joint inversion of the GPS, InSAR, and leveling data, is illustrated in Fig. 6a. We illustrate the fitness to the GPS, InSAR and leveling observations in Fig. 7. The mean misfits of the GPS horizontal and vertical components are 1.38 and 1.47 mm/year, respectively, exhibiting small velocity residuals. The mean misfits of the InSAR and leveling data are 0.7, and 0.4 mm/year, respectively. The reduction in the data variance is ∼ 91% for the GPS horizontal measurements, ∼ 82% for GPS vertical velocities, ∼ 86% for the leveling data, and ∼ 92% for the InSAR observations, suggesting that the geodetic observations can be satisfactorily explained by the model.

In general, the MHT is fully locked along the strike of the Himalaya, with no obvious shallow creeping zone identified, and exhibits a homogeneous coupling pattern. Unlike the coupling characteristics along strike, the coupling distribution along the downdip direction exhibits obvious spatial variations. To illustrate the details of the coupling variation, three arc-normal profiles of the coupling coefficient were projected (Fig. 6b). These profiles show the

term convergence rate, so that the coupling coefficient can be constrained between 0 and 1. A coupling value of 0 means that the MHT is creeping at the long-term convergence rate, and a coupling of 1 means that there is no creep on the MHT during the interseismic period (i.e., it is fully locked). A coupling value between 0 and 1 means that the fault is partially locked.

The optimal slip rate deficits can be solved through the linear equation

$$\begin{bmatrix} d^p \\ 0 \end{bmatrix} = \begin{bmatrix} G^{e,p} \\ k^2 \nabla^2 \end{bmatrix} s^e. \qquad (3)$$

The optimal objective function can be described as follows:

$$\|W(Gs - d)\|^2 + k^2 \|\nabla^2 s\|^2 = \min, \qquad (4)$$

where d^p represents the displacement at point (p) on the surface caused by the slip (s^e) on the triangular elements (e) of a buried fault, $G^{e,p}$ denotes the Green's function matrix, ∇^2 represents the smoothing operator to avoid a slip anomaly between neighboring elements, k^2 is the smoothing factor based on the visual inspection of the trade-off curve between the model roughness and the weighted residual sum of squares (WRSS) (Fig. 5), and W is the weight matrix combining the GPS (including the horizontal and vertical components), InSAR, and leveling

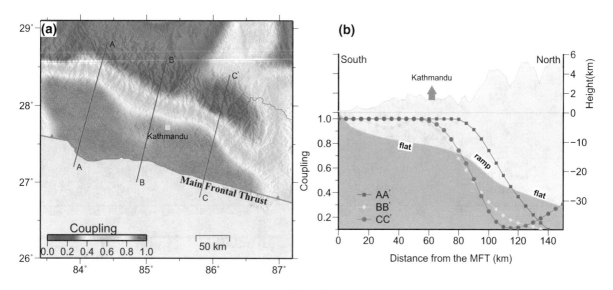

Figure 6
a Interseismic coupling model in central Nepalese Himalaya. **b** Three coupling profiles perpendicular to the mountain range

smooth distribution of the coupling coefficient with respect to the downdip width. From the coupling profiles, we can observe the following. First, the MHT under the sub-Himalaya and Lesser Himalaya is fully coupled. Nevertheless, the mid-crust ramp beneath the front of the Higher Himalaya is partly coupled, suggesting that this ramp is affected by aseismic creeping. This result is different from the 100% coupling of the mid-crust ramp found by a previous study (Ader et al. 2012). Second, maximal creeping (with coupling as low as 0.2) is observed north of the mid-crust ramp. The free creeping zones always correspond to rate-strengthening areas, which probably act as barriers resisting fault ruptures. Third, if we consider a coupling coefficient of 0.5 as the boundary value to distinguish fault locking or creeping, then the locking width west of Kathmandu (profile AA') is ∼ 110 km, which is slighter longer than the locking width of ∼ 90 km east of Kathmandu (profile CC'). The average locking width of ∼ 100 km in the central Nepalese Himalaya is reasonably consistent with previous estimations (Ader et al. 2012; Stevens and Avouac 2015; Jouanne et al. 2017).

A sharp transition zone where the fault interface gradually decouples from full locking to free creeping ($0.2 < \phi < 1.0$) can be identified beneath the front of

the Higher Himalaya. The width of this transition zone at the longitude of Kathmandu (profile BB') is approximately 50 km, which is considerably shorter than the typical range of subduction zones (Lay et al. 2012). The sharpest transition zone corresponds to an area where the resolution is enhanced by the leveling data. This transition zone coincides spatially with the location of the mid-crust ramp (Fig. 6b), suggesting that the geometric structure of the MHT influences the coupling status at the interface. The modeled maximal uplift rates reach 6 mm/year in the high chain of the central Nepalese Himalaya, which is slightly less than the maximal observed uplift of 8 mm/year from leveling measurements (Fig. 7c). This discrepancy probably indicates that the real dip of the mid-crust ramp might be slightly greater than 20° (which was adopted in this study). In addition, the ongoing glacial isostatic adjustment induced by the current melting of the Himalayan glaciers might also contribute to the larger observed uplift rate than that of the modeling (Duputel et al. 2016).

4.2. Resolution Test

A series of checkerboard tests were performed to evaluate the spatial resolution of the coupling image. We imposed alternating slip rates of 20 mm/year (full

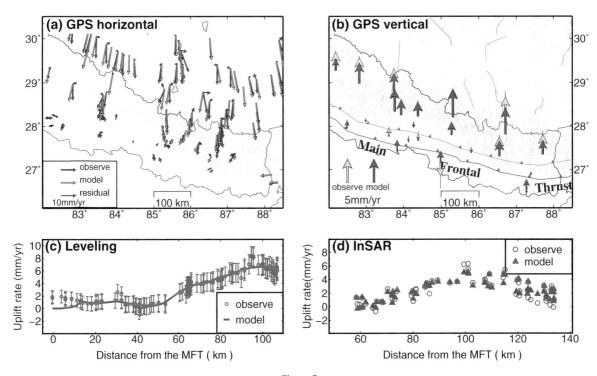

Figure 7
Post-fit residual of the GPS data and the misfit to the InSAR and leveling observations. **a** The fitness to the GPS horizontal velocities. The red, green, and blue arrows represent the observed, predicted, and residual GPS horizontal velocities, respectively. **b** The fitness to the GPS vertical velocities. The yellow and blue arrows indicate the observed and predicted velocities, respectively. **c** Leveling velocities along the profile perpendicular to the MFT. The blue circles with error bars show the observed leveling velocities, and the red solid line represents the model prediction. **d** InSAR velocities along the transect perpendicular to the MFT. The blue circles and red triangles denote the observed and simulated InSAR velocities, respectively

locking) and 0 mm/year (free creeping) on the input checkerboard patches. The surface velocities (GPS, InSAR, and leveling) were synthesized by the input model. The synthetic data were disturbed by Gaussian noise, which followed a normal distribution with zero mean and standard deviation for the observed data. We inverted the noisy velocities using the same strategy as for the real data. The results for the checkerboard tests are illustrated in Fig. 8. In general, the resolution decays as the checkerboard dimension decreases. When the patch of the checkerboard is approximately 50 km, the input coupling distribution can be retrieved effectively except for the part from the western portion to Kathmandu ($84° - 85° E$), on account of the sparse data in this region. A patch size of ~ 30 km along the downdip direction near Kathmandu can be resolved owing to the incorporation of leveling data, which is smaller than the width of the coupling transition zone in central Nepal (~ 50 km),

suggesting that this transition zone can be retrieved by surface geodetic measurements. In general, the main features of fault coupling can be resolved in terms of both size and pattern using geodetic data.

5. Discussion

5.1. Viscoelastic Effects on the Convergence Rate and Coupling

Previous studies have indicated that postseismic deformation induced by great earthquakes M > 8 in Mongolia and Chile may persist for several decades, or even a century (Vergnolle et al. 2003; Lorenzo-Martín et al. 2006). In the Himalaya, the geodetic measurements were all obtained in recent decades, with some of them located close to the epicentral area of the 1934 M $\sim$ 8.2 earthquake. This large

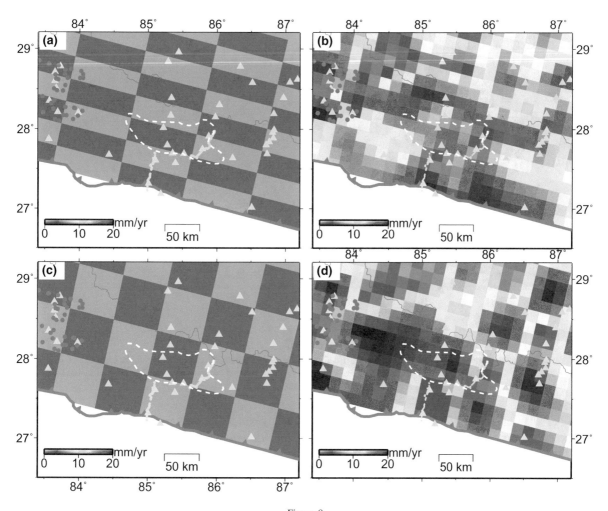

Figure 8

Checkerboard tests showing the spatial resolution of the fault coupling based on the joint inversion of the GPS, leveling, and InSAR data. **a** and **c**: Synthetic input slip distributions used to create the synthetic deformation field. **b** and **d**: Recovered slip patterns inverted from the synthetic surface deformation. The yellow triangles denote the locations of GPS sites. Green and pink points show the locations of leveling and InSAR data, respectively. The white dashed line outlines the rupture zone of the 2015 Gorkha earthquake

megathrust earthquake may still be contributing a small postseismic signal, which could affect the interseismic velocities and the estimated convergence rates. In order to test the possible magnitude of this effect, we adopt a simple source model according to the seismic moments (Chen and Molnar 1977) and a scaling law for the relation between the rupture size and seismic moment (Feldl and Bilham 2006; Sapkota et al. 2013). A 2D rheologic structure for the Himalaya and southern Tibet is considered in our viscoelastic model (Fig. 9a). We select the steady-

state viscosities for the lower crust and upper mantle from two recent studies on the postseismic deformation of the 2015 Gorkha earthquake (Zhao et al. 2017; Wang and Fialko 2018). We employ the spectral element method code VISCO2.5D to calculate the deformation resulting from viscoelastic relaxation in the lower crust and upper mantle of Tibet and India caused by the 1934 event (Pollitz 2014). The results are illustrated in Fig. 9b. For sites to the north of the earthquake source area, notable southward motions can be identified with a maximum value of 2.7 mm/

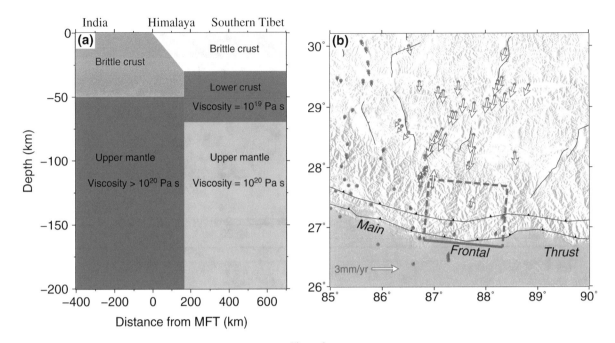

Figure 9
a Rheological model structure across the central Nepalese Himalaya. **b** Simulated postseismic deformation at the locations of GPS sites in eastern Nepal induced by the viscoelastic relaxation 80 years after the 1934 Bihar–Nepal M ∼ 8.2 earthquake. The red square indicates the earthquake source area

year. For sites just above the source area, ongoing northward motions of 1–2 mm/year are suggested. This kind of contraction suggests that the continued relaxation caused by the 1934 earthquakes would increase the interseismic convergence rate across the eastern Nepalese Himalaya. In fact, the convergence rate across the western Nepalese Himalaya, where no large earthquakes have occurred for several centuries, is 3–4 mm/year larger than that across the eastern Nepalese Himalaya, while the opposite would be expected according to the modeling. Thus, the model predication cannot account for the apparent differences in convergence rates between the eastern and western Nepal Himalaya. If we removed the postseismic contributions from GPS velocities, then the along-strike differences in the long-term convergence rates between the eastern and western Nepalese Himalaya might be even larger. We note that the coupling pattern in the central Nepalese Himalaya is barely modified after removing the postseismic effects caused by the 1934 earthquake, although the transition zone of the coupling is slightly broadened compared to the original coupling map.

5.2. Spatial Correlation Between Interseismic Coupling and Coseismic Slip

Only two large earthquakes (Mw > 7.5) have occurred along the Nepal Himalaya during the past century: the 1934 Bihar–Nepal earthquake (M ∼ 8.2) in eastern Nepal and the 2015 Gorkha earthquake (Mw 7.9) in central Nepal. The detailed coseismic slip distribution for the 1934 earthquake is particularly uncertain, thus preventing a comparison with the interseismic coupling (Sapkota et al. 2013). The 2015 Gorkha earthquake nucleated ∼ 80 km northwest of Kathmandu, and failed a 150 km long segment of the MHT (Avouac et al. 2015). The rupture image of the Gorkha earthquake is now well documented by seismic and geodetic observations (Galetzka et al. 2015; Grandin et al. 2015), allowing a detailed comparison with the pattern of fault locking.

Before assessing the spatial correlation between the coseismic slip distribution and interseismic coupling, we developed a geodetically constrained slip distribution of the Gorkha earthquake on the same fault geometry as used in the coupling estimation (Fig. 10a). The detailed process for the slip inversion

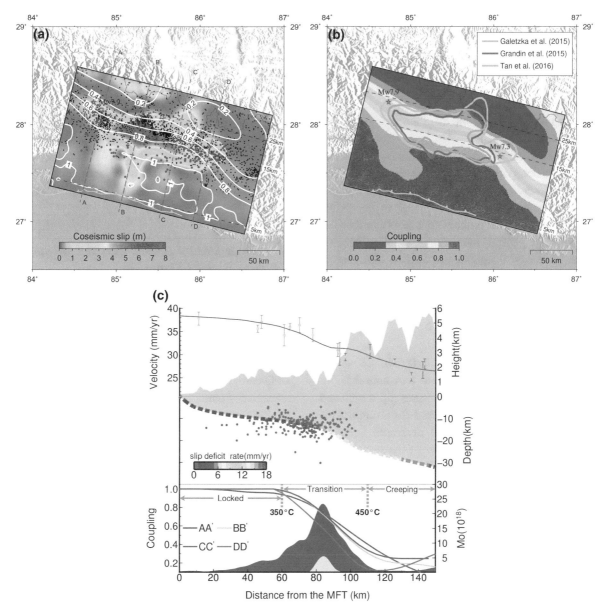

Figure 10

Map view illustrating the spatial correlation between the interseismic coupling and coseismic slip distribution of the 2015 Mw 7.9 Gorkha earthquake. **a** The colored basemap shows the coseismic fault slip. White contour lines denote the interseismic coupling distribution. Black dots show small background earthquakes derived by Pandey et al. (1995). Blue circles mark the locations of aftershocks according to Bai et al. (2016). The four red dashed lines indicate the locations of four coupling profiles (AA', BB', CC', DD'). **b** The blue-to-red basemap indicates the coupling distribution. The blue, red, and green solid lines outline the spatial extent of the rupture of the Gorkha earthquake based on three different studies (Galetzka et al. 2015; Grandin et al. 2015; Tan et al. 2016). **c** Fault profile perpendicular to the MFT. Top: The green dots with error bars represent horizontal GPS velocities, and the red line denotes the model prediction. Middle: The thick dashed line denotes the interface of the MHT and its color represents the slip deficit rate calculated by the coupling coefficient and convergence rate. Red circles indicate the relocated aftershocks. Bottom: Blue and yellow histograms indicate the distributions of the moment release of the Mw 7.9 mainshock and the Mw 7.3 aftershock. The four solid lines (red, blue, green, and pink) show the coupling profiles along the downdip direction

is reported in the auxiliary material (Text S1, Figs. S1–S5). The coseismic slip model exhibits an elongated unilateral rupture, with a maximal slip of

7.8 m. In addition, three previous rupture models derived from the inversion of geodetic and seismological data are also included for comparison with the interseismic coupling (Fig. 10b). Comparing Figs. 10a and b, we conclude that the spatial extents of the four slip models behave consistently, although their maximal slip magnitudes exhibit slight differences.

The spatial correlation between the interseismic coupling and coseismic slip distribution suggests that the Gorkha earthquake unzipped the lower edge of the locked portion of the MHT, in accordance with the conclusion proposed by Avouac et al. (2015). As shown in Fig. 10a, the Gorkha earthquake initiated from the epicenter, where the coupling coefficient was ~ 0.5, and then propagated laterally in the ESE direction. In general, interseismic coupling behaves homogeneously along the propagation direction of the rupture. The fault patches are estimated to have slipped ≥ 4 m, corresponding to an area of the interface in which $0.5 <$ coupling < 0.8, with a peak slip near a region coupled at ≥ 0.7. The updip end of the rupture stopped at the leading edge of the Lesser Himalaya, with a coupling coefficient of ~ 1.0, leaving the shallow fully locked part unbroken. In the downdip direction, the coseismic slip terminated almost at the bottom of the mid-crust ramp. This result closely agrees with the decreased coupling (< 0.2) limiting the propagation of the rupture to a greater depth. This characteristic indicates that the ruptures of large megathrust earthquakes in the Himalaya can fail patches that are inferred to be less strongly coupled (as low as 0.2).

In more detail, our modeling suggests that the Gorkha earthquake appears to have mainly ruptured the transition zone of coupling on the MHT. This is demonstrated by the distributions of the moment release along the downdip direction (Fig. 10c). For the mainshock, approximately 85% of the released moment was concentrated on the partially locked transition zone, where the coupling decreases from 1.0 to less than 0.2. Only $\sim 15\%$ of the released moment was in the upper fully locked portion. For the Mw 7.3 aftershock, all the released energy was confined to the transition zone of the coupling (Fig. 10c). In the map view, this brittle/ductile transition zone is followed by a narrow belt of background seismicity (Fig. 10a), reflecting a high stress loading rate (~ 10 kPa/year) during the interseismic period (Pandey et al. 1995). Laboratory

experiments on quartzo-feldspathic rocks also show that the transition zone of coupling is characterized by temperature boundaries of 350 °C and 450 °C, implying that there is a thermally controlled downdip change from brittle–seismic to ductile–aseismic behavior (Hyndman 2013). The width of the transition zone is believed to be controlled by the geothermal gradient along the MHT, which in turn depends on the dip of the fault. A sharp transition zone usually indicates a relatively large dip angle for the MHT below the Higher Himalaya (Bilham et al. 2017).

5.3. Seismic Moment Accumulation and Release since the 1833 Earthquake

The moments released by known large earthquakes should equal the seismic moment deficit since the latest historical earthquake, assuming that only large earthquakes have contributed to strain release (Stevens and Avouac 2015). In central Nepal, the 1833 earthquake has been suggested to have been the latest large earthquake to occur prior to the 2015 Gorkha earthquake (Bilham 1995). It has been proposed that the epicenter of the 1833 earthquake was located northeast of Kathmandu, based on the orientations of the fractures and dikes that developed during the event (Mugnier et al. 2013). Bilham (1995) reported that the epicenter was probably north or northeast of Kathmandu, adjacent to the rupture zone of the 1934 earthquake. In contrast, Ambraseys and Douglas (2004) calculated an epicenter location for the 1833 earthquake nearly 40 km east of Kathmandu based on available macroseismic data. The uncertainty concerning the epicenter location for this event is in great part owing to the lack of records from eastern Nepal, precluding a reliable determination of the isoseismals (Mugnier et al. 2017). The magnitude of the 1833 earthquake is also poorly constrained, owing partly to the large uncertainty in the extent of the 1833 rupture. Bilham (1995) suggested that the extent of the 1833 earthquake reached to the northern part of the Kathmandu basin, with a maximal slip of 5–6 m, corresponding to a Mw 7.7 ± 0.2 event, which is somewhat greater than the value (M 7.3 ± 0.1) determined by Szeliga et al. (2010) using new intensity versus attenuation relations for the Himalayan region. In addition, in the shallow part the rupture of the 1833 earthquake did not reach the surface, exhibiting similar characteristics to the 2015 Gorkha earthquake (Mugnier et al. 2011). These characteristics suggest that the 2015 earthquake may have occurred on the same segment of the décollement as the 1833 event. The 2015 Gorkha earthquake may have reruptured at least part, and perhaps all, of the portion of the décollement that slipped during the 1833 earthquake. Thus, the 1833 earthquake provides a critical window to understand the balance of the seismic moment accumulation and release.

Given the long-term slip rate and pattern of the interseismic coupling derived from this study, we estimate a seismic moment buildup rate $M_0 = 3.15 \pm 0.5 \times 10^{18}$ Nm/year (assuming a shear modulus of 30 GPa) in the central Nepalese Himalaya. This quantity is calculated from the area of the fault plane and the slip deficit rate on each patch. Based on the seismic moment buildup rate calculated above, we estimate an earthquake with an average slip of ~ 2.5 m is required to balance the strain accumulated since the 1833 earthquake. However, according to the coseismic slip distribution, the average slip during the 2015 Gorkha earthquake was greater than 4 m, and reached 7.8 m locally (Fig. 10a). This simple comparison suggests that the strain released in the 2015 Gorkha earthquake was considerably greater than the strain accumulated over the 182 years from 1833, which implies that the 1833 earthquake only partially released the strain that accumulated prior to that event. A fraction of the elastic energy that was not released during the 1833 event was finally released by the 2015 Gorkha earthquake. The partial strain released during an earthquake has been reported in western Nepal, where the 1505 M 8.5 earthquake occurred. The transient slip for the 1505 earthquake was over 9 m, while the accumulated displacement since the 1255 earthquake was approximately 5 m assuming a long-term shortening rate of ~ 20 mm/year, suggesting that nearly half of the strain accumulated before 1255 was not released by the Mw > 8 1255 earthquake (Mugnier et al. 2013).

The unbalance between the seismic moment accumulation and the release by large known earthquakes in central Nepal has important implications concerning the recurrence intervals of large Himalayan earthquakes. Assuming that the next earthquake in central Nepal has

the same spatial extent (150 km $\times$ 60 km) and magnitude (Mw 7.9) as the Gorkha earthquake, a recurrence interval of ~ 200 years is estimated considering the average slip deficit rate of 14 mm/year from our coupling model. However, if the 2015 Gorkha earthquake also did not release all of the accumulated elastic strain, similar to the 1833 event, then the residual strain would reduce the recurrence interval of the next large earthquake near Kathmandu, and a realistic recurrence interval might be less than 200 years.

5.4. Implications for Earthquake Hazards in Central Nepal

The seismic risk in the central Nepalese Himalaya remains high, although the 2015 Gorkha earthquake partially released the accumulated strain. We calculated the coseismic static Coulomb stress change at a depth of 15 km triggered by the Gorkha earthquake using the expressions for calculating the stress field associated with angular dislocations provided by Meade (2007) (Fig. 11). This result clearly shows that the stress status adjacent to the rupture zone was perturbed by the Gorkha earthquake. To the south of the rupture zone, the coulomb stress calculations exhibited loading at a depth of 0–10 km, pushing this region ($\sim 150 \times 70$ km^2) closer to failure. Geodetic measurements over the past 20 years suggest that this shallow décollement has been minimally affected by aseismic slip (e.g., slow slip events) and postseismic afterslip (Ader et al. 2012; Mencin et al. 2016; Sreejith et al. 2016; Gualandi et al. 2017). Thus, we believe that the strain accumulated in this portion will be released via large seismic events. Our coupling model suggests a slip deficit rate of 18 mm/year in this segment. The slip deficit in this shallow décollement could have exceeded 3.2 m since the 1833 event, which is sufficient to trigger an Mw ~ 8.0 earthquake. In addition, a 300 km long section to the west of the Gorkha earthquake has not undergone any earthquakes for 500 years, since the 1505 M ~ 8.5 earthquake. However, our model shows that this section is fully locked, with a downdip width of 110 km. The slip deficit could have exceeded 7.5 m based on a slip rate of 15 mm/year on the MHT, which is sufficient to fuel an Mw ~ 8.3 earthquake if a 150×110 km^2 zone is ruptured.

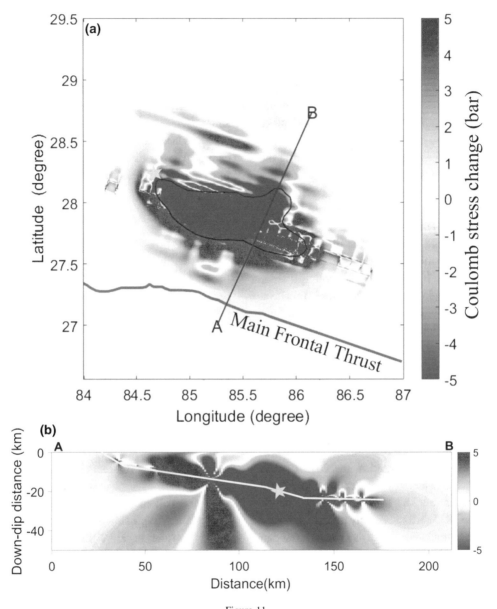

Figure 11
a Static Coulomb stress change at a depth of 15 km triggered by the Gorkha earthquake. Red areas denote stress loading and blue areas denote stress release. The black line outlines the coseismic slip of the mainshock. The yellow star shows the epicenter of the mainshock. **b** Cross-section with the Coulomb stress change calculated by the coseismic mainshock slip model. The white solid line denotes the interface of the MHT

6. Conclusions

The 2015 Gorkha Mw 7.9 earthquake highlights the urgent need to re-analyze the strain accumulation and release in Central Nepal. In this study, we incorporated GPS observations in southern Tibet with those publicly available for geodetic imaging of interseismic coupling in the central Nepalese Himalaya. Our improved GPS measurements yield a convergence rate of 20.6 ± 1.6 mm/year in west-central Nepal, 17.4 ± 0.3 mm/year in central Nepal, and 16.9 ± 0.9 mm/year in eastern Nepal, exhibiting an along-strike variation of ~ 3 mm/year from west to east. The refined coupling model confirms the finding of previous studies that the locking is essentially continuous throughout the MHT. In addition, we identified a sharp transition zone of the coupling $(0.2 < \phi < 1.0)$ with an average width of ~ 50 km in the central Nepalese Himalaya. This transition zone coincides spatially with the location of the mid-crust ramp, suggesting that this ramp might control the deep transition from locked to creeping. We found that $\sim 85\%$ of the total moment released by the 2015 Gorkha earthquake was confined to the partially coupled transition portion of the MHT. This spatial correlation indicates that the earthquake mainly ruptured the brittle/ductile transition zone at the lower edge of the locked portion of the MHT.

Acknowledgements

We acknowledge the Crustal Movement Observation Network of China (CMONOC) for providing us with the GPS data in southern Tibet. We thank the editor Carla F. Braitenberg and two anonymous reviewers for their constructive and helpful comments, which greatly helped in improving our manuscript. This work is supported by the National Natural Science Foundation of China (41674015, 41731071, 41574012, 41674017, 41274037, 41541030), China postdoctoral science foundation (2016M592408), the Fundamental Research Funds for National Universities (CUG160225), Hubei Subsurface Multi-scale Imaging Key Laboratory (SMIL-2017-02) in China University of Geosciences, Wuhan, and Key Laboratory of Geospace Environment and Geodesy, Ministry of Education, Wuhan University (16-01-06). The figures are plotted using the Generic Mapping Tool (GMT) software.

References

Ader, T., Avouac, J.-P., Liu-Zeng, J., Lyon-Caen, H., Bollinger, L., Galetzka, J., et al. (2012). Convergence rate across the Nepal Himalaya and interseismic coupling on the Main Himalayan Thrust: Implications for seismic hazard. *Journal of Geophysical Research, 117*(B4), B04403. https://doi.org/10.1029/2011jb009071.

Ambraseys, N. N., & Douglas, J. (2004). Magnitude calibration of north Indian earthquakes. *Geophysical Journal International, 159,* 165–206.

Arora, B. R., Bansal, B. K., Prajapati, S. K., Sutar, A. K., & Nayak, S. (2017). Seismotectonics and seismogenesis of Mw7.8 Gorkha earthquake and its aftershocks. *Journal of Asian Earth Sciences, 133,* 2–11.

Avouac, J. P. (2003). Mountain building, erosion, and the seismic cycle in the Nepal Himalaya. *Advances in Geophysics, 46*(03), 1–80.

Avouac, J. P., Meng, L., Wei, S., Wang, T., & Ampuero, J.-P. (2015). Lower edge of locked Main Himalayan Thrust unzipped by the 2015 Gorkha earthquake. *Nature Geoscience, 8*(9), 708–711. https://doi.org/10.1038/ngeo2518.

Bai, L., Liu, H., Ritsema, J., Mori, J., Zhang, T., Ishikawa, Y., et al. (2016). Faulting structure above the Main Himalayan Thrust as shown by relocated aftershocks of the 2015 Mw 7.8 Gorkha, Nepal, earthquake. *Geophysical Research Letters, 43*(2), 637–642.

Banerjee, P., Bürgmann, R., Nagarajan, B., & Apel, E. (2008). Intraplate deformation of the Indian subcontinent. *Geophysical Research Letters, 35*(18), 7–12.

Bettinelli, P., Avouac, J. P., Flouzat, M., Jouanne, F., Bollinger, L., Willis, P., et al. (2006). Plate motion of India and interseismic strain in the Nepal Himalaya from GPS and DORIS measurements. *Journal of Geodesy, 80*(8), 567–589. https://doi.org/10.1007/s00190-006-0030-3.

Bilham, R. (1995). Location and magnitude of the 1833 Nepal earthquake and its relation to the rupture zones of contiguous great Himalayan earthquakes. *Current Science, 69*(2), 155–187.

Bilham, R., & Ambraseys, N. (2005). Apparent Himalayan slip deficit from the summation of seismic moments for Himalayan earthquakes, 1500–2000. *Current Science, 88*(10), 1658–1663.

Bilham, R., Larson, K., & Freymueller, J. (1997). GPS measurements of present-day convergence across the Nepal Himalaya. *Nature, 386*(6620), 61–64.

Bilham, R., Mencin, D., Bendick, R., & Bürgmann, R. (2017). Implications for elastic energy storage in the Himalaya from the Gorkha 2015 earthquake and other incomplete ruptures of the Main Himalayan Thrust. *Quaternary International, 462,* 3–21.

Bollinger, L., Avouac, J. P., Cattin, R., & Pandey, M. R. (2004). Stress buildup in the Himalaya. *Journal of Geophysical Research: Atmospheres, 109*(B11), 179–204.

Cattin, R., & Avouac, J. P. (2000). Modeling mountain building and the seismic cycle in the Himalaya Nepal. *Journal of Geophysical Research: Solid Earth, 105*(B6), 13389–13407.

Cheloni, D., D'Agostino, N., & Selvaggi, G. (2014). Interseismic coupling, seismic potential and earthquake recurrence on the southern front of the Eastern Alps (NE Italy). *Journal of Geophysical Research, 119*(5), 4448–4468.

Chen, W. P., & Molnar, P. (1977). Seismic moments of major earthquakes and the average rate of slip in central Asia. *Journal of Geophysical Research: Atmospheres, 82*(20), 2945–2970.

Chlieh, M., Perfettini, H., Tavera, H., Avouac, J. P., Remy, D., Nocquet, J. M., et al. (2011). Interseismic coupling and seismic potential along the Central Andes subduction zone. *Journal of Geophysical Research: Solid Earth, 116*(B12), 12405.

Duputel, Z., Vergne, J., Rivera, L., Wittlinger, G., Farra, V., & Hetényi, G. (2016). The 2015 Gorkha earthquake: A large event illuminating the Main Himalayan Thrust fault. *Geophysical Research Letters, 43*(6), 2517–2525. https://doi.org/10.1002/2016gl068083.

Elliott, J. R., Jolivet, R., González, P. J., Avouac, J. P., Hollingsworth, J., Searle, M. P., et al. (2016). Himalayan megathrust geometry and relation to topography revealed by the Gorkha earthquake. *Nature Geoscience, 9*(2), 174–180. https://doi.org/10.1038/ngeo2623.

Feldl, N., & Bilham, R. (2006). Great Himalayan earthquakes and the Tibetan plateau. *Nature, 444*(7116), 165–170.

Freund, L. B., & Barnett, D. M. (1976). A two-dimensional analysis of surface deformation due to dip-slip faulting. *Bulletin of the Seismological Society of America, 66*(3), 667–675.

Fu, Y., & Freymueller, J. T. (2012). Seasonal and long-term vertical deformation in the Nepal Himalaya constrained by GPS and GRACE measurements. *Journal of Geophysical Research: Solid Earth, 117*, B03407. https://doi.org/10.1029/2011JB008925.

Galetzka, J., Melgar, D., Genrich, J. F., Geng, J., Owen, S., Lindsey, E. O., et al. (2015). Slip pulse and resonance of the Kathmandu basin during the 2015 Gorkha earthquake. *Nepal. Science, 349*(6252), 1091–1095. https://doi.org/10.1126/science.aac6383.

Grandin, R., Doin, M.-P., Bollinger, L., Pinel-Puysségur, B., Ducret, G., Jolivet, R., et al. (2012). Long-term growth of the Himalaya inferred from interseismic InSAR measurement. *Geology, 40*(12), 1059–1062. https://doi.org/10.1130/g33154.1.

Grandin, R., Vallée, M., Satriano, C., Lacassin, R., Klinger, Y., Simoes, M., et al. (2015). Rupture process of the Mw = 7.9 2015 Gorkha earthquake (Nepal): Insights into Himalayan megathrust segmentation. *Geophysical Research Letters, 42*(20), 8373–8382.

Gualandi, A., Avouac, J.-P., Galetzka, J., Genrich, J. F., Blewitt, G., Adhikari, L. B., et al. (2017). Pre- and post-seismic deformation related to the 2015, Mw 7.8 Gorkha earthquake. *Nepal. Tectonophysics, 714–715*, 90–106.

He, P., Wang, Q., Ding, K., Wang, M., Qiao, X., Li, J., et al. (2016). Source model of the 2015 Mw 6.4 Pishan earthquake constrained by interferometric synthetic aperture radar and GPS: Insight into blind rupture in the western Kunlun Shan. *Geophysical Research Letters, 43*(4), 1511–1519.

Hyndman, R. D. (2013). Downdip landward limit of Cascadia great earthquake rupture. *Journal of Geophysical Research: Solid Earth, 118*(10), 5530–5549.

Jackson, M., & Bilham, R. (1994). Constraints on Himalayan deformation inferred from vertical velocity fields in Nepal and Tibet. *Journal of Geophysical Research: Solid Earth, 99*(B7), 13897–13912. https://doi.org/10.1029/94JB00714.

Jouanne, F., Mugnier, J. L., Gamond, J. F., Fort, P. L., Pandey, M. R., Bollinger, L., et al. (2004). Current shortening across the Himalayas of Nepal. *Geophysical Journal International, 157*(1), 1–14. https://doi.org/10.1111/j.1365-246X.2004.02180.x.

Jouanne, F., Mugnier, J. L., Sapkota, S. N., Bascou, P., & Pecher, A. (2017). Estimation of coupling along the Main Himalayan Thrust in the central Himalaya. *Journal of Asian Earth Sciences, 133*, 62–71.

Kumar, S., Wesnousky, S. G., Jayangondaperumal, R., Nakata, T., Kumahara, Y., & Singh, V. (2010). Paleoseismological evidence of surface faulting along the northeastern Himalayan front, India: Timing, size, and spatial extent of great earthquakes. *Journal of*

Geophysical Research: Solid Earth, 115, B12422. https://doi.org/10.1029/2009JB006789.

Lavé, J., & Avouac, J. P. (2000). Active folding of fluvial terraces across the Siwaliks Hills, Himalayas of central Nepal. *Journal of Geophysical Research: Solid Earth, 105*(B3), 5735–5770.

Lavé, J., & Avouac, J. P. (2001). Fluvial incision and tectonic uplift across the Himalayas of central Nepal. *Journal of Geophysical Research: Solid Earth, 106*(B11), 26561–26591.

Lawson, C. L., & Hanson, R. J. (1974). *Solving least squares problems. Prentice-Hall, 77*(1), 673–682.

Lay, T., Kanamori, H., Ammon, C. J., Koper, K. D., Hutko, A. R., Ye, L., et al. (2012). Depth-varying rupture properties of subduction zone megathrust faults. *Journal of Geophysical Research: Solid Earth, 117*, B04311. https://doi.org/10.1029/2011JB009133.

Liang, S., Gan, W., Shen, C., Xiao, G., Liu, J., Chen, W., et al. (2013). Three-dimensional velocity field of present-day crustal motion of the Tibetan Plateau derived from GPS measurements. *Journal of Geophysical Research: Solid Earth, 118*(10), 5722–5732.

Lorenzo-Martín, F., Roth, F., & Wang, R. (2006). Inversion for rheological parameters from post-seismic surface deformation associated with the 1960 Valdivia earthquake, Chile. *Geophysical Journal International, 164*(1), 75–87.

Maerten, F. (2005). Inverting for slip on three-dimensional fault surfaces using angular dislocations. *Bulletin of the Seismological Society of America, 95*(5), 1654–1665.

McNamara, D. E., Yeck, W. L., Barnhart, W. D., Schulte-Pelkum, V., Bergman, E., Adhikari, L. B., et al. (2017). Source modeling of the 2015 Mw 7.8 Nepal (Gorkha) earthquake sequence: Implications for geodynamics and earthquake hazards. *Tectonophysics, 714–715*, 21–30. https://doi.org/10.1016/j.tecto.2016.08.004.

Meade, B. J. (2007). Algorithms for the calculation of exact displacements, strains, and stresses for triangular dislocation elements in a uniform elastic half space. *Computers & Geosciences, 33*(8), 1064–1075.

Mencin, D., Bendick, R., Upreti, B. N., Adhikari, D. P., Gajurel, Ananta P., Bhattarai, R. R., et al. (2016). Himalayan strain reservoir inferred from limited afterslip following the Gorkha earthquake. *Nature Geoscience, 9*, 533–537. https://doi.org/10.1038/ngeo2734.

Métois, M., Socquet, A., & Vigny, C. (2012). Interseismic coupling, segmentation and mechanical behavior of the central Chile subduction zone. *Journal of Geophysical Research: Solid Earth, 117*(B3), B03406. https://doi.org/10.1029/2011JB008736.

Molnar, P., & Chen, W. P. (1984). S–P wave travel time residuals and lateral inhomogeneity in the mantle beneath Tibet and the Himalaya. *Journal of Geophysical Research: Atmospheres, 89*(B8), 6911–6917.

Molnar, P., & Tapponnier, P. (1975). Cenozoic tectonics of Asia: Effects of a continental collision: Features of recent continental tectonics in Asia can be interpreted as results of the India–Eurasia collision. *Science, 189*(4201), 419–426.

Morsut, F., Pivetta, T., Braitenberg, C., & Poretti, G. (2017). Strain Accumulation and Release of the Gorkha, Nepal, Earthquake (Mw 7.8, 25 April 2015). *Pure and Applied Geophysics, 175*(5), 1909–1923.

Mugnier, J. L., Gajurel, A., Huyghe, P., Jayangondaperumal, R., Jouanne, F., & Upreti, B. (2013). Structural interpretation of the great earthquakes of the last millennium in the central Himalaya. *Earth-Science Reviews, 127*(1), 30–47.

Mugnier, J. L., Huyghe, P., Gajurel, A. P., Upreti, B. N., & Jouanne, F. (2011). Seismites in the Kathmandu basin and seismic hazard in central Himalaya. *Tectonophysics, 509*(1), 33–49.

Mugnier, J. L., Jouanne, F., Bhattarai, R., Cortes-Aranda, J., Gajurel, A., Leturmy, P., et al. (2017). Segmentation of the Himalayan megathrust around the Gorkha earthquake (25 April 2015) in Nepal. *Journal of Asian Earth Science, 141,* 236–252. https://doi.org/10.1016/j.jseaes.2017.01.015.

Nábělek, J., Hetényi, G., Vergne, J., Sapkota, S., Kafle, B., Jiang, M., et al. (2009). Underplating in the Himalaya-Tibet collision zone revealed by the Hi-CLIMB experiment. *Science, 325*(5946), 1371–1374.

Pandey, M. R., Tandukar, R. P., Avouac, J. P., Lave, J., & Massot, J. P. (1995). Interseismic strain accumulation on the Himalaya crustal ramp (Nepal). *Geophysical Research Letters, 22,* 751–754. https://doi.org/10.1029/94GL02971.

Pollitz, F. F. (2014). Post-earthquake relaxation using a spectral element method: 2.5-D case. *Geophysical Journal International, 198*(1), 308–326.

Prawirodirdjo, L., Mccaffrey, R., Chadwell, C. D., Bock, Y., & Subarya, C. (2010). Geodetic observations of an earthquake cycle at the Sumatra subduction zone: Role of interseismic strain segmentation. *Journal of Geophysical Research: Solid Earth, 115*(B3), 153–164.

Sapkota, S. N., Bollinger, L., Klinger, Y., Tapponnier, P., Gaudemer, Y., & Tiwari, D. (2013). Primary surface ruptures of the great Himalayan earthquakes in 1934 and 1255. *Nature Geoscience, 6*(1), 71–76. https://doi.org/10.1038/ngeo1669.

Savage, J. C. (1983). A dislocation model of strain accumulation and release at a subduction zone. *Journal of Geophysical Research: Solid Earth, 88*(B6), 4984–4996.

Sreejith, K. M., Sunil, P. S., Agrawal, R., Saji, A. P., Ramesh, D. S., & Rajawat, A. S. (2016). Coseismic and early postseismic deformation due to the 25 April 2015, Mw 7.8 Gorkha, Nepal, earthquake from InSAR and GPS measurements. *Geophysical Research Letters, 43*(7), 3160–3168.

Stevens, V. L., & Avouac, J. P. (2015). Interseismic coupling on the main Himalayan thrust. *Geophysical Research Letters, 42*(14), 5828–5837. https://doi.org/10.1002/2015gl064845.

Stevens, V. L., & Avouac, J. P. (2016). Millenary Mw > 9.0 earthquakes required by geodetic strain in the Himalaya. *Geophysical Research Letters, 43*(3), 1118–1123.

Suwa, Y., Miura, S., Hasegawa, A., Sato, T., & Tachibana, K. (2006). Interplate coupling beneath NE Japan inferred from three-dimensional displacement field. *Journal of Geophysical Research: Atmospheres, 111*(B4), 258–273.

Szeliga, W., Hough, S., Martin, S., & Bilham, R. (2010). Intensity, magnitude, location, and attenuation in India for felt earthquakes since 1762. *Bulletin of the Seismological Society of America, 100*(2), 570–584.

Tan, K., Zhao, B., Zhang, C. H., Du, R. L., Wang, Q., Huang, Y., et al. (2016). Rupture models of the Nepal Mw 7.9 earthquake and Mw 7.3 aftershock constrained by GPS and InSAR coseismic deformations. *Chinese Journal of Geophysics, 59*(6), 2080–2093. **(in Chinese)**.

Vergne, J., Cattin, R., & Avouac, J. P. (2001). On the use of dislocations to model interseismic strain and stress build-up at intracontinental thrust faults. *Geophysical Journal International, 147*(1), 155–162. https://doi.org/10.1046/j.1365-246X.2001.00524.x.

Vergnolle, M., Pollitz, F., & Calais, E. (2003). Constraints on the viscosity of the continental crust and mantle from GPS measurements and postseismic deformation models in western Mongolia. *Journal of Geophysical Research: Solid Earth, 108*(B10), 2502. https://doi.org/10.1029/2002JB002374.

Wang, K., & Fialko, Y. (2018). Observations and modeling of coseismic and postseismic deformation due to the 2015 Mw 7.8 Gorkha (Nepal) earthquake. *Journal of Geophysical Research: Solid Earth, 123*(1), 761–779.

Wang, W., Qiao, X., Yang, S., & Wang, D. (2017). Present-day velocity field and block kinematics of Tibetan Plateau from GPS measurements. *Geophysical Journal International, 208,* 1088–1102.

Wang, Q., Zhang, P. Z., Freymueller, J. T., Bilham, R., Larson, K. M., Lai, X., et al. (2001). Present-day crustal deformation in China constrained by global positioning system measurements. *Science, 294*(5542), 574–577.

Xiong, W., Tan, K., Qiao, X., Liu, G., Nie, Z., & Yang, S. (2017). Coseismic, postseismic and interseismic coulomb stress evolution along the himalayan main frontal thrust since 1803. *Pure and Applied Geophysics, 174*(5), 1889–1905.

Xu, C., Ding, K., Cai, J., & Grafarend, E. W. (2009). Methods of determining weight scaling factors for geodetic–geophysical joint inversion. *Journal of Geodynamics, 47*(1), 39–46.

Yi, L., Xu, C., Zhang, X., Wen, Y., Jiang, G., Li, M., et al. (2017). Joint inversion of GPS, InSAR and teleseismic data sets for the rupture process of the 2015 Gorkha, Nepal, earthquake using a generalized ABIC method. *Journal of Asian Earth Sciences, 148,* 121–130.

Zhao, B., Bürgmann, R., Wang, D., Tan, K., Du, R., & Zhang, R. (2017). Dominant controls of downdip afterslip and viscous relaxation on the postseismic displacements following the Mw 7.9 Gorkha, Nepal, earthquake. *Journal of Geophysical Research: Solid Earth, 122*(10), 8376–8401.

Zhao, W., Nelson, K. D., Che, J., Quo, J., Lu, D., Wu, C., et al. (1993). Deep seismic reflection evidence for continental underthrusting beneath southern Tibet. *Nature, 366*(6455), 557–559.

Zumberge, J. F., Heflin, M. B., Jefferson, D. C., Watkins, M. M., & Webb, F. H. (1997). Precise point positioning for the efficient and robust analysis of GPS data from large networks. *Journal of Geophysical Research: Solid Earth, 102*(B3), 5005–5017.

(Received April 9, 2018, revised January 27, 2019, accepted January 30, 2019, Published online February 5, 2019)

Pure Appl. Geophys. 176 (2019), 3913–3928
© 2019 Springer Nature Switzerland AG
https://doi.org/10.1007/s00024-018-02090-3

| **Pure and Applied Geophysics** |

Role of Lower Crust in the Postseismic Deformation of the 2010 Maule Earthquake: Insights from a Model with Power-Law Rheology

Carlos Peña,[1,2] (iD) Oliver Heidbach,[1] Marcos Moreno,[1,3] Jonathan Bedford,[1] Moritz Ziegler,[1,4]
Andrés Tassara,[5,6] and Onno Oncken[1,2]

Abstract—The surface deformation associated with the 2010 M_w 8.8 Maule earthquake in Chile was recorded in great detail before, during and after the event. The high data quality of the continuous GPS (cGPS) observations has facilitated a number of studies that model the postseismic deformation signal with a combination of relocking, afterslip and viscoelastic relaxation using linear rheology for the upper mantle. Here, we investigate the impact of using linear Maxwell or power-law rheology with a 2D geomechanical-numerical model to better understand the relative importance of the different processes that control the postseismic deformation signal. Our model results reveal that, in particular, the modeled cumulative vertical postseismic deformation pattern in the near field (< 300 km from the trench) is very sensitive to the location of maximum afterslip and choice of rheology. In the model with power-law rheology, the afterslip maximum is located at 20–35 km rather than > 50 km depth as suggested in previous studies. The explanation for this difference is that in the model with power-law rheology the relaxation of coseismically imposed differential stresses occurs mainly in the lower crust. However, even though the model with power-law rheology probably has more potential to explain the vertical postseismic signal in the near field, the uncertainty of the applied temperature field is substantial, and this needs further investigations and improvements.

1. Introduction

At subduction zones, the sudden release of strain that has accumulated over tens to hundreds of years repeatedly produces the failure of large areas of the boundary interface, resulting in great ($M_w > 8.5$) or even giant ($M_w > 9.0$) earthquakes (Barrientos and Ward 1990; Chlieh et al. 2008; Moreno et al. 2012; Schurr et al. 2014). This sudden slip is followed by postseismic deformation that gradually relaxes the coseismically induced stress perturbations. The rate of postseismic deformation is time-dependent and has been attributed to three primary processes: (1) afterslip (Bedford et al. 2013; Hsu et al. 2006; Perfettini et al. 2010; Tsang et al. 2016), (2) poro-elastic rebound (Hu et al. 2014; Hughes et al. 2010) and (3) viscoelastic relaxation (Hu et al. 2004; Pollitz et al. 2006; Qiu et al. 2018; Rundle, 1978; Wang et al. 2012). Interseismic relocking or simply relocking is another process that may occur shortly after megathrust events. Bedford et al. (2016) inferred that the fault interface relocked within the first year after the 2010 Maule earthquake. A similar finding was obtained by Remy et al. (2016) after the 2007 Pisco, Peru, earthquake. In the past decade, the increased spatial density of continuous GPS (cGPS) instrumentation at subduction zones together with the implementation of geomechanical-numerical models has allowed us to test the relative importance of these processes in time and space (Bedford et al. 2016; Govers et al. 2017; Klein et al. 2016; Li et al. 2017; 2018; Sun et al. 2014). In these studies, linear viscoelastic relaxation has been used to infer the viscosity structure of the upper mantle and to understand the postseismic deformation signal in the

[1] Helmholtz Centre Potsdam, GFZ German Research Centre for Geosciences, Potsdam, Germany. E-mail: carlosp@gfz-potsdam.de
[2] Freie Universität Berlin, Berlin, Germany.
[3] Departamento de Geofísica, Universidad de Concepción, Concepción, Chile.
[4] Institute of Earth and Environmental Science, University of Potsdam, Potsdam, Germany.
[5] Departmento de Ciencias de la Tierra, Universidad de Concepción, Concepción, Chile.
[6] Millenium Nucleus CYCLO "The Seismic Cycle along Subduction Zones", Valdivia, Chile.

near, middle and far field. These models assume that the crust is purely elastic and that the relaxation in the upper mantle can be described with a linear viscoelastic rheology using either the Maxwell (Govers et al. 2017; Hu et al. 2004; Li et al. 2017, 2018) or Burgers body (Klein et al. 2016; Sun et al. 2014). Furthermore, most of these models consider an inversion scheme to estimate the location and magnitude of afterslip as well as the viscosity structure of the mantle that results in a best fit of the observed cumulative postseismic deformation signal derived from GPS observations. Alternatively, in their 2D geomechanical-numerical forward model Hergert and Heidbach (2006) showed that a power-law rheology with dislocation creep can also fit the vertical and horizontal time series of the postseismic relaxation after the 2001 Arequipa earthquake. However, for their study only one cGPS station at 225 km distance from the trench was available and no afterslip was considered.

The 2010 M_w 8.8 Maule earthquake that struck south-central Chile was one of the first great events to be captured by modern space-geodetic monitoring networks (Vigny et al. 2011; Moreno et al. 2012). Through a rapid international collaborative effort, a dense cGPS network of 67 stations (Bedford et al. 2013; Bevis et al. 2010; Vigny et al. 2011) was installed to monitor the postseismic surface deformation (Fig. 1). Recent analyses of the postseismic deformation signal from the Maule earthquake have drawn attention to the limits posed by using a linear viscoelastic relaxation with homogeneous viscosity distribution in the mantle (Klein et al. 2016; Li et al. 2017, 2018) to explain the heterogeneity of the vertical postseismic signal, showing that a simple process is not a candidate to explain the postseismic signal associated with the 2010 Maule case. The best-fit model from Klein et al. (2016) results in a heterogeneous viscosity structure with a deep viscoelastic channel up to 135 km depth along the fault interface and afterslip at regions close to the up- and down-dip limits to explain in particular the pattern of the observed vertical displacement and the displacement over time in the north, east and vertical components recorded by the cGPS time series. On the other hand, Li et al. (2017, 2018) showed how lateral viscosity variations improve the fit of the observed

cumulative postseismic vertical deformation while having less effect on the horizontal predictions. Furthermore, they speculate that a power-law rheology could also explain the postseismic relaxation, in agreement with results from laboratory experiments (Bürgmann and Dresen 2008; Hirth and Tullis 1992; Karato and Wu 1993; Kirby and Kronenberg 1987).

In this article, we investigate the general differences that result from the use of a power-law rheology compared with a linear viscoelastic relaxation in a Maxwell body for the purpose of better understanding the processes controlling the spatio-temporal patterns of the postseismic deformation signal. We construct a 2D geomechanical-numerical model along a cross section perpendicular to the strike of the subduction zone at 36°S sub-parallel to the maximum of the coseismic slip of the Maule earthquake (Fig. 1). We model the first 6 years of postseismic deformation and compare our model results with the vertical and horizontal components of the cumulative and time series displacements of cGPS sites as a function of distance from the trench. The primary focus of this study is not to achieve a best-fit solution of the cGPS signal using an inversion scheme; instead, we use forward models to study the principal differences between a linear Maxwell and power-law rheology. However, the results of our test series to study the sensitivity due to linear Maxwell versus power-law rheology as well as due to the location and magnitude of afterslip partly show a remarkably good fit to the observed postseismic signals.

Our model results indicate that the overall contribution of relocking to the cumulative postseismic deformation signal is small compared with the impact of afterslip and viscoelastic relaxation. Our model results confirm previous studies (Klein et al. 2016; Li et al. 2017, 2018; Qiu et al. 2018) that showed that the vertical postseismic deformation signal is the key to better assess the relative importance of the involved processes, i.e., the viscosity, effective viscosity, maximum magnitude and location of afterslip. We show that in particular the predicted cumulative vertical postseismic signal in the near field (distance < 300 km from the trench) is very sensitive to the choice of model rheology as well as the afterslip location and maximum. The model with power-law

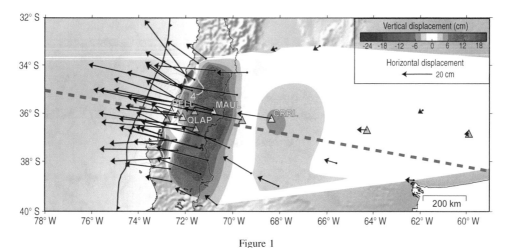

Figure 1
Study area and cumulative postseismic displacement after 6 years of the Maule event derived from cGPS observations in the stable South American reference frame. Horizontal (black arrows) and interpolated vertical displacements (color coded) show the cumulative postseismic deformation in the first 6 years after the M_w 8.8 Maule earthquake. Green and yellow triangles display the 11 cGPS sites used in this study. Yellow triangles show the four cGPS sites considered for the time series analysis. Yellow contour lines depict the 2010 Maule earthquake coseismic slip from Moreno et al. (2012). Blue dotted line represents the 2D model cross section oriented parallel to the horizontal postseismic deformation

rheology favors afterslip at depths of 20–35 km rather than at the down-dip limit of the seismogenic zone > 50 km. This shift of afterslip location is explained with the dislocation creep process that occurs in the deeper part of the lower crust and the uppermost mantle.

2. Model Description

2.1. Model Setup

In the first 6 years following the Maule event, the postseismic surface displacement is almost perpendicular to the strike of the trench. We thus choose a 2D model cross section oriented parallel to the direction of the observed horizontal cumulative postseismic displacement vector. The model geometry is derived from the model of Li et al. (2017). The cross section is almost perpendicular to the trench and cuts through the center of the coseismic rupture where the key postseismic deformation processes take place (Fig. 1). The model geometry takes into account the geometry of the slab (Hayes et al. 2012) and extends 3800 km in the horizontal and 400 km in the vertical direction to avoid boundary effects (Fig. 2a).

The model is discretized with 112,000 finite elements with a high resolution close to the slab interface where the coseismic displacement occurs and a significantly coarser resolution at the model boundaries where no deformation is expected. We assign to each element the rock properties presented in Table 1 differentiating the continental crust, oceanic crust/slab and upper mantle. At the lower and lateral model boundaries, the model cannot displace in the normal direction, but it is free to move parallel to the model boundaries; the model surface is free of constraints (Fig. 2a).

The temperature field of the model is taken from Springer (1999) by interpolating the temperature contours and assigning the according temperature to each node of the finite elements (Fig. 2b). The temperature field is assumed to be time-independent as no significant changes are expected within 6 years. Coseismic slip models for the Maule earthquake (Bedford et al. 2013; Klein et al. 2016; Moreno et al. 2012; Vigny et al. 2011; Yue et al. 2014) show some differences, mainly in magnitude and location of maximum slip. This is most probably due to the use of different data sets and regularization methods in the inversion process. Postseismic deformation modeled with power-law rheology depends on the

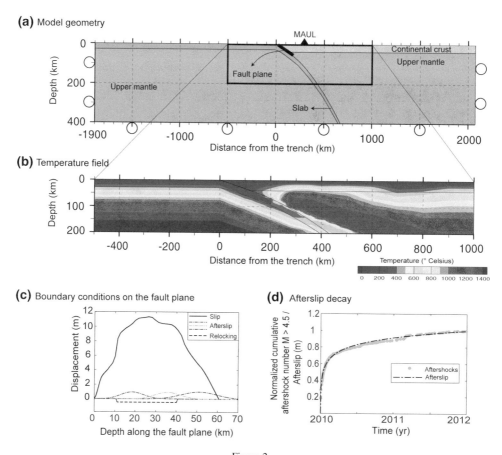

Figure 2
Model setup. **a** The 2D model geometry along the cross section is shown in Fig. 1. Circles indicate that no displacement is allowed perpendicular to the model boundary. **a** Exaggerated in the vertical by a factor of two. **b** The implemented temperature field according to Springer (1999) in the area of key interest. **c** Distribution of coseismic slip taken from the inversion of Moreno et al. (2012) and afterslip distributions. **d** Afterslip decay law used in this study. The aftershocks seismicity corresponds to $M_w > 4.5$ taken from the NEIC catalogue (http://www.usgs.gov)

Table 1

Elastic and creep parameters

Layer	Rock type[b]	Young's module E (MPa)[a]	Poisson's ratio v[a]	Pre-exponent A (MPa^{-n} s^{-1})[b]	Stress exponent n[b]	Activation enthalpy Q (kJ mol^{-1})[b]
Continental crust	Wet quartzite	1×10^5	0.265	3.2×10^{-4}	2.3	154
Oceanic crust/slab	Diabase	1.2×10^5	0.3	2.0×10^{-4}	3.4	260
Upper mantle	Olivine	1.6×10^5	0.25	2.0	3.0	433

[a]Reference source from Christensen (1996) and Khazaradze et al. (2002)

[b]Reference source from Ranalli (1997) and Karato and Wu (1993)

coseismic stress changes, and therefore may vary depending on the coseismic slip distribution. In this study, we decided to implement the coseismic slip distribution from the inversion of Moreno et al. (2012) as a displacement boundary condition on the fault plane (Fig. 2c), because our study shares the same numerical approach (FEM), margin geometry (slab and Moho discontinuities) and elastic material parameters as Moreno et al. (2012). To fit the observed coseismic displacement from previous studies (Moreno et al. 2012; Vigny et al. 2011), we assign 70% of the coseismic slip to the upper side of the fault plane toward the up-dip direction and 30% to the bottom side toward the down-dip direction (Govers et al. 2017; Hergert and Heidbach 2006; Sun and Wang 2015). The same ratio is applied to simulate afterslip and relocking.

The afterslip is modeled with a Gaussian distribution curve and decays exponentially to the 2nd year as explained by Marone et al. (1991). The afterslip decay law also is in agreement with the aftershock seismicity (Fig. 2d), which is a first-order approximation for the afterslip decay law for the 2010 Maule case (Bedford et al. 2016; Lange et al. 2014). Klein et al. (2016) found cumulated afterslip values on the order of 100 cm at 45 km depth between 2011 and 2012 for the postseismic deformation associated with the Maule event. Thus, we start with 100 cm of maximum afterslip centered at 48 km depth, but vary these values in different model scenarios. Different afterslip decay laws may achieve a better fit to the data; however, we do not explore this parameter since the main focus of this study is to investigate the first-order differences between the models that use linear Maxwell or power-law rheology instead of perfectly fitting the observations. Relocking is assumed as backslip on the rupture plane with a convergence velocity of 6 cm $year^{-1}$ and takes place linearly up to the 6th year. With these kinematic boundary conditions, i.e., the coseismic rupture, afterslip distribution and relocking, the model simulates the postseismic relaxation of stresses during 6 years. The resulting numerical problem is solved using the commercial finite element code ABAQUS^{TM}, version 6.11.

2.2. Model Rheology

We implement the dislocation creep law for models with power-law rheology using the expression stated in Kirby and Kronenberg (1987)

$$\dot{\varepsilon} = A\sigma^n \exp\left(\frac{-Q}{RT}\right), \qquad (1)$$

where $\dot{\varepsilon}$ is the strain rate, A is a pre-exponent parameter, σ the differential stress, n the stress exponent, Q the activation enthalpy for creep, R the gas constant and T the absolute temperature. The key control is the stress exponent n and the temperature field. In particular, the latter controls were in the

continental crust where the brittle-ductile transition (BDT) zone is located (Brace and Kohlstedt 1980; Ranalli 1997). Below the BDT the differential stress is relaxed by dislocation creep processes. Our models with linear Maxwell rheology use a viscosity of 1.3×10^{19} Pa s for the uppermost mantle and elastic parameters for the crust and oceanic/slab. This value is in agreement with previous studies on the Chilean subduction zone (Bedford et al. 2016; Hu et al. 2004) that found viscosity values on the order of 10^{19} Pa s. We emphasize that the main difference is the fact that in our model with linear Maxwell rheology the whole crust is considered as an elastic material above a viscous mantle, while in the model with power-law rheology the viscosity distribution is controlled by the implemented temperature field. Elastic and creep parameters used in the model area are listed in Table 1.

2.3. GPS Observations

The cGPS observations in the Maule region show trench-ward motion in the horizontal component and different patterns of deformation in the vertical component along longitude, with a pronounced uplift in the Andean region (Fig. 1). We use the first 6 years of postseismic surface displacements observed by cGPS as reported by Li et al. (2017). In this data set, the effect of aftershocks was removed by applying the trajectory model of Bevis and Brown (2014). To compare with the prediction of our 2D model, we selected 11 cGPS sites distributed in the near, middle and far field for comparison with our model (yellow triangles in Fig. 1).

3. Results

Based on the model described in the previous section, we set up three different model groups to test the general difference when using linear Maxwell or power-law rheology in the model. An overview of different model parameters is provided in Table 2. In the first test group we focus on models with power-law rheology and investigate the relative impact of relocking and afterslip on the postseismic deformation pattern (Sect. 3.1 and Fig. 3). In the second test

Table 2

Description of the model parameters (rheology, afterslip and relocking) used in this study

Model	Maximum of afterslip (cm)	Depth of maximum afterslip (km)	Relocking (cm year^{-1})	Temperature (°C)	Graph color and type
NLA100D48R	100	48	6	T	Figures 3, 4 and 5: solid blue
NLA100D35R	100	35	6	T	Figure 5: solid orange
NLA100D20R	100	20	6	T	Figures 5 and 6: solid red
NLA100D48	100	48	–	T	Figure 3: solid thin blue
NLA20D48R	20	48	6	T	Figures 3 and 4: solid cyan
NLA20D48	20	48	–	T	Figure 3: solid thin cyan
NLA0R	0	–	6	T	Figures 3 and 4: solid green
NLA0	0	–	–	T	Figures 3, 6 and 8: solid thin green
NLA0T + 100	0	–	–	T + 100	Figure 8: solid dark red
NLA0T-100	0	–	–	T − 100	Figure 8: solid pink
LA100D48R	100	48	6	T	Figures 4 and 5: dashed blue
LA100D35R	100	35	6	T	Figure 5: dashed orange
LA100D20R	100	20	6	T	Figures 5 and 6: dashed red
LA20D48R	20	48	6	T	Figure 4: dashed cyan
LA0R	0	–	6	T	Figure 4: dashed green
LA0	0	–	–	T	Figure 6: dashed pink

The rheology, linear (L, Maxwell) and non-linear (NL, power-law), maximum afterslip (A), relocking (R) and changes in the initial temperature field from the Springer model (T) are indicated in the model name. If relocking is considered, it is always with a rate of 6 cm year^{-1}

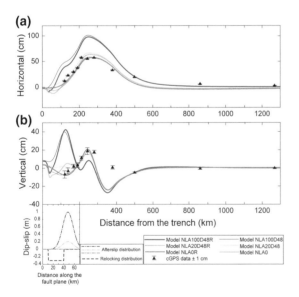

Figure 3

Relative impact of afterslip and relocking for the cumulative surface displacement 6 years after the Maule event compared with cGPS observations. Afterslip and relocking distributions for the six models are shown below the figures at the location relative to the trench. **a** Horizontal displacement: positive values represent trench-ward motion and negative landward motion. cGPS displacements are projected onto the model cross section. **b** Vertical displacement

group we focus on differences when using linear Maxwell or power-law model rheology and different afterslip magnitudes (Sect. 3.2 and Fig. 4), and in the third test group we investigate the differences when using linear Maxwell or power-law model rheology and different depth locations of the maximum after-slip (Sect. 3.3 and Fig. 5).

3.1. Relative Impact of Relocking and Afterslip in Models with Power-Law Rheology

Figure 3 shows the comparison of the cumulative postseismic surface displacement after 6 years between the model results and the data from the cGPS stations. We used three different maximum amplitudes of afterslip at 48 km depth. To evaluate the relative contribution of relocking, we fully and uniformly locked the fault interface as backslip between 10 and 40 km depth (Govers et al. 2017; Tichelaar and Ruff 1993). We also perform tests without relocking to assess its relative impact on the cumulative vertical and horizontal postseismic displacement signal (Fig. 3). The models with and without relocking produce landward motion in the

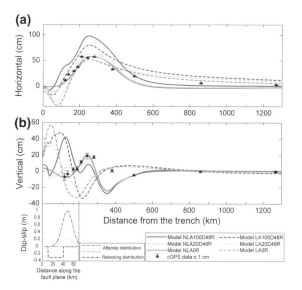

Figure 4

Impact of rheology and afterslip maximum on the cumulative surface displacement 6 years after the Maule earthquake compared with cGPS observations. Afterslip and relocking distributions for the six models are shown below the figures at the location relative to the trench. **a** Horizontal displacement. GPS velocities are projected onto the model cross section. **b** Vertical displacement

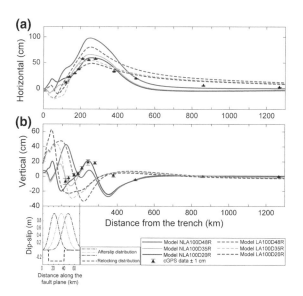

Figure 5

Impact of rheology and location of the afterslip maximum on the cumulative surface displacement 6 years after the Maule earthquake compared with cGPS observations. Afterslip and relocking distributions for the six models are shown below the figures at the location relative to the trench. **a** Horizontal displacement. GPS velocities are projected onto the model cross section. **b** Vertical displacement

very near field (< 50 km from the trench). In general, our results indicate that relocking does not affect the deformation field significantly (see continuous versus dashed lines in Fig. 3). A small signal is seen close to the trench (< 80 km from the trench), and it vanishes at distances > 200 km from the trench for both the horizontal and vertical displacements. Changing the maximum of the afterslip does not change the pattern of the horizontal surface deformation at distances > 600 km from the trench, but it changes the magnitude of trench-ward motion at distances between 150 and 400 km from the trench. Beyond distances of 600 km from the trench, the results show trench-ward motion when 100 cm of maximum afterslip is used, but small landward motion when it is reduced to 20 and 0 cm, respectively. Interestingly, our results show that the vertical deformation is the component most sensitive to the afterslip maximum. The afterslip centered at the down-dip limit of the seismogenic zone produces maximum uplift around 100 km from the trench. When 100 cm afterslip is applied, an uplift of 40 cm after 6 years is accumulated. This number is considerably reduced when only 20 cm maximum afterslip is used; without any afterslip it changes to subsidence. These results are in agreement with Wang and Fialko (2014, 2018), who found afterslip at the down-dip limit produces uplift at that region, while subsidence is controlled by viscoelastic relaxation. Beyond distances of 400 km, the impact of different afterslip magnitudes is negligible.

The overall pattern of the horizontal cGPS signal is better explained by models with small afterslip at the down-dip limit of the seismogenic zone than when 100 cm of afterslip is considered, in particular in the area of largest deformation between 200 and 400 km from the trench. An increase in maximum afterslip results in an increase in surface deformation that leads to an overestimation of the horizontal component in the near field.

The observed patterns in the vertical signal are also in better agreement with models when a smaller afterslip is applied. Adding afterslip shifts the higher uplift signal toward the trench in a different pattern, as observed by the cGPS observations. All models are in a good agreement with the cGPS observations in the far field (> 500 km from the trench). However, none of the models can explain the wavelength of the declining uplift signal observed between 300 and 500 km from the trench (Fig. 3b). In general, the geomechanical-numerical model with power-law rheology results qualitatively in a good fit to the overall surface deformation pattern observed at the cGPS sites.

3.2. Impact of Afterslip Maximum in Models with Linear Maxwell and Power-Law Rheology

In the second model group, we model the cumulative surface deformation 6 years after the 2010 Maule event using models with linear Maxwell or power-law rheology and different afterslip magnitudes of 100, 20 and 0 cm located at the down-dip limit of the seismogenic zone (Fig. 4). We use the same three models with power-law rheology (as in Fig. 3), where the afterslip maximum is at 48 km depth, and compare these with models that have the same setup, but considering linear Maxwell rheology. Furthermore, despite the results presented in Fig. 3 that show a minor contribution from relocking on the cumulative surface deformation, in Fig. 4 we consider all models with relocking after 2 years.

Similar to the results presented in Sect. 3.1, the maximum of the afterslip also has an impact on the horizontal and vertical deformation signal for the models with linear Maxwell rheology, but it is smaller than the magnitude inferred using the models with power-law rheology, in particular for the vertical component (Fig. 4b). The horizontal component shows the largest differences between models with linear Maxwell and power-law rheology in amplitude and patterns in the near field among the models, but the difference in the overall pattern is small (Fig. 4a). In the far field all models with linear Maxwell rheology overestimate the horizontal displacement compared with the ones with power-law rheology. Significant differences between the models with linear and non-linear rheology are found in particular in the near field for the vertical component and to a lesser extent in the middle and far field (Fig. 4b). While models with power-law rheology show uplift at about 200–300 km and subsidence at about 300–700 km from the trench, models with linear rheology show the opposite surface displacement pattern.

Compared with the horizontal cGPS signal, the overall pattern from the models with linear Maxwell and power-law rheology agrees with the observations equally well in the area of key postseismic deformation, in the Andean region (Fig. 4a). However, for the vertical cGPS signal the models with linear Maxwell rheology reveal larger differences from the observed patterns than models with power-law rheology. This holds especially for the area 150–300 km from the trench.

3.3. Impact of Afterslip Location on Models with Linear Maxwell and Power-Law Rheology

In the third model group we shift the location of the maximum afterslip of 100 cm from 48 to 35 km and 20 km depth to investigate the impact on the surface deformation in models with linear Maxwell and power-law rheology. The choice of the maximum afterslip location has important effects on the surface deformation. In particular, for the horizontal component, models with linear Maxwell or power-law rheology and shallow afterslip result in a larger surface deformation than those using moderate deep afterslip for distances closer to 100 km from the trench (Fig. 5a). Beyond distances of 200 km from the trench, the surface deformation is smaller as shallow afterslip takes place, and it is also in the same fashion as the results from models without afterslip. These differences also apply to the vertical component, mainly in models with power-law rheology (Fig. 5b). For models with power-law rheology, the impact is much larger for distances closer to 200 km from the trench than the effect observed in the horizontal component. There, the differences are both in magnitude and patterns. This effect is less pronounced in models with linear Maxwell rheology. These models show a similar pattern of deformation, where the maximum uplift and subsidence are shifted around 40 km toward the trench as afterslip moves to closer distances from the trench on the fault plane.

The different patterns of deformation shown by these models can be compared with the cGPS signal to evaluate the relative impact of afterslip on the surface deformation signal. From models with power-law rheology, our results indicate that they can better explain the overall pattern observed by cGPS where shallow afterslip is considered. In particular, the vertical component gives clear insight to evaluate the relative impact of afterslip location for surface regions closer to 300 km from the trench. Here, the remarkable uplift at about 250 km and small subsidence at about 140 km from the trench can be just explained by the power-law rheology model with maximum afterslip at either 35 km or 20 km depth. None of these models result in very small uplift as shown by one cGPS site about 400 km from the trench. However, beyond these distances, power-law rheology models explain the cGPS displacement pattern.

In summary, the key findings from previous sections are: (1) relocking is not contributing significantly to the cumulative postseismic deformation signal along the chosen model profile; (2) models with linear Maxwell rheology without adaptation of the viscosity structure at depth fail to reproduce the pattern of the observed cumulative vertical postseismic deformation signal regardless of where the maximum afterslip is located and the amplitude of the afterslip; finally, (c) the general patterns of the cGPS observations are better explained by models with power-law rheology when small values of afterslip at the down-dip limit are considered and/or when afterslip is occurring at shallower regions.

3.4. Model Results Versus Time Series of the cGPS Stations

In this section we analyze the time series for 6 years after the Maule earthquake from four cGPS stations at different distances from the trench and compare these with the models with linear Maxwell and power-law rheology (Fig. 6). For this comparison we choose the models with 100 cm maximum afterslip at a depth of 20 km and 0 cm afterslip (Fig. 6). We selected the cGPS time series of the stations PELL, QLAP, MAUL and CRRL for comparison, which are located in the near, middle and far field (yellow triangles in Fig. 1) at about 130 km, 190 km, 270 km and 500 km distance from the trench, respectively.

The largest differences from models with and without afterslip are found in the near field (cGPS site PELL). As expected, models with afterslip

(NLA100D20R and LA100D20R for the power-law and linear Maxwell case, respectively) result in larger deformation than when afterslip is assumed to be zero in particular in the near field. It is also observed that for the two cGPS sites at larger distance from the trench (MAUL and CRRL), the power-law rheology models with afterslip have very close deformation patterns and magnitudes but linear Maxwell rheology models keep small differences after 6 years. For sites at 190 km and 270 km from the trench, models with linear Maxwell and power-law rheology show very similar surface cumulative deformation for the horizontal component; however, there are large differences in the early part of the postseismic phase. In this period, the transient deformation of models with power-law rheology is much faster than linear Maxwell model scenarios, especially at 270 km from the trench where the cGPS MAUL site is located.

By comparing with the cGPS PELL site in the near field, it can be shown that the effect of afterslip is larger than that of viscous relaxation, in agreement with previous studies (Bedford et al. 2013; Hsu et al. 2006). A combination of afterslip and viscous relaxation can resemble the deformation patterns, in particular in the first 2 years. However, after the 2nd year, the model with power-law rheology can better explain the observed horizontal and vertical postseismic deformation pattern than models with linear Maxwell rheology. Compared with cGPS sites further from the trench, our results indicate that the preferred model also is a combination of power-law rheology and afterslip for both the horizontal and vertical component. Even though the models with Maxwell rheology and afterslip can produce good agreement with the cumulative surface deformation signal, they cannot produce the transient deformation in the early postseismic phase, as observations show. In the far field, at the cGPS CRRL site, no model is in agreement with the early postseismic deformation during the first years for the horizontal component. The vertical component is in very good agreement with models considering power-law rheology. In general, compared with the selected cGPS sites, models with power-law rheology show a better agreement with the overall deformation pattern signal than models with linear Maxwell rheology.

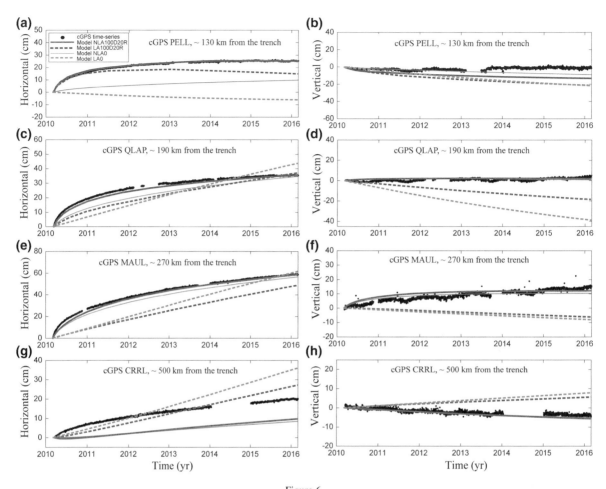

Figure 6
Time series of four cGPS stations versus model results from four models with linear Maxwell and power-law rheology for 6 years after the
Maule event. Black dots are daily solutions of the cGPS observations; distance from the trench is given in km next to the station names. Left
row (**a, c, e, g**) shows the horizontal displacement. GPS velocities are projected onto the model cross section. Right row (**b, d, f, h**) shows the
vertical displacement

4. Discussion

4.1. Location of the Viscous Relaxation Process

The largest deformation for models with power-law rheology is produced in a region about 280 km landward from the trench (Fig. 7a). Interestingly, most of the viscoelastic relaxation occurs in the lower continental crust. This is in contrast to previous studies in the Chilean subduction zone, since these assumed that the whole crust is an elastic medium above a viscoelastic mantle (Hu et al. 2004; Klein et al. 2016; Li et al. 2017, 2018), resulting in relaxation mainly occurring in the mantle wedge, in

agreement with our model results with linear Maxwell rheology (Fig. 7b).

Below the cGPS station MAUL, at 36 km depth, we infer a creep strain after 6 years of 7.9×10^{-5} and an effective viscosity of 1.1×10^{18} Pa s from the power-law model with 1 m of afterslip at 20 km depth. The creep strain and effective viscosity values are very similar for all models with power-law rheology. For the same region but at a shallower depth of only 10 km in the continental crust, we infer after 6 years a creep strain and effective viscosity on the order of 1×10^{-10} and 1×10^{22} Pa s, respectively. The model results using power-law rheology are in good agreement with a brittle upper crust and a

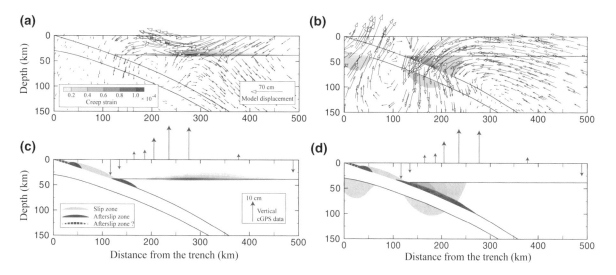

Figure 7
Modeled accumulated displacement field and creep strain 6 years after the Maule earthquake compared with the accumulated observed vertical displacement from nine cGPS stations along the model profile as shown in Fig. 1. **a** Modeled cumulative creep strain (second invariant of the creep strain tensor) and displacement vectors from model NLA0 (power-law rheology, no afterslip and no relocking). **b** Same as **a** but with linear model rheology (model LA0). **c** Schematic representation of where the afterslip occurs in case of the model shown in **a**. **d** Same as **c** using the linear Maxwell model rheology

ductile lower crust shown by laboratory extrapolation of the rock strength with depth (Brace and Kohlstedt 1980; Ranalli 1997). The high creep strain rate in the lower crust predicted by our model may be a result of the vertical geothermal gradient and rock composition at the boundary between the continental lower crust and the upper mantle. These results support the conclusion from Griggs and Blacic (1965) who raised the possibility of great stress relaxation in the deeper crust and uppermost mantle at temperatures far below the melting point. The latter is in agreement with other studies of postseismic relaxation that also consider rock viscosity below the solidus (Barbot 2018; Klein et al. 2016; Wang et al. 2012). Hence, this rheologic boundary likely affects geodetic observations of the postseismic deformation at the earth's surface.

4.2. Implication of Linear Maxwell and Power-Law Model Rheology on Afterslip Location

Uplift deformation observed by cGPS sites at distances between 200 and 300 km from the trench is also found for the postseismic deformation after the great 1960 Valdivia, Chile; 2011 Tohoku-Oki, Japan; great 2004 Sumatra-Andaman, Indonesia and 2015

Gorkha, Nepal, earthquakes (Hu et al. 2004; Muto et al. 2016; Qiu et al. 2018; Wang and Fialko 2018; Zhao et al. 2017), suggesting that postseismic surface deformation is driven by common relaxation processes. To explain this deformation pattern, our preferred model scenarios are those with power-law rheology and afterslip at the upper part of the fault plane (< 30 km depth) or at the down-dip limit less than 20 cm. Our model results suggest that such a remarkable uplift is mainly the result of stress relaxation in the lower crust due to dislocation creep (Fig. 7a), showing that afterslip in a deeper region of the megathrust fault plays a secondary role to explain the uplift pattern at those distances (Fig. 7c). The dislocation creep process occurs at distances relatively close to the surface; thus, the deformation produced by this process does not need to be high to explain this pattern. Previous studies showed that this pattern can be explained by using linear viscoelastic rheology in the uppermost mantle in combination with afterslip, especially at the down-dip limit at about 55 km depth or deeper regions (Govers et al. 2017; Klein et al. 2016; Noda et al. 2017; Yamagiwa et al. 2015). In the same fashion, our model results from linear Maxwell model rheology suggest that deeper afterslip is required to explain this pattern

(Fig. 7d). However, evidence from interseismic locking obtained from GPS velocities (Moreno et al. 2010) or friction laws (Scholz 1998) along megathrust faults suggests that below approximately 55 km depth the megathrust is probably fully unlocked and no strain is built up to be released as frictional slip after the earthquake. Such a deep aseismic slip may not be only due to frictional processes, but may also occur as strain localization within ductile shear zones. Montési and Hirth (2003) proposed a theoretical model to investigate the impact of dislocation and diffusion creep processes on the transient behavior of ductile shear zones considering grain size evolution. They found that a ductile shear zone resembles frictional afterslip on a deep extension of the fault. This result is also supported by Takeuchi and Fialko (2013). Nevertheless, they found that thermally activated shear zones have little effect of postseismic relaxation. Diffusion creep processes depend strongly on grain size evolution. Here, we have considered the dominance of dislocation creep over diffusion creep processes; therefore, we have not considered grain size evolution. However, further experiments are required to investigate its impact on postseismic deformation, in particular on ductile shear zones along the megathrust fault.

In the very near field (< 50 km from the trench), our results show important differences in the cumulative surface displacement between models with linear Maxwell and power-law rheology, providing a key discriminant between the predominant rheology (linear or non-linear) and the magnitude and location of afterslip. Observations from the postseismic phase of the 2011 Tohoku-Oki earthquake indicated that the impact of afterslip is much smaller than was previously assumed when near-trench time series of GPS stations are used (Sun et al. 2014). Such stations observe a landward motion, which is not in agreement with substantial afterslip at the up-dip limit, which results in a seaward motion. Recently, Barbot (2018) used a power-law rheology in a 2D model to show that landward motion above the rupture area of the main shock can be produced by transient deformation in the oceanic asthenosphere. Our model with power-law rheology (Model LNA20D48R), in fact, results in a landward motion of ∼ 10 cm at 50 km distance from the trench, but since near-trench

observations are missing in Chile, it remains a speculation whether landward motion would be observed or not.

4.3. Uncertainties of the Temperature Field

The largest uncertainty of the models with power-law rheology originates from the incorporated temperature model since this, besides the stress exponent, is the key control of the effective viscosity and thus the stress relaxation process induced by the coseismic slip and afterslip. Unfortunately, no temperature model exists for the entire cross section of the model, and we thus adopt the model from Springer (1999) that is located in the central Andes at 21°S. There, the age of the oceanic crust is older (∼ 50 Ma) in contrast to the younger plate at 36°S (∼ 35 Ma). Other temperature models closer to the Maule area (Oleskevich et al. 1999; Völker et al. 2011) only provide a temperature field 300 km landward from the trench not covering our model area. In contrast, the Springer model is covering the entire E–W extent of the modeled plate boundary system. Furthermore, Oleskevich et al. (1999) showed that in the fore arc and arc regions at 21°S and 34° the temperature contours have a very similar pattern, but absolute values can vary by 100 °C and more (Lamontagne and Ranalli 1996).

To show the model sensitivity due to the initial temperature field T, we increased (Model NLA0T + 100) and decreased (Model NLA0T − 100) the temperatures by 100 °C, respectively (Fig. 8). Since we would like to investigate only the impact of viscoelastic relaxation due to temperature changes on the deformation, we considered the model with power-law rheology and without afterslip. The results display a strong impact of the temperature field on the surface deformation, undergoing a maximum surface displacement change by a factor of about two, in the region of largest deformation at the Andean region (Fig. 8c, d). Thus, the mismatch of patterns of the slight uplift at about 350 km from the trench and the trench-ward motion in the far field (> 570 km) shown by cGPS observations and our model results, but also obtaining the afterslip, might be due to the temperature uncertainties.

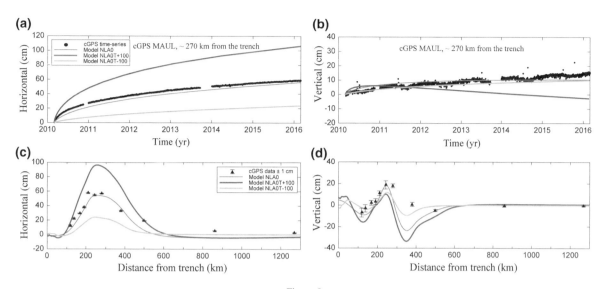

Figure 8
Results of the temperature sensitivity test for the model with power-law rheology. **a** Time series of the horizontal displacement of the cGPS station MAUL projected onto the model profile compared with model results for the temperature test. **b** Same as **a** for the vertical displacement. **c** Cumulative horizontal displacement of the cGPS stations indicated in Fig. 1 after 6 years compared with model results for the temperature test. **d** Same as **c** for the cumulative vertical displacement

5. Conclusion

We used a 2D geomechanical-numerical model to study the relative impact of afterslip, relocking and viscoelastic relaxation on the observed postseismic deformation 6 years after the 2010 Maule earthquake. In particular, we tested the general difference of using linear Maxwell or power-law rheology. The overall impact of relocking is only visible at distances < 200 km from the trench, but small compared with afterslip and viscoelastic relaxation. For the cumulative horizontal displacement the overall pattern from models with linear Maxwell and power-law rheology is similar. However, for the cumulative vertical displacement this is different. Here the used afterslip magnitudes as well as its depth location have a different expression in the modeled cumulative vertical displacement. To reproduce the pattern of the cGPS observations, the model with power-law rheology requires afterslip in shallower regions at 20–30 km depth rather than afterslip at depth > 50 km as suggested by models with linear rheology (Bedford et al. 2016; Klein et al. 2016). It also seems that less afterslip is needed at shallow depths. This difference is due to the different processes that are induced. In the models with power-law rheology the coseismically induced differential stresses in the lower crust and upper mantle are relaxed in shallower regions, i.e., the lower crust, whereas the models with linear Maxwell rheology assume that the crust is elastic. To produce the same vertical postseismic displacement these models require a relatively high afterslip at greater depth. To discriminate which model assumption is ultimately controlling the postseismic relaxation processes, cGPS stations near the trench are needed, and these turning points between subsidence and uplift as well as the change in direction of the horizontal displacement toward or away from the trench could be used as a proxy for the location and amount of afterslip as well as for the depth where differential stresses are relaxed by linear or non-linear viscoelastic processes.

Acknowledgements

Carlos Peña appreciates the scholarship granted to him by both the German Academic Exchange Service (DAAD) and the National Commission for Scientific and Technological Research (CONICYT-Becas

Chile). Jonathan Bedford is grateful to the German Science Foundation (DFG, MO-2310/3-1). Marcos Moreno acknowledges support from the Chilean National Fund for Development of Science and Technology (FONDECYT) grants 1181479, Millennium Scientific Initiative (ICM) grant NC160025, and National Research Center for Integrated Natural Disaster Management (CIGIDEN), CONICYT/FONDAP/15110017. Andrés Tassara is grafetul to the National Fund for Scientific and Technological Development, FONDECYT 1151175. The authors thank Shaoyang Li for the discussion in the early stage of the manuscript. All data used are properly cited in the reference list, figures, and tables.

Publisher's Note   Springer Nature remains neutral with regard to jurisdictional claims in published maps and institutional affiliations.

References

Barbot, S. (2018). Asthenosphere flow modulated by megathrust earthquake cycles. *Geophysical Research Letters*, *45*, 6018–6031. https://doi.org/10.1029/2018GL078197.

Barrientos, S., & Ward, S. (1990). The 1960 Chile earthquake: Inversion for slip distribution from surface deformation. *Geophysical Journal International*, *103*(3), 589–598. https://doi.org/10.1111/j.1365-246X.1990.tb05673.x.

Bedford, J., Moreno, M., Baez, J. C., Lange, D., Tilmann, F., Rosenau, M., et al. (2013). A high-resolution, time-variable after slip model for the 2010 Maule M_w = 8.8, Chile megathrust earthquake. *Earth and Planetary Science Letters, 383*, 26–36. https://doi.org/10.1016/j.epsl.2013.09.020.

Bedford, J., Moreno, M., Li, S., Oncken, O., Baez, J. C., Bevis, M., et al. (2016). Separating rapid relocking, afterslip, and viscoelastic relaxation: An application of the postseismic straightening method to the Maule 2010 cGPS. *Journal of Geophysical Research Solid Earth, 121*, 7618–7638. https://doi.org/10.1002/2016JB013093.

Bevis, M., & Brown, A. (2014). Trayectory models and reference frames for crustal motion geodesy. *Journal of Geodynamics, 88*(3), 283–311.

Bevis, B.A., Brooks, M.G., Smalley, R., Baez, J.C., Parra, H., Kendrick, E.C., Foster, J.H., Blanco, M., Simons, M., Caccamise, I., Genrich, D.A., Sladen, J.F., Melnick, M., Moreno, D., Cimbaro, S., Ryder, I.M., Wang, K., Bataille, K., Cassasa, G., Klotz, A., Folguera, J., Tong, X., Sandwell, D.T. (2010). The 2010 (M 8.8) Maule, Chile Earthquake: an overview of the emergency geodetic response and some of its early findings. Presented at 2010 Fall Meeting, AGU, U21B–04, San Francisco, Calif., 13–17 Dec.

Brace, W. F., & Kohlstedt, D. L. (1980). Limits on lithospheric stress imposed by laboratory experiments. *Journal of Geophysical Research, 85*(B11), 6248–6252. https://doi.org/10.1029/JB085iB11p06248.

Bürgmann, R., & Dresen, G. (2008). Rheology of the lower crust and upper mantle: Evidence from rock mechanics, geodesy and field observations. *Annual Review of Earth Planetary Sciences., 36*(1), 531–567. https://doi.org/10.1146/annurev.earth.36.031207.124326.

Chlieh, M., Avouac, J., Sieh, K., Natawidjaja, D., & Galetzka, J. (2008). Heterogeneous coupling of the Sumatran megathrust constrained by geodetic and paleogeodetic measurements. *Journal of Geophysical Research, 113*, B5. https://doi.org/10.1029/2007JB004981.

Christensen, N. (1996). Poisson's ratio and crustal seismology. *Journal of Geophysical Research, 101*(B2), 3139–3156. https://doi.org/10.1029/95JB03446.

Govers, R., Furlong, K., van de Wiel, L., Herman, M., & Broerse, T. (2017). The geodetic signature of the earthquake cycle at subduction zones: Model constraints on the deep processes. *Reviews of Geophysics, 56*(1), 6–49. https://doi.org/10.1002/2017RG000586.

Griggs, D. T., & Blacic, D. J. (1965). Quartz: Anomalous weakness of synthetic crystals. *Siences, 147*(3755), 292–295. https://doi.org/10.1126/science.147.3655.292.

Hayes, G., Wald, D., & Johnson, R. (2012). Slab1.0: A three-dimensional model of global subduction zone geometries. *Journal of Geophysical Research Solid Earth, 117*(B1), B01302. https://doi.org/10.1029/2011JB008524.

Hergert, T., & Heidbach, O. (2006). New insights into the mechanism of the postseismic stress relaxation exemplified by the 23 June M_w = 8.4 earthquake in southern Peru. *Geophysical Research Letters, 30*, 02307. https://doi.org/10.1029/2005GL024858.

Hirth, G., & Tullis, J. (1992). Dislocation creep regimes in quartz aggregates. *Journal of Structural Geology, 14*(2), 145–159. https://doi.org/10.1016/0191-8141(92)90053-Y.

Hsu, Y. J., Simons, M., Avouac, J. P., Galeteka, J., Sieh, K., Chlieh, M., et al. (2006). Frictional afterslip following the 2005 Nias-Simeulue earthquake, Sumatra. *Science, 312*(5782), 1921–1926. https://doi.org/10.1126/science.1126960.

Hu, Y., Bürgmann, R., Freymueller, J., Banerjee, P., & Wang, K. (2014). Contributions of poroelastic rebound and a weak volcanic arc to the postseismic deformation of the 2011 Tohoku earthquake. *Earth Planets and Space, 66*(1), 106. https://doi.org/10.1186/1880-5981-66-106.

Hu, Y., Wang, K., He, J., Klotz, J., & Khazaradze, G. (2004). Three-dimensional viscoelastic finite element model for postseismic deformation of the great 1960 Chile earthquake. *Journal of Geophysical Research, 109*(B12), B12403. https://doi.org/10.1029/2004JB003163.

Hughes, K., Masterlark, T., & Mooney, W. (2010). Poroelastic stress-triggering of the 2005 M8.7 Nias earthquake by the 2004 M9.2 Sumatra-Andaman earthquake. *Earth and Planetary Science Letters, 293*(3–4), 289–299. https://doi.org/10.1016/j.epsl.2010.02.043.

Karato, S., & Wu, P. (1993). Rheology of the upper mantle: A synthesis. *Science, 260*, 771–778. https://doi.org/10.1126/science.260.5109.771.

Kirby, S., & Kronenberg, A. (1987). Rheology of the lithosphere: Selected topics. *Reviews of Geophysics, 25*, 1219–1244. https://doi.org/10.1029/RG025i006p01219.

Klein, E., Fleitout, L., Vigny, C., & Garaud, J. D. (2016). Afterslip and viscoelastic relaxation model inferred from the large-scale postseismic deformation following the 2010 M_w 8.8 Maule earthquake (Chile). *Geophysical Journal International, 205*(3), 1455–1472. https://doi.org/10.1093/gji/ggw086.

Lamontagne, M., & Ranalli, G. (1996). Thermal and rheological constraints on the earthquake depth distribution in the Charlevoix, Canada, intraplate seismic zone. *Tectonophysics, 257*(1), 55–69. https://doi.org/10.1016/0040-1951(95)00120-4.

Lange, D., Bedford, J., Moreno, M., Tilmann, F., Baez, J., Bevis, M., et al. (2014). Comparison of postseismic afterslip models with aftershock seismicity for three subduction-zone earthquakes: Nias 2005, Maule 2010 and Tohoku 2011. *Geophysical Journal International, 199*(2), 784–799. https://doi.org/10.1093/gji/ggu292.

Li, S., Bedford, J., Moreno, M., Barnhart, W. D., Rosenau, M., & Oncken, O. (2018). Spatiotemporal variation of mantle viscosity and the presence of cratonic mantle inferred from 8 years of postseismic deformation following the 2010 Maule, Chile, earthquake. *Geochemistry Geophysics Geosystems*. https://doi.org/10.1029/2018GC007645.

Li, S., Moreno, M., Bedford, J., Rosenau, M., Heidbach, O., Melnick, D., et al. (2017). Postseismic uplift of the Andes following the 2010 Maule earthquake: Implications for the mantle rheology. *Geophysical Research Letters, 44*(4), 1768–1776. https://doi.org/10.1002/2016GL071995.

Marone, C., Scholtz, C., & Bilham, R. (1991). On the mechanics of earthquake afterslip. *Journal of Geophysical Research, 96*(B5), 8441. https://doi.org/10.1029/91JB00275.

Montési, L., & Hirth, G. (2003). Grain size evolution and the rheology of ductile shear zones: From laboratory experiments to postseismic creep. *Earth and Planetary Science Letters, 211*(1–2), 97–110. https://doi.org/10.1016/S0012-821X(03)00196-1.

Moreno, M., Melnick, D., Rosenau, M., Baez, J., Klotz, J., Oncken, O., et al. (2012). Toward understanding tectonic control on the M_w 8.8 2010 Maule Chile earthquake. *Earth and Planetary Science Letters, 321–322*, 152–165. https://doi.org/10.1016/j.epsl.2012.01.006.

Moreno, M., Rosenau, M., & Oncken, O. (2010). 2010 Maule earthquake slip correlates with pre-seismic locking of Andean subduction zone. *Nature, 467*(7312), 198–202.

Muto, J., Shibazaki, B., Iinuma, T., Ito, Y., Ohta, Y., Miura, S., et al. (2016). Heterogeneous rheology controlled postseismic deformation of the 2011 Tohoku-Oki earthquake. *Geophysical Research Letters, 43*(10), 4971–4978. https://doi.org/10.1002/2016GL068113.

Noda, A., Takahama, T., Kawasato, T., & Matsu'ura, M. (2017). Interpretation of offshore crustal movements following the 2011 Tohoku-Oki earthquake by the combined effect of afterslip and viscoelastic stress relaxation. *Pure and Applied Geophysics, 175*(2), 559–572.

Oleskevich, D., Hyndman, R., & Wang, K. (1999). The updip and downdip limits to great subduction earthquakes: Thermal and structural models of Cascadia, south Alaska, SW Japan, and Chile. *Journal of Geophysical Research Solid Earth, 104*(B7), 14965–14991. https://doi.org/10.1029/1999JB900060.

Perfettini, H., Avouac, J.-P., Tavera, H., Kositsky, A., Nocquet, J.-M., Bondoux, F., et al. (2010). Seismic and aseismic slip on the central Peru megathrust. *Nature, 465*(7294), 78–81. https://doi.org/10.1038/nature09062.

Pollitz, F. F., Bürgmann, R., & Banerjee, P. (2006). Post-seismic relaxation following the great 2004 Sumatra–Andaman earthquake on a compressible self-gravitating Earth. *Geophysical Journal International, 167*, 397–420. https://doi.org/10.1111/j.1365-246X.2006.03018.x.

Qiu, Q., Moore, J. D. P., Barbot, S., Feng, L., & Hill, E. M. (2018). Transient rheology of the Sumatran mantle wedge revealed by a decade of great earthquakes. *Nature Communications, 9*, 995. https://doi.org/10.1038/s41467-018-03298-6.

Ranalli, G. (1997). Rheology and deep tectonics. *Annali di Geofisica XL, 3*, 671–780. https://doi.org/10.4401/ag-3893.

Remy, D., Perfettini, H., Cotte, N., Avouac, J. P., Chlieh, M., Bondoux, F., et al. (2016). Postseismic relocking of the subduction megathrust following the 2007 Pisco, Peru, earthquake. *Journal of Geophysical Research Solid Earth, 121*, 3978–3995. https://doi.org/10.1002/2015JB012417.

Rundle, J. B. (1978). Viscoelastic crustal deformation by finite quasi-static sources. *Journal of Geophysical Research, 83*(B12), 5937–5946. https://doi.org/10.1029/JB083iB12p05937.

Scholz, C. (1998). Earthquakes and friction laws. *Nature, 391*(6662), 37–42.

Schurr, B., Asch, G., Hainzl, S., Bedford, J., Hoechner, A., Palo, M., et al. (2014). Gradual unlocking of plate boundary controlled initiation of the 2014 Iquique earthquake. *Nature, 512*(7514), 299–302. https://doi.org/10.1038/nature13681.

Springer, M. (1999). Interpretation of heat-flow density in the central Andes. *Tectonophysics, 306*(3), 377–395. https://doi.org/10.1016/S0040-1951(99)00067-0.

Sun, T., & Wang, K. (2015). Viscoelastic relaxation following subduction earthquakes and its effects on afterslip determination. *Journal of Geophysical Research Solid Earth, 120*, 1329–1344. https://doi.org/10.1002/2014JB011707.

Sun, T., Wang, K., Iinuma, T., Hino, R., He, J., Fujimoto, H., et al. (2014). Prevalence of viscoelastic relaxation after the 2011 Tohoku-Oki earthquake. *Nature, 514*(7520), 84–87. https://doi.org/10.1038/nature13778.

Takeuchi, C. S., & Fialko, Y. (2013). On the effects of thermally weakened ductile shear zones on postseismic deformation. *Journal of Geophysical Research Solid Earth, 118*(12), 6295–6310. https://doi.org/10.1002/2013JB010215.

Tichelaar, B. W., & Ruff, L. J. (1993). Depth of seismic coupling along subduction zones. *Journal of Geophysical Research, 98*(B2), 2017–2037. https://doi.org/10.1029/92JB02045.

Tsang, L. L. H., Hill, E. M., Barbot, S., Qiu, Q., Feng, L., Hermawan, I., et al. (2016). Afterslip following the 2007 M_w 8.4 Bengkulu earthquake in Sumatra loaded the 2010 M_w 7.8 Mentawai tsunami earthquake rupture zone. *Journal of Geophysical Research Solid Earth, 121*, 9034–9049. https://doi.org/10.1002/2016JB013432.

Vigny, C., Socquet, A., Peyrat, S., Ruegg, J.-C., Métois, M., Madariaga, R., et al. (2011). The M_w 8.8 Maule megathrust earthquake of central Chile, monitored by GPS. *Science, 332*, 1417–1421. https://doi.org/10.1126/science.1204132.

Völker, D., Grevemeyer, I., Stipp, M., Wang, K., & He, J. (2011). Thermal control of the seismogenic zone of southern central Chile. *Journal of Geophysical Research*. https://doi.org/10.1029/2011JB008247.

Wang, K., & Fialko, Y. (2014). Space geodetic observations and models of postseismic deformation due to the 2005 M7.6 Kashmir (Pakistan) earthquake. *Journal of Geophysical Research*

Solid Earth, 119(9), 7306–7318. https://doi.org/10.1002/2014JB011122.

Wang, K., & Fialko, Y. (2018). Observations and modeling of coseismic and postseismic deformation due to the 2015 M_w 7.8 Gorkha (Nepal) earthquake. *Journal of Geophysical Research Solid Earth, 123*(1), 761–779. https://doi.org/10.1002/2017JB014620.

Wang, K., Hu, Y., & He, J. (2012). Deformation cycles of subduction earthquakes in a viscoelastic Earth. *Nature, 484*(7394), 327–332. https://doi.org/10.1038/nature11032.

Yamagiwa, S., Miyazaki, S., Hirahara, K., & Fukahata, Y. (2015). Afterslip and viscoelastic relaxation following the 2011 Tohoku-oki earthquake (M_w 9.0) inferred from inland GPS and seafloor

GPS/Acoustic data. *Geophysical Research Letters, 42*(1), 66–73. https://doi.org/10.1002/2014GL061735.

Yue, H., Lay, T., Rivera, L., An, C., Vigny, C., Tong, X., & Báez Soto, J.C. (2014). Localized fault slip to the trench in the 2010 Maule, Chile Mw = 8.8 earthquake. from joint inversion of high-rate GPS, teleseismic body waves, InSAR, campaign GPS, and tsunami observations. *J. Geophys. Res. Solid Earth, 119*, 7786–7804. https://doi.org/10.1002/2014JB011340.

Zhao, B., Bürgmann, R., Wang, D., Tan, K., Du, R., & Zhang, R. (2017). Dominant Controls of Downdip Afterslip and Viscous Relaxation on the Postseismic Displacements Following the M_w 7.9 Gorkha, Nepal, Earthquake. *Journal of Geophysical Research Solid Earth, 122*(10), 8376–8401. https://doi.org/10.1002/2017JB014366.

(Received July 25, 2018, revised November 2, 2018, accepted December 29, 2018, Published online January 22, 2019)

Pure Appl. Geophys. 176 (2019), 3929–3949
© 2018 Springer Nature Switzerland AG
https://doi.org/10.1007/s00024-018-2054-z

Green's Functions for Post-seismic Strain Changes in a Realistic Earth Model and Their Application to the Tohoku-Oki M_w 9.0 Earthquake

Tai Liu,[1] Guangyu Fu,[2] Yawen She,[3] and Cuiping Zhao[2]

Abstract—Based on a spherically symmetric, self-gravitating viscoelastic Earth model, we derive a complete set of Green's functions for the post-seismic surface strain changes for four independent dislocation sources: strike-slip, dip-slip, and horizontal and vertical tensile point sources. The post-seismic surface strain changes caused by an arbitrary earthquake can be obtained by a combination of the above Green's functions. The post-seismic surface strain changes in the near field agree well with the results calculated by the method in a half-space Earth model (Wang et al. in Comput Geosci 32:527–541, 2006), which verifies our Green's functions. With an increase in the epicentral distance, the effect of the curvature on both the co- and post-seismic strain changes clearly increases, revealing the importance of our spherical theory for far-field calculations. Next, we use our Green's functions to simulate the post-seismic surface strain changes that were caused by the viscoelastic relaxation of the mantle over the 6-year period after the Tohoku-Oki M_w 9.0 earthquake. Based on continuous Global Positioning System (GPS) observations around Honshu Island of Japan, Northeastern China, South Korea and the Russian Far East, we also deduce the post-seismic strain changes caused by the Tohoku-Oki M_w 9.0 earthquake. Overall, the distributions of the calculated and GPS-derived strain changes agree well each other. Finally, we compare the relative error between the observed and simulated strain changes over the 3.0–4.5-year period after the earthquake in both the near and far field. We find that the relative errors decrease as the epicentral distance increases, which validates our Green's functions for research in the far field.

Key words: Post-seismic strain changes, spherical dislocation theory, Green's functions, preliminary reference Earth model (PREM), Tohoku-Oki M_w 9.0 earthquake.

1. Introduction

Because of developments in modern geodetic techniques, post-seismic deformations caused by large earthquakes can be detected by Global Positioning System (GPS), interferometric synthetic-aperture radar (InSAR) and other measurements. Such data make it possible to study the Earth's internal structure and seismogenic mechanisms. In particular, the GPS provides efficient, stable and accurate results for monitoring of large-scale crustal movements, plays an important role in inversion of fault slip distributions (Ozawa et al. 2011; Zhou et al. 2014) and supplements detection of stress and strain changes (Savage et al. 1986, 2001; Tape et al. 2009). Using geodetic data, the strain changes induced by earthquakes can be continuously monitored (Araya et al. 2010, 2017; Ohzono et al. 2012). Strain is a more direct indicator than displacements to represent deformation (Takahashi 2011). To explain these strain data, theoretical calculations of co- and post-seismic strain changes are necessary.

Scientists have developed dislocation theories to compute co- and post-seismic deformations. Based on a half-space Earth model, Steketee (1958) first introduced dislocation theory into the field of seismology. Since then, many researchers have studied co-seismic deformations caused by earthquakes (Chinnery 1961, 1963; Maruyama 1964). Okada (1985) and Okubo (1991, 1992) presented several sets of concise formulae to calculate co-seismic deformations (gravity, geoid, displacement and strain), which have been widely applied in seismology and geodesy. Fukahata and Matsu'ura (2005, 2006), Hashima et al. (2008, 2014) and Wang et al. (2003, 2006) provided dislocation theories to

[1] Institute of Geophysics, China Earthquake Administration, Beijing 100081, China.

[2] Institute of Earthquake Forecasting, China Earthquake Administration, Beijing 100036, China. E-mail: fugy@cea-ies.ac.cn

[3] Hebei Seismological Bureau, Shijiazhuang 050021, China.

compute the co- and post-seismic strain changes in a multilayered half-space model. Their theories ignore the effect of the curvature of the Earth (Sun and Okubo 2002; Dong et al. 2014, 2016), and the validity of their theories is limited to the near field. Taking the curvature of the Earth into account, Sun et al. (2006, 2009) presented a complete theory to calculate the co-seismic strain changes in a layered spherical model. Takagi and Okubo (2017) and Tang and Sun (2017) presented asymptotic expressions for the co-seismic surface strain changes in a homogeneous Earth model, which improves the efficiency of spherical dislocation theories greatly. However, their methods can only be used to study co-seismic strain changes. They ignore the viscosity structure of the Earth.

Using the normal mode method, Piersanti et al. (1995) solved the post-seismic deformation equations in a spherically symmetrical Earth model. They considered the curvature and the viscosity structure of the Earth but not the issue of compressibility. Pollitz (1997) took into account the compressibility of the Earth, whereas his model consists of limited layers, in which the allowable radial structure of density and gravity in the unperturbed state is limited, viz. $\rho(r)g(r) = \text{const}/r$). To bypass this difficulty following the normal mode method, Tanaka et al. (2006, 2007) used numerical inverse Laplace integration along a rectangular path to obtain the post-seismic gravity, geoid and displacement changes. However, they did not consider the strain changes following an earthquake.

In this paper, we present a new method to calculate the post-seismic surface strain changes in a self-gravitating, compressible and realistically stratified Earth model. In the following section, we explain the theory of displacements provided by Tanaka et al. (2006, 2007), on which our work is based. In Sect. 3, we derive the expressions for the Green's functions for the post-seismic surface strain changes for four independent point sources. In Sect. 4, we verify our method by comparing the surface strain changes induced by particular dislocation sources using our methods and previous theory (Sun et al. 2006; Wang et al. 2006) to confirm the validity of our method. In Sect. 5, we simulate the post-seismic strain changes induced by the Tohoku-Oki M_w 9.0 earthquake. By

comparing the simulated strain changes with those derived from GPS observations, we further show the importance of our Green's functions.

2. Basic Theory of Post-seismic Displacements

In this section, we give an outline of the numerical inverse Laplace integration along a rectangular path presented by Tanaka et al. (2006, 2007), a technique that will be further considered when we discuss the post-seismic strain changes. We assume a dislocation model as shown in Fig. 1. Here, n and v are a unit vector normal to the infinitesimal fault surface dS and a unit slip vector, respectively. Axes 1 and 2 are within the horizontal surface, and axis 3 is the vertical axis.

The viscoelastic global deformations caused by an earthquake can be expressed in the same form as the elastic ones provided by Sun et al. (1996). Specifically, we can get the solution for the equivalent elastic medium in the Laplace domain and then obtain the viscoelastic solution by taking the inverse Laplace transform (Lee 1955; Radok 1957). The displacement $\mathbf{u}$, stress tensor $\mathbf{\tau}$ and potential ψ are expanded using the vector spherical harmonics in spherical coordinates (r, θ, φ) as follows (Tanaka et al. 2007):

$$
\begin{aligned}
\mathbf{u}(r,\theta,\varphi,t) &= \sum_{n,m} \left[y_1(r,t;n,m)\mathbf{R}_n^m(\theta,\varphi) + y_3(r,t;n,m)\mathbf{S}_n^m(\theta,\varphi) \right. \\
&\quad \left. + y_1^\mathrm{T}(r,t;n,m)\mathbf{T}_n^m(\theta,\varphi) \right], \mathbf{\tau}(r,\theta,\varphi,t) \cdot \mathbf{e}_r \\
&= \sum_{n,m} \left[y_2(r,t;n,m)\mathbf{R}_n^m(\theta,\varphi) + y_4(r,t;n,m)\mathbf{S}_n^m(\theta,\varphi) \right. \\
&\quad \left. + y_2^\mathrm{T}(r,t;n,m)\mathbf{T}_n^m(\theta,\varphi) \right], \\
\psi(r,\theta,\varphi,t) &= \sum_{n,m} y_5(r,t;n,m)Y_n^m(\theta,\varphi),
\end{aligned}
$$

$$(1)$$

where

$$
\mathbf{R}_n^m(\theta,\varphi) = \mathbf{e}_r Y_n^m(\theta,\varphi), \tag{2}
$$

$$
\mathbf{S}_n^m(\theta,\varphi) = \left[\mathbf{e}_\theta \frac{\partial}{\partial\theta} + \mathbf{e}_\varphi \frac{1}{\sin\theta}\frac{\partial}{\partial\varphi} \right] Y_n^m(\theta,\varphi), \tag{3}
$$

$$
\mathbf{T}_n^m(\theta,\varphi) = \left[\mathbf{e}_\theta \frac{1}{\sin\theta}\frac{\partial}{\partial\varphi} - \mathbf{e}_\varphi \frac{\partial}{\partial\theta} \right] Y_n^m(\theta,\varphi), \tag{4}
$$

$$
Y_n^m(\theta,\varphi) = P_n^m(\cos\theta)\mathrm{e}^{im\varphi}. \tag{5}
$$

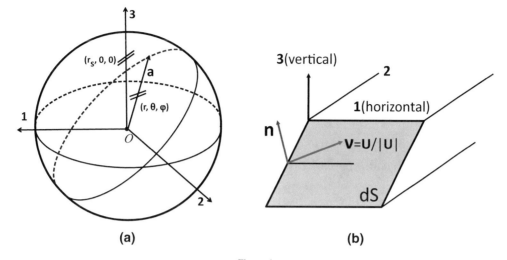

Figure 1

Diagram of the seismic source. Axes 1 and 2 are the two horizontal axes, and axis 3 is the vertical axis. $(r_S, 0, 0)$ is the location of the source. (r, θ, φ) is an arbitrary point. **a** is the radius of the Earth. dS denotes the fault surface with a normal vector n. v is a unit slip vector parallel to U

Herein, (r, θ, φ) and e_i are the conventional polar coordinates and unit base vectors, respectively. $P_n^m(\cos\theta)$ is the unnormalized Legendre's function and $P_n^{-|m|}(\cos\theta) = (-1)^m P_n^{|m|}(\cos\theta)$. $y_1(r, t; n, m)$ and $y_3(r, t; n, m)$ are the radial functions representing the vertical and horizontal displacement, respectively. $y_1^{\mathrm{T}}(r, t; n, m)$ denotes the displacement for the toroidal mode. $y_5(r, t; n, m)$ denotes the potential change. $y_2(r, t; n, m)$ and $y_4(r, t; n, m)$ are the radial functions corresponding to the vertical and horizontal stress, respectively. $y_2^{\mathrm{T}}(r, t; n, m)$ denotes the stress for the toroidal mode. An additional $y_6(r, t; n, m)$ function is introduced for mathematical convenience.

The solution for a Maxwell solid in the Laplace domain in an equivalent elastic medium can be found in Peltier (1974), and the apparent elastic deformation with a Laplace variable s satisfies the following differential equations:

$$\frac{d\tilde{y}_i(r, s; n, m)}{dr} = \sum_j A_{i,j}(r, s; n)\tilde{y}_j(r, s; n, m) + \tilde{S}_i(s; n, m), \qquad (6)$$

$$\frac{d\tilde{y}_i^{\mathrm{T}}(r, s; n, m)}{dr} = \sum_j B_{i,j}(r, s; n)\tilde{y}_j^{\mathrm{T}}(r, s; n, m) + \tilde{S}_i^{\mathrm{T}}(s; n, m), \qquad (7)$$

where explicit expressions for the homogeneous terms are given by Takeuchi and Saito (1972) and are listed in Appendix 1. The inhomogeneous terms

$$\tilde{S}_i = \tilde{y}_i(r_s + 0, s; n, m) - \tilde{y}_i(r_s - 0, s; n, m), \qquad (8)$$

$$\tilde{S}_i^{\mathrm{T}} = \tilde{y}_i^{\mathrm{T}}(r_s + 0, s; n, m) - \tilde{y}_i^{\mathrm{T}}(r_s - 0, s; n, m) \qquad (9)$$

are the source functions with a point dislocation at $r = r_s$ (Saito 1967; Takeuchi and Saito 1972), and the boundary condition on the free surface is

$$\tilde{y}_2(a) = \tilde{y}_4(a) = \tilde{y}_6(a) = \tilde{y}_2^{\mathrm{T}}(a) = 0. \qquad (10)$$

We can obtain the solutions $\tilde{y}_i(r, s; n, m)$ and $\tilde{y}_i^{\mathrm{T}}(r, s; n, m)$ for the equivalent elastic problem by solving Eqs. (6) and (7). The viscoelastic solution in the time domain can be obtained by taking the inverse Laplace transform of these equations. However, it is not feasible to obtain the solutions directly when innumerable layers and compressibility are considered simultaneously (Fang and Hager 1994; Piersanti et al. 1995; Vermeersen and Sabadini 1997; Wang 1999). It is extremely difficult to obtain the solutions for the y-variables by summing the normal modes of all the poles in the Laplace domain. This is because the normal mode method cannot evaluate an infinite set of poles when compressibility and continuous radial structures are considered. Tanaka et al.

(2006, 2007) carried out the numerical inverse Laplace integration on a closed path to include all poles, which evaluates the contributions from all the eigenmodes and avoids the intrinsic numerical difficulties of previous theories. Inverting the Laplace integration in Eqs. (11, 12), we obtain expressions for the y-variables $y_i(r, t; n, m)$ and $y_i^{\mathrm{T}}(r, t; n, m)$ (Eqs. 13–20).

$$y_i(r, t; n, m) = \frac{1}{2\pi i} \oint \tilde{y}_i(r, s; n, m) \frac{e^{st}}{s} ds, \quad (11)$$

$$y_i^{\mathrm{T}}(r, t; n, m) = \frac{1}{2\pi i} \oint \tilde{y}_i^{\mathrm{T}}(r, s; n, m) \frac{e^{st}}{s} ds, \quad (12)$$

Finally, we obtain the expressions for the post-seismic displacements by substituting Eqs. (13–20) into Eq. (1):

$$\begin{aligned} y_1(a, t; n, 0) = & \left[(n_1 v_1 + n_2 v_2) F_u^1(t; n) \right. \\ & \left. + n_3 v_3 F_u^2(t; n) \right] U\, dS, \end{aligned} \quad (13)$$

$$\begin{aligned} y_1(a, t; n, \pm 1) = & [\pm(n_3 v_1 + n_1 v_3) \\ & -i(n_2 v_3 + n_3 v_2)] F_u^3(t; n) U\, dS, \end{aligned}$$
$$(14)$$

$$\begin{aligned} y_1(a, t; n, \pm 2) = & [(-n_1 v_1 + n_2 v_2) \\ & \pm i(n_1 v_2 + n_2 v_1)] F_u^4(t; n) U\, dS, \end{aligned}$$
$$(15)$$

$$\begin{aligned} y_3(a, t; n, 0) = & \left[(n_1 v_1 + n_2 v_2) F_v^1(t; n) \right. \\ & \left. + n_3 v_3 F_v^2(t; n) \right] U\, dS, \end{aligned} \quad (16)$$

$$\begin{aligned} y_3(a, t; n, \pm 1) = & [\pm(n_3 v_1 + n_1 v_3) \\ & -i(n_2 v_3 + n_3 v_2)] F_v^3(t; n) U\, dS, \end{aligned}$$
$$(17)$$

$$\begin{aligned} y_3(a, t; n, \pm 2) = & [(-n_1 v_1 + n_2 v_2) \\ & \pm i(n_1 v_2 + n_2 v_1)] F_v^4(t; n) U\, dS, \end{aligned}$$
$$(18)$$

$$\begin{aligned} y_1^{\mathrm{T}}(a, t; n, \pm 1) = & [\pm(n_2 v_3 + n_3 v_2) \\ & -i(n_3 v_1 + n_1 v_3)] F_t^1(t; n) U\, dS, \end{aligned}$$
$$(19)$$

$$\begin{aligned} y_1^{\mathrm{T}}(a, t; n, \pm 2) = & [(n_1 v_2 + n_2 v_1) \\ & \pm i(n_1 v_1 - n_2 v_2)] F_t^2(t; n) U\, dS. \end{aligned} \quad (20)$$

The expressions for F_v^i ($i = 1, \ldots, 4$), F_u^i ($i = 1, \ldots, 4$) and F_t^i ($i = 1, 2$) can be found in Appendix 2.

3. Green's Functions for Post-seismic Strain Changes

Sun et al. (2006) presented a set of Green's functions for co-seismic strain changes in a spherically symmetric Earth model. Here, we derive a set of Green's functions for the post-seismic strain changes based on the same Earth model. According to the conventional theory of elasticity, we express the components of the strain tensor in terms of the displacements as follows (Takeuchi and Saito 1972):

$$\begin{aligned} e_{rr} &= \frac{\partial u_r}{\partial r}, \\ e_{\theta\theta} &= \frac{1}{r} \frac{\partial u_\theta}{\partial \theta} + \frac{1}{r} u_r, \\ e_{\varphi\varphi} &= \frac{1}{r \sin\theta} \frac{\partial u_\varphi}{\partial \varphi} + \frac{1}{r} u_\theta \cot\theta + \frac{1}{r} u_r, \\ e_{\theta\varphi} &= \frac{1}{r} \frac{\partial u_\varphi}{\partial \theta} - \frac{1}{r} u_\varphi \cot\theta + \frac{1}{r \sin\theta} \frac{\partial u_\theta}{\partial \varphi}, \\ e_{r\theta} &= \frac{\partial u_\theta}{\partial r} - \frac{1}{r} u_\theta + \frac{1}{r} \frac{\partial u_r}{\partial \theta}, \\ e_{r\varphi} &= \frac{1}{r \sin\theta} \frac{\partial u_r}{\partial \varphi} + \frac{\partial u_\varphi}{\partial r} - \frac{1}{r} u_\varphi. \end{aligned} \quad (21)$$

After inserting the displacement components that were defined in Sect. 2 into Eq. (21), we obtain the strain components in Eqs. (22–45). Because the last two components $e_{r\theta}$ and $e_{r\varphi}$ vanish on the Earth's free surface, the total strain change on the Earth's surface can be completely represented by the remaining four strain components.

The strain changes in vertical direction e_{rr} can be expressed as

$$\begin{aligned} e_{rr}(a, \theta, \varphi, t) = & \{(n_1 v_1 + n_2 v_2) G_{rr}^1(\theta, t) + n_3 v_3 G_{rr}^2(\theta, t) \\ & + [(n_1 v_3 + n_3 v_1) \cos\varphi \\ & + (n_2 v_3 + n_3 v_2) \sin\varphi] G_{rr}^3(\theta, t) \\ & + [(n_1 v_1 - n_2 v_2) \cos 2\varphi \\ & + (n_1 v_2 + n_2 v_1) \sin 2\varphi] G_{rr}^4(\theta, t)\} \frac{U\, dS}{a^3}. \end{aligned}$$
$$(22)$$

Herein, $G_{rr}^i(\theta, t)$ $(i = 1,\ldots,4)$ denotes the Green's function for strain changes in vertical direction. $a = 6371$ km, U is the magnitude of the dislocation. dS denotes the area of the dislocation.

$$G_{rr}^1(\theta, t) = \frac{\lambda}{\lambda + 2\mu} \sum_{n=0}^{\infty} \big[-2a^2 F_u^1(t; n) + n(n+1)a^2 F_v^1(t; n)\big] P_n^0(\cos\theta), \qquad (23)$$

$$G_{rr}^2(\theta, t) = \frac{\lambda}{\lambda + 2\mu} \sum_{n=0}^{\infty} \big[-2a^2 F_u^2(t; n) + n(n+1)a^2 F_v^2(t; n)\big] P_n^0(\cos\theta), \qquad (24)$$

$$G_{rr}^3(\theta, t) = \frac{\lambda}{\lambda + 2\mu} \sum_{n=1}^{\infty} \big[-4a^2 F_u^3(t; n) + n(n+1)a^2 F_v^3(t; n)\big] P_n^1(\cos\theta), \qquad (25)$$

$$G_{rr}^4(\theta, t) = \frac{\lambda}{\lambda + 2\mu} \sum_{n=2}^{\infty} \big[-4a^2 F_u^4(t; n) + n(n+1)a^2 F_v^4(t; n)\big] P_n^2(\cos\theta). \qquad (26)$$

The expressions for F_v^i $(i = 1,\ldots,4)$ and F_u^i $(i = 1,\ldots,4)$ can be found in Appendix 2.

Similarly, the Green's functions for strain changes in horizontal direction, viz. $G_{\theta\theta}^i(\theta, t), G_{\varphi\varphi}^i(\theta, t),$ $G_{\theta\varphi}^i(\theta, t)$, $(i = 1,\ldots,4)$, can be expressed as follows:

$$e_{\theta\theta}(a, \theta, \varphi, t) = \{(n_1 v_1 + n_2 v_2)G_{\theta\theta}^1(\theta, t) + n_3 v_3 G_{\theta\theta}^2(\theta, t)$$
$$+ \big[(n_1 v_3 + n_3 v_1)\cos\varphi$$
$$+ (n_2 v_3 + n_3 v_2)\sin\varphi\big]G_{\theta\theta}^3(\theta, t)$$
$$+ \big[(n_1 v_1 - n_2 v_2)\cos 2\varphi$$
$$+ (n_1 v_2 + n_2 v_1)\sin 2\varphi\big]G_{\theta\theta}^4(\theta, t)$$
$$+ \big[-(n_2 v_3 + n_3 v_2)\sin\varphi$$
$$+ (n_1 v_3 + n_3 v_1)\cos\varphi\big]G_{\theta\theta}^{t1}(\theta, t)$$
$$+ \big[(n_1 v_2 + n_2 v_1)\sin 2\varphi$$
$$+ (n_1 v_1 + n_2 v_2)\cos 2\varphi\big]G_{\theta\theta}^{t2}(\theta, t)\}\frac{U dS}{a^3}. \qquad (27)$$

Herein,

$$G_{\theta\theta}^1(\theta, t) = \sum_{n=0}^{\infty}\Big[a^2 F_u^1(t; n)P_n^0(\cos\theta) + a^2 F_v^1(t; n)\frac{d^2 P_n^0(\cos\theta)}{d\theta^2}\Big], \qquad (28)$$

$$G_{\theta\theta}^2(\theta, t) = \sum_{n=0}^{\infty}\Big[a^2 F_u^2(t; n)P_n^0(\cos\theta) + a^2 F_v^2(t; n)\frac{d^2 P_n^0(\cos\theta)}{d\theta^2}\Big], \qquad (29)$$

$$G_{\theta\theta}^3(\theta, t) = \sum_{n=1}^{\infty}\Big[2a^2 F_u^3(t; n)P_n^0(\cos\theta) + 2a^2 F_v^3(t; n)\frac{d^2 P_n^1(\cos\theta)}{d\theta^2}\Big], \qquad (30)$$

$$G_{\theta\theta}^4(\theta, t) = \sum_{n=2}^{\infty}\Big[2a^2 F_u^4(t; n)P_n^0(\cos\theta) + 2a^2 F_v^4(t; n)\frac{d^2 P_n^2(\cos\theta)}{d\theta^2}\Big], \qquad (31)$$

$$G_{\theta\theta}^{t1}(\theta, t) = \sum_{n=1}^{\infty}2a^2 F_t^1(t; n)\Big[\frac{1}{\sin\theta}\frac{dP_n^1(\cos\theta)}{d\theta} - \frac{\cos\theta}{\sin^2\theta}P_n^1(\cos\theta)\Big], \qquad (32)$$

$$G_{\theta\theta}^{t2}(\theta, t) = \sum_{n=2}^{\infty}4a^2 F_t^2(t; n)\Big[\frac{1}{\sin\theta}\frac{dP_n^2(\cos\theta)}{d\theta} - \frac{\cos\theta}{\sin^2\theta}P_n^2(\cos\theta)\Big]. \qquad (33)$$

The expressions for F_t^i $(i = 1, 2)$ can be found in Appendix 2.

Again,

$$e_{\varphi\varphi}(a, \theta, \varphi, t) = \{(n_1 v_1 + n_2 v_2).G_{\varphi\varphi}^1(\theta, t) + n_3 v_3 G_{\varphi\varphi}^2(\theta, t)$$
$$+ \big[(n_1 v_3 + n_3 v_1)\cos\varphi$$
$$+ (n_2 v_3 + n_3 v_2)\sin\varphi\big]G_{\varphi\varphi}^3(\theta, t)$$
$$+ \big[(n_1 v_1 - n_2 v_2)\cos 2\varphi$$
$$+ (n_1 v_2 + n_2 v_1)\sin 2\varphi\big]G_{\varphi\varphi}^4(\theta, t)$$
$$+ \big[-(n_2 v_3 + n_3 v_2)\sin\varphi$$
$$+ (n_3 v_1 + n_1 v_3)\cos\varphi\big]G_{\varphi\varphi}^{t1}(\theta, t)$$
$$+ \big[(n_1 v_2 + n_2 v_1)\sin 2\varphi$$
$$+ (n_1 v_1 + n_2 v_2)\cos 2\varphi\big]G_{\varphi\varphi}^{t2}(\theta, t)\}\frac{U dS}{a^3}. \qquad (34)$$

Herein,

$$G_{\varphi\varphi}^1(\theta, t) = \sum_{n=0}^{\infty}\Big[a^2 F_u^1(t; n)P_n^0(\cos\theta) + a^2 F_v^1(t; n)\cot\theta\frac{dP_n^0(\cos\theta)}{d\theta}\Big], \qquad (35)$$

$$G_{\varphi\varphi}^2(\theta, t) = \sum_{n=0}^{\infty}\Big[a^2 F_u^2(t; n)P_n^0(\cos\theta) + a^2 F_v^2(t; n)\cot\theta\frac{dP_n^0(\cos\theta)}{d\theta}\Big], \qquad (36)$$

$$G^3_{\varphi\varphi}(\theta, t) = \sum_{n=1}^{\infty} \left[2a^2 F^3_u(t; n) P^1_n(\cos\theta) \right.$$
$$\left. + 2a^2 F^3_v(t; n) \cot\theta \frac{dP^1_n(\cos\theta)}{d\theta} \right], \quad (37)$$

$$G^4_{\varphi\varphi}(\theta, t) = \sum_{n=2}^{\infty} \left[2a^2 F^4_u(t; n) P^2_n(\cos\theta) \right.$$
$$\left. + 2a^2 F^4_v(t; n) \cot\theta \frac{dP^2_n(\cos\theta)}{d\theta} \right], \quad (38)$$

$$G^{t1}_{\varphi\varphi}(\theta, t) = \sum_{n=1}^{\infty} \frac{2a^2 F^1_t(t; n)}{\sin\theta} \left[\cot\theta P^1_n(\cos\theta) - \frac{dP^1_n(\cos\theta)}{d\theta} \right], \quad (39)$$

$$G^{t2}_{\varphi\varphi}(\theta, t) = \sum_{n=2}^{\infty} \frac{4a^2 F^2_t(t; n)}{\sin\theta} \left[\cot\theta P^2_n(\cos\theta) - \frac{dP^2_n(\cos\theta)}{d\theta} \right], \quad (40)$$

and

$$e_{\theta\theta}(a, \theta, \varphi, t)$$
$$= \{ [-(n_1 v_3 + n_3 v_1)\sin\varphi + (n_2 v_3 + n_3 v_2)\cos\varphi] G^1_{\theta\varphi}(\theta, t)$$
$$+ [-(n_1 v_1 + n_2 v_2)\sin 2\varphi + (n_1 v_2 + n_2 v_1)\cos 2\varphi] G^2_{\theta\varphi}(\theta, t)$$
$$- [(n_2 v_3 + n_3 v_2)\cos\varphi + (n_3 v_1 - n_1 v_3)\sin\varphi] G^{t1}_{\theta\varphi}(\theta, t)$$
$$+ [(n_1 v_2 + n_2 v_1)\cos 2\varphi - (n_1 v_1 - n_2 v_2)\sin 2\varphi] G^{t2}_{\theta\varphi}(\theta, t) \}$$
$$\times \frac{U dS}{a^3}. \quad (41)$$

Herein,

$$G^1_{\theta\varphi}(\theta, t) = \sum_{n=1}^{\infty} \frac{4a^2 F^3_v(t; n)}{\sin\theta} \left[\frac{dP^1_n(\cos\theta)}{d\theta} - \cot\theta P^1_n(\cos\theta) \right], \quad (42)$$

$$G^2_{\theta\varphi}(\theta, t) = \sum_{n=2}^{\infty} \frac{8a^2 F^4_v(t; n)}{\sin\theta} \left[\frac{dP^2_n(\cos\theta)}{d\theta} - \cot\theta P^2_n(\cos\theta) \right], \quad (43)$$

$$G^{t1}_{\theta\varphi}(\theta, t) = \sum_{n=1}^{\infty} 2a^2 F^1_t(t; n) \left[\frac{P^1_n(\cos\theta)}{\sin^2\theta} \right.$$
$$\left. - \cot\theta \frac{dP^1_n(\cos\theta)}{d\theta} + \frac{d^2 P^1_n(\cos\theta)}{d\theta^2} \right], \quad (44)$$

$$G^{t2}_{\theta\varphi}(\theta, t) = \sum_{n=2}^{\infty} 2a^2 F^2_t(t; n) \left[\frac{P^2_n(\cos\theta)}{\sin^2\theta} \right.$$
$$\left. - \cot\theta \frac{dP^2_n(\cos\theta)}{d\theta} + \frac{d^2 P^2_n(\cos\theta)}{d\theta^2} \right]. \quad (45)$$

In the equations above, the expressions for F^i_v ($i = 1,...,4$), F^i_u ($i = 1,...,4$) and F^i_t ($i = 1,2$) are given in Appendix 2 and can also be found in Tanaka et al. (2006, 2007). We can calculate the four components of the strain changes excited by an arbitrary shear and tensile dislocation using the above Eqs. (22–45). The physical meaning of $G^i_{rr,\theta\theta,\varphi\varphi,\theta\varphi}$ ($i = 1,...,4$) is related to the geometry of the four independent point sources (n_i, v_i) that are shown in Table 1. In particular, for the vertical tensile case, the expressions for strain have both phi-independent (isotropic) and phi-dependent terms. The phi-dependent term can be expressed by a vertical strike-slip source (Tanaka et al. 2006; 2007), but we display only the isotropic component of the vertical tensile source in Table 1 to avoid redundant definitions.

4. Characteristics of the Green's Functions for Post-seismic Strain Changes

4.1. Behaviour of the Green's Functions for Post-seismic Strain Changes

The Green's functions for the strain change are a function of source depth, source distance and the time passed since the earthquake. In this section, we compute the strain Green's functions for the strain change for the four independent sources in Table 1. We use the PREM (Dziewonski and Anderson 1981) as an Earth model for the density and the elastic constants (Fig. 2). We assume the thickness of the elastic lithosphere and the viscosity of the mantle to be 40 km and 1×10^{19} Pa s, respectively. Figure 3 shows the time variation in the strain Green's functions component $G_{\varphi\varphi}(\theta, t)$ for a source depth (D_s) of 32 km, at $t = 0+$ year (immediately after an event), 2, 4, 6, 8 and 10 years. The magnitude of the deformations is normalized using $U dS = a^2$, which is the same operation as carried out by Sun et al. (2006).

Table 1

Relationship between Green's functions (e_{rr}, $e_{\theta\theta}$, $e_{\varphi\varphi}$, $e_{\theta\varphi}$) of strain changes and geometry of point sources

Green's function	(n_1, n_2, n_3) (v_1, v_2, v_3)	Dislocation type
$G^1_{rr}(\theta, t)$ $G^1_{\theta\theta}(\theta, t)$ $G^1_{\varphi\varphi}(\theta, t)$ 0	(1, 0, 0) (1, 0, 0)	Isotropic component of vertical tensile
$G^2_{rr}(\theta, t)$ $G^2_{\theta\theta}(\theta, t)$ $G^2_{\varphi\varphi}(\theta, t)$ 0	(1, 0, 0) (1, 0, 0)	Horizontal tensile
$G^3_{rr}(\theta, t)\ \cos\varphi$ $\left[G^3_{\theta\theta}(\theta, t) + G'^1_{\theta\theta}(\theta, t)\right]\cos\varphi$ $\left[G^3_{\varphi\varphi}(\theta, t) + G'^1_{\varphi\varphi}(\theta, t)\right]\cos\varphi$ $-\left[G^1_{\theta\varphi}(\theta, t) + G'^1_{\theta\varphi}(\theta, t)\right]\sin\varphi$	(1, 0, 0) (1, 0, 0)	Vertical dip-slip
$G^4_{rr}(\theta, t)\ \sin 2\varphi$ $\left[G^4_{\theta\theta}(\theta, t) + G'^2_{\theta\theta}(\theta, t)\right]\sin 2\varphi$ $\left[G^4_{\varphi\varphi}(\theta, t) + G'^2_{\varphi\varphi}(\theta, t)\right]\sin 2\varphi$ $-\left[G^2_{\theta\varphi}(\theta, t) + G'^2_{\theta\varphi}(\theta, t)\right]\cos 2\varphi$	(1, 0, 0) (0, 1, 0)	Vertical strike-slip

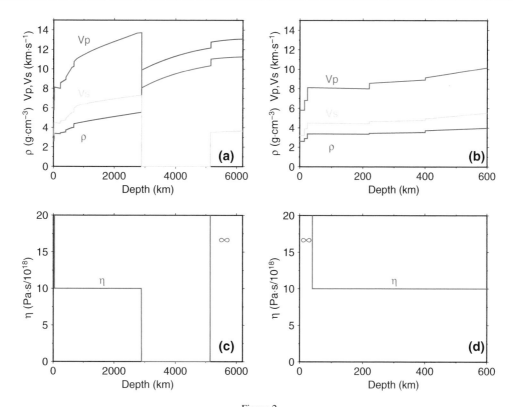

Figure 2

Viscoelastic Earth model used to compute the strain Green's functions. Density ρ and *P*- and *S*-wave velocities are shown in **a** and **b**, respectively. The viscosity profiles are shown in **c**, **d**. The viscosity is infinite at a depth of 0–40 km and in the inner core. The shallower part of the profile is magnified and shown in **b**, **d**

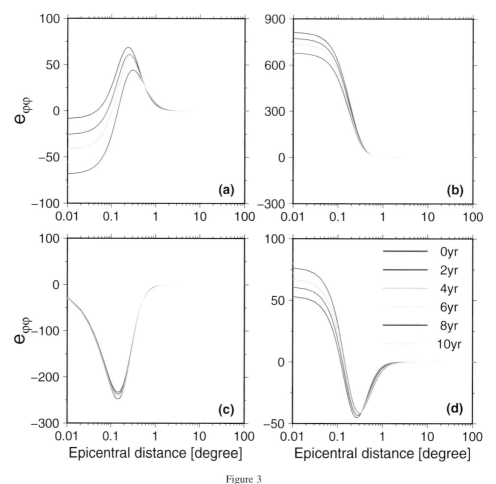

Figure 3
Strain component of $e_{\varphi\varphi}$ for four sources in Table 1. **a** Isotropic component of vertical tensile, **b** horizontal tensile, **c** vertical dip-slip, **d** vertical strike-slip. The results are normalized using $Uds = a^2$

The strain Green's functions in Fig. 3 show a similar tendency in the time domain to those of the displacements in Tanaka et al. (2006, 2007). The shorter the time since earthquake occurrence, the closer the post-seismic Green's function is to the co-seismic Green's function. As the time elapsed after the earthquake increases, the Green's functions increase with the same trend of convergence. This reflects the consistency and gradual change of the co- and post-seismic Green's functions (Gao et al. 2017). Assembling the strain Green's functions for the four independent seismic sources mentioned above, we can calculate the post-seismic strain changes raised by an arbitrary point source.

4.2. Verification of the Green's Functions for Post-seismic Strain Changes

To prove the validity of our results, we compare our results with those of Sun et al. (2006) and Wang et al. (2006), which are based on the same layered Earth model (Fig. 2). For simplicity, we study the co- and post-seismic strain changes caused by a point event with a dip-slip angle of $45°$ and a seismic depth of 10 km. In this case, n and v are $\left(-\frac{1}{\sqrt{2}}, 0, \frac{1}{\sqrt{2}}\right)\left(\frac{1}{\sqrt{2}}, 0, \frac{1}{\sqrt{2}}\right)$, respectively, and the strain components $e_{\varphi\varphi}(\theta, t)$ in Eq. (34) can be written as $-\frac{1}{2}G^1_{\varphi\varphi}(\theta, t) + \frac{1}{2}G^2_{\varphi\varphi}(\theta, t) - \frac{1}{2}G^4_{\varphi\varphi}(\theta, t)\cos(2\varphi)$. As an example, we calculate the co-seismic strain changes in the $e_{\varphi\varphi}$ direction ($\varphi = 0°$) caused by a simple point event using the above three methods

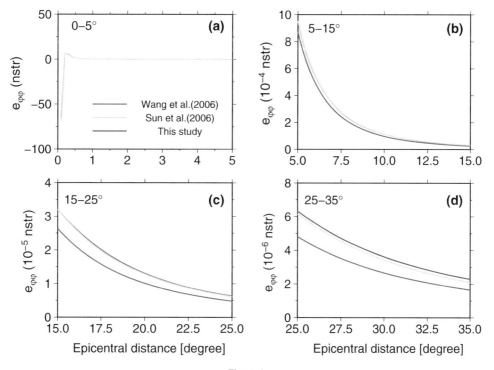

Figure 4

Comparison of our co-seismic strain changes in $e_{\varphi\varphi}$ direction ($\varphi = 0°$) caused by a point event (a dip-slip angle of 45° and a depth of 10 km) with those calculated by Sun et al. (2006) and Wang et al. (2006). **a–d** the results in the four epicentral intervals: 0–5°, 5–15°, 15–25° and 25–35°, respectively

(Fig. 4), using the same layered Earth model as that shown in Fig. 2. The magnitude of the deformations is normalized using $UdS = 1\,\text{m} \times (1\,\text{km})^2$. The blue, green and red lines denote the co-seismic strain change results obtained from the half-space dislocation theory of Wang et al. (2006), the spherical dislocation theory of Sun et al. (2006) and the method presented in this study. We find excellent agreement between the results of Sun et al. (2006) and our method in both the near and far fields, which verifies the validity of our results for the co-seismic responses. However, the agreement between the results of Wang et al. (2006) and ours is limited to the near field. The differences between our results and theirs reach 17.8% for epicentral distance larger than 15°. This reflects the curvature effect.

Next, we examine the viscoelastic responses of our method with the same operation. We calculate the post-seismic strain changes a period of 100 years after a dip-slip event using the method of Wang et al. (2006) and our method. To show the overall trend in the two sets of

results, we display the time-dependent variation in the post-seismic strain changes at four epicentral distances (5°, 15°, 25° and 35°) in Fig. 5. We find that the results have similar relaxation characteristics, and the differences between the post-seismic strain change results in the far field are more obvious than those in the near field, which indicates the necessity of using spherical dislocation theory for studying the viscoelastic relaxation in the far field. The relaxation characteristics of the two sets of results are similar, especially in the near field (Fig. 5a), which again verifies our method.

Next, we display the effect of the curvature at each of the epicentral distances shown in Fig. 6, which was calculated using the following equation:

$$\varepsilon = \frac{|e_s - e_h|}{|e_s|}. \tag{46}$$

Herein, e_s denotes the strain change corresponding to a spherical model (in this study), and e_h denotes the strain change corresponding to the half-space model used by Wang et al. (2006). Figure 6 shows the rapid

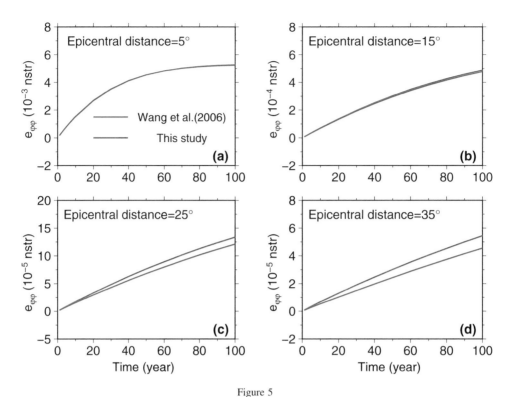

Figure 5
Comparison of our post-seismic strain changes caused by a point event (a dip-slip angle of 45° and a seismic depth of 10 km) with those obtained using the half-space theory of Wang et al. (2006). The post-seismic strain changes from 1 to 100 years after the seismic event. **a–d** The results at four epicentral distances: 5°, 15°, 25° and 35°, respectively

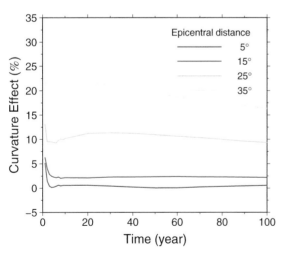

Figure 6
Time-dependent variation in the curvature effect at each epicentral distance. Red, blue, green and cyan lines show the results at four epicentral distances: 5°, 15°, 25° and 35°, respectively

increase in the curvature effect at an epicentral distance larger than 15°, while the effect is less than 5% as a whole for distance smaller than 15°.

Assuming an epicentral distance of 15°, we study the effect of curvature over the first 10,000 years after the seismic event and show the results in Fig. 7. The effect of curvature is approximately 6.9% about 1 year after the event, and then it becomes as small as 2.0% about 10 years after the event. As a whole, the effect of the viscoelastic relaxation becomes steady after 1000 years following the event. As time goes on, the effect of curvature decreases to 0.5% eventually.

5. Application to the Tohoku-Oki M_w 9.0 Earthquake

5.1. GPS Data around the Tohoku-Oki M_w 9.0 Earthquake

As an actual example, we estimate the post-seismic strain changes due to the Tohoku-Oki M_w 9.0 earthquake using our new Green's functions. The earthquake occurred near the east coast of Honshu Island on 11 March 2011, generated a huge tsunami, and caused huge loss of property and lives. The GPS

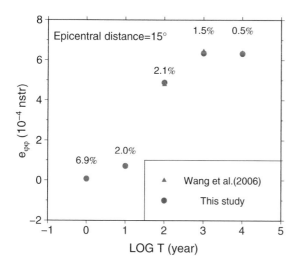

Figure 7

Comparison of our post-seismic strain changes caused by a point event (dip-slip angle of 45° and seismic depth of 10 km) with the results obtained from the half-space theory of Wang et al. (2006). The post-seismic strain changes occur at 1, 10, 100, 1000 and 10,000 years after the seismic event. The number above the red dots and blue triangle displays the curvature effect

network in Japan is of high quality and spatial resolution and provides valuable data to study the post-seismic processes that were associated with this earthquake (Fig. 8). We select the data of 298 continuous GPS stations in Fig. 8 that are evenly distributed around Honshu Island, with a date range between 11 March 2009 and 11 March 2017. The GPS data were processed by the Nevada Geodetic Laboratory (http://geodesy.unr.edu/) using GIPSY software in the global reference frame of IGS08. We remove the pre-earthquake velocities, and the annual and semiannual trends from the GPS data, using data from the 2 years before the event. The detailed process of GPS processing can be found in our previous study (Liu et al. 2017). We then divide the GPS data from the first 6 years after the earthquake into four periods (0–1.5, 1.5–3, 3–4.5 and 4.5–6 years after the earthquake). We show the velocities recorded by the GPS in Fig. 9.

5.2. Post-seismic Strain Changes on Honshu Island Deduced from the GPS Data

It is possible to calculate strain changes using GPS data. Some researchers approximate the surface of the Earth by a flat surface and use rectangular coordinates to calculate the strain changes; this approximation neglects the curvature of the Earth. Based on the above GPS data, in this study, we use the expressions in spherical coordinates developed by Savage et al. (2001) to calculate the strain changes induced by the Tohoku-Oki earthquake.

$$u_\theta = -\omega_\theta r_0 \cos \theta_0 \Delta\varphi + \omega_\phi r_0 + e_{\theta\varphi} r_0 \sin \theta_0 \Delta\varphi \\ + r_0 e_{\theta\theta} \Delta\theta - \omega_r r_0 \sin \theta_0 \Delta\varphi, \tag{47}$$

$$u_\varphi = -\omega_\theta r_0 - \omega_\varphi r_0 \cos \theta_0 \Delta\varphi + e_{\varphi\varphi} r_0 \sin \theta_0 \Delta\theta \\ + r_0 e_{\theta\varphi} \Delta\theta + r_0 \omega_r \Delta\theta. \tag{48}$$

Herein, θ corresponds to colatitude, φ to longitude, and r_0 to radial distance from the centre of the Earth. u_φ and u_θ represent the GPS-derived displacements in the east–west and south–north directions, respectively. $\Delta\varphi$, $\Delta\theta$ and Δr represent the distances of the GPS stations to the points that are chosen to calculate the strain changes. Given the displacements u_φ and u_θ at several stations, one can use the least-squares method to solve Eqs. (47, 48) and obtain the three components of the strain changes ($e_{\varphi\varphi}, e_{\theta\theta}, e_{\theta\varphi}$). Based on the three components of the strain, we can calculate the principal angle (α), the surface strain ($\varepsilon_{\text{area}}$) and the maximum and minimum principal strains ($\varepsilon_{1,2}$) using Eqs. (49–51).

$$\varepsilon_{1,2} = \frac{e_{\theta\theta} + e_{\varphi\varphi}}{2} \pm \sqrt{\left(\frac{e_{\theta\theta} - e_{\varphi\varphi}}{2}\right)^2 + \left(\frac{e_{\theta\varphi}}{2}\right)^2}, \tag{49}$$

$$\tan 2\alpha = \frac{e_{\theta\varphi}}{e_{\theta\theta} - e_{\varphi\varphi}}, \tag{50}$$

$$\varepsilon_{\text{area}} = e_{\theta\theta} + e_{\varphi\varphi}. \tag{51}$$

We show the surface strain changes on Honshu Island, including the directions and magnitudes of the principal strain changes, in Fig. 10, showing that significant strain changes [several thousand nstr (nano-strain)] occurred within 1.5 years after the earthquake. On the east coast of Honshu Island, near the earthquake, the surface shows a contraction state. However, most parts of Honshu Island are in a dilatation state. In addition, the strain changes in the northern part of the island are larger than those in the southern part, because the large ruptures were located in the north of the fault plane (Wei et al. 2012).

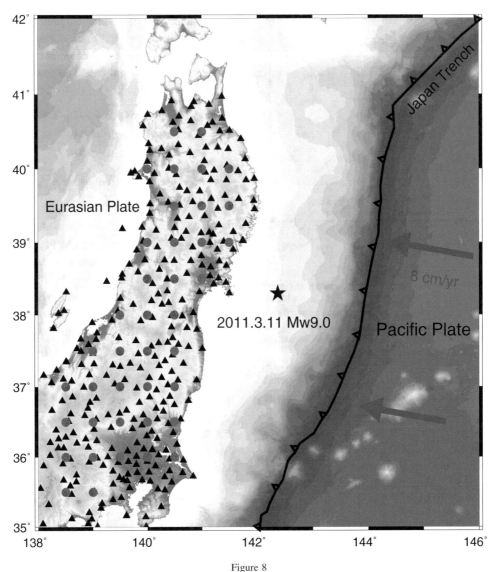

Figure 8
Configuration of the plate boundary and the locations of the GPS stations. The Pacific Plate is subducting westward beneath the Eurasian Plate with a velocity of 8 cm/year. The black triangles indicate the distribution of the 298 continuous GPS stations. The red circles indicate the points at which we deduce the post-seismic strain changes. The black star indicates the epicentre of the Tohoku-Oki M_w 9.0 earthquake

5.3. Post-seismic Strain Changes on Honshu Island Calculated by Our Method

In this subsection, we use our strain Green's functions to simulate the post-seismic strain changes induced by the Tohoku-Oki M_w 9.0 earthquake. In our calculations, we use the slip model proposed by Wei et al. (2012), which is inverted from both strong-motion waves and GPS observations. As in our previous study (Liu et al. 2017), we set the viscosity

of the mantle and the thickness of the lithosphere to be about 6×10^{18} Pa s and 30 km, respectively. Finally, we obtain the post-seismic strain changes in the 6 years after the earthquake (Fig. 11).

Remember that our method is built on the point dislocation theory. Our Green's functions are calculated under the assumption that the size of the fault is negligible. However, real faults, such as the one that caused the Tohoku-Oki M_w 9.0 earthquake, always have finite size and dimensions. According to Sun

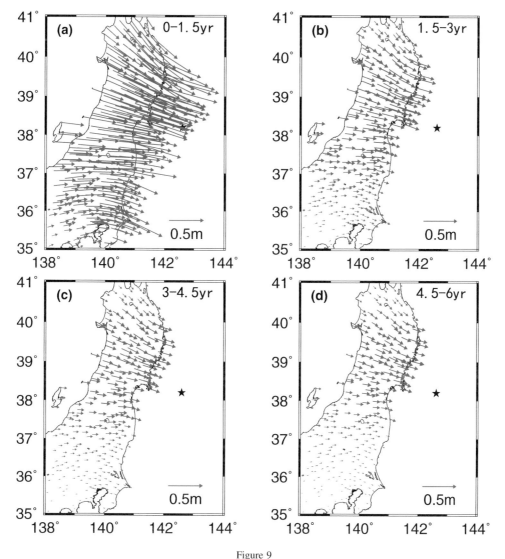

Figure 9
The GPS-derived post-seismic displacements on Honshu Island after the Tohoku-Oki M_w 9.0 earthquake

and Okubo (1998), the determination of whether the size of a fault is negligible is dependent on the relative position between the fault and the observation point. Without careful treatment of this distance, a calculation based on the point dislocation theory will produce large errors (Fu and Sun 2004). In this study, we regard a fault as a point source when the distance between the observation point and the centre of the fault is ten times larger than the longest side of the fault. Otherwise, we subdivide the subfault into smaller cells to ensure that they can be regarded as point sources, and the operational process becomes

the same as Gao et al. (2017). Using this segment-summation scheme, we can calculate the near-field strain changes caused by the Tohoku-Oki M_w 9.0 earthquake with high accuracy.

After the Tohoku-Oki M_w 9.0 earthquake, many authors studied the mechanisms of the post-seismic deformations based on GPS data (Ozawa et al. 2012; Diao et al. 2014; Sun et al. 2014; Yamagiwa et al. 2015; Freed et al. 2017; Noda et al. 2018). They found that post-seismic deformation was primarily caused by after-slip over the initial 2 or 3 years after the event. The post-seismic deformations at longer

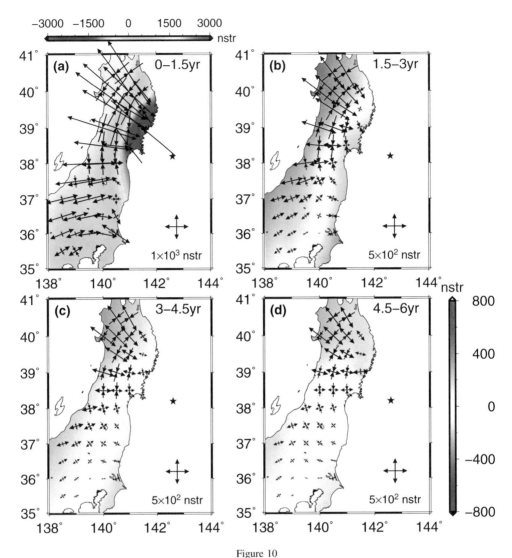

Figure 10
The post-seismic strain changes on Honshu Island deduced from the GPS data. The color indicates either contraction or dilatation fields.
Orthogonal double arrows represent the horizontal directions and magnitudes of the principal strain changes

times following this great event result from vis-coelastic relaxation. For this reason, we compare the strain changes deduced from the GPS displacement data (Fig. 10c, d) with the results obtained using our method (Fig. 11c, d) during 3–6 years after the main shock of the Tohoku-Oki M_w 9.0 earthquake, which was primarily caused by viscoelastic relaxation. In general, the two sets of results agree with each other.

The viscosity of the mantle plays an important role in simulations of post-seismic deformations. We change the viscosity of the mantle from 6×10^{18} to 6×10^{19} Pa s, and show the post-seismic strain changes within 6 years after the earthquake in Fig. 12. Comparing the strain changes in Figs. 11 and 12, we find that the amplitudes are significantly different between the two figures. The amplitudes of the deformations become much smaller when the viscosity becomes 10 times larger, although the patterns of the strain changes are overall consistent. Therefore, we can invert the viscosity of the mantle around the epicentre using the GPS data, the fault

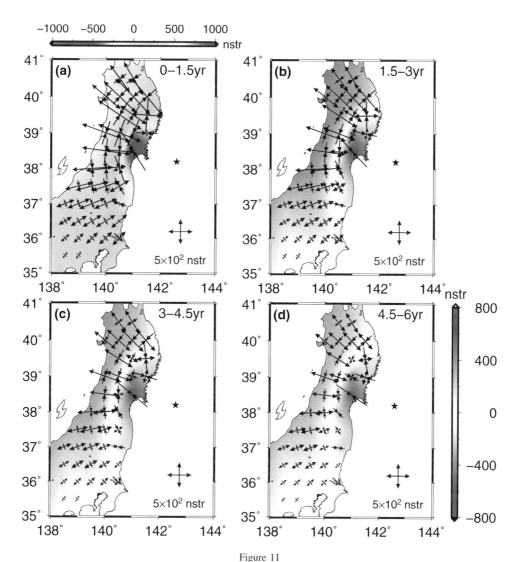

Figure 11
The post-seismic strain changes caused by the Tohoku-Oki M_w 9.0 earthquake on Honshu island of Japan, calculated using our new method. Symbols and colors as in Fig. 10

model of the great earthquake, as well as our post-seismic strain Green's functions.

5.4. Far-Field Post-seismic Strain Changes Following the 2011 Tohoku-Oki M_w 9.0 Earthquake

This section discusses the far-field post-seismic strain changes following the 2011 Tohoku-Oki M_w 9.0 earthquake. Zhao et al. (2018) extracted the post-seismic displacements in Northeastern China, South Korea and the Russian Far East following the

earthquake. Northeast Asia is more than 1000 km away from the epicentre of the Tohoku-Oki M_w 9.0 earthquake. When we calculate the post-seismic strain changes induced by the earthquake in Northeast Asia using a dislocation theory, it is necessary to consider the curvature effect to obtain more accurate results.

We first divide the displacements presented by Zhao et al. (2018) over the first 4.5 years after the earthquake into three periods (0–1.5, 1.5–3.0 and 3.0–4.5 years after the earthquake). We find that the far-field post-seismic displacements decay more

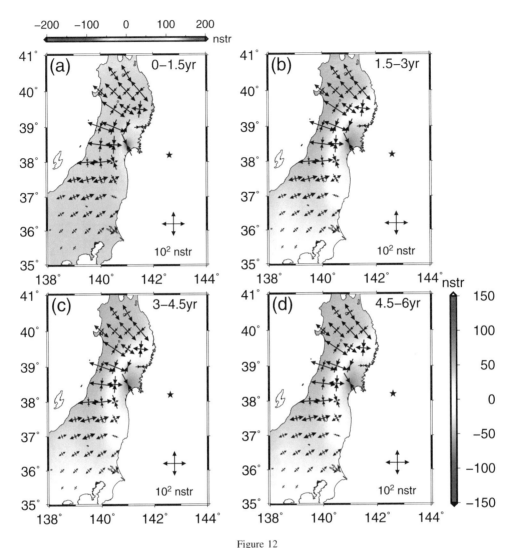

Figure 12
Post-seismic strain changes following the Tohoku-Oki M_w 9.0 earthquake calculated using our method when the viscosity of the mantle is 6×10^{19} Pa s. Symbols and colors as in Fig. 10

slowly than the near-field deformations do (Fig. 13). We then deduce the post-seismic strain changes over the above three periods using the least-squares method and GPS data and show them in Fig. 14. Limited by the sparse GPS stations used by Zhao et al. (2018), we calculate the strain changes at five points in the far field.

We also calculate the post-seismic strain changes following the Tohoku-Oki M_w 9.0 earthquake theoretically using the fault model of Wei et al. (2012) and our post-seismic Green's functions. In our calculation, we set the viscosity of the mantle to be 6×10^{18} Pa s, the same value as in the simulation in

the near field (Fig. 11). From both the observations and simulations (Fig. 14), we find extensional strain changes in roughly east–west direction and compressional in roughly south–north direction in the far field. The observed and simulated strain changes agree with each other overall. The strain changes in the east–west direction are much larger than the corresponding ones in the south–north direction. This is because the Tohoku-Oki M_w 9.0 earthquake was a thrust-dominated event and the surface displacements along the dip of the seismic fault plane were significant, which has verified by GPS observations (Zhao et al. 2018). The GPS-deduced strain changes

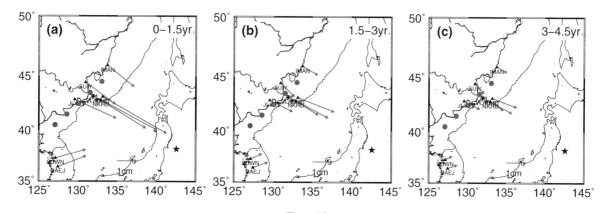

Figure 13
Post-seismic displacements in Northeastern China, South Korea and the Russian Far East. The red circles are the points at which the post-seismic strain changes are calculated. The black star indicates the epicentre of the Tohoku-Oki M_w 9.0 earthquake

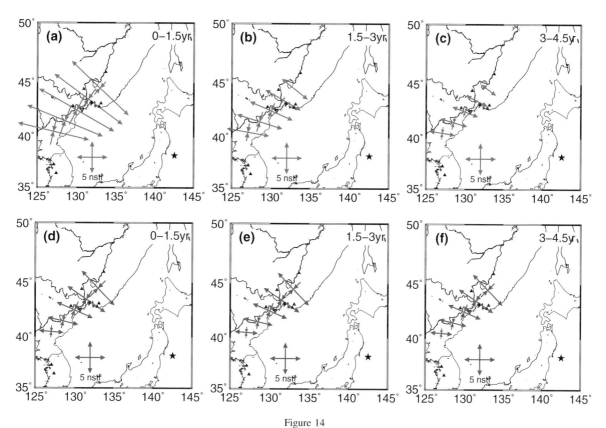

Figure 14
Far-field post-seismic strain changes following the Tohoku-Oki M_w 9.0 earthquake. **a–c** are the GPS-deduced post-seismic strain changes over the periods 0–1.5, 1.5–3.0 and 3.0–4.5 years after the earthquake, respectively. **d–f** are the corresponding theoretical strain changes calculated by our method

within the first 1.5 years after the earthquake (Fig. 14a) are much larger than those from the simulations (Fig. 14d). This is because the

observations of this period include the effects of after-slips, which are large soon after the earthquake but decline quickly as time goes on (Liu et al. 2017).

Next, we compare the computed and observed strain changes in both the near and far field to display the extent to which the theoretical computations agree with the observed strain changes. We first calculate the epicentral distances of each point to deduce the post-seismic strain changes. We then divide these points into seven groups according to epicentral distances (100–150 km, 150–200 km, 200–250 km, 250–300 km, 300–350 km, 350–400 km and > 400 km). Next, we calculate the average error between observations and expectations in each group using Eq. (52). Note that we focused on the surface strains (ε_{area}) between 3.0 and 4.5 years after the great earthquake, because the deformation in this period is primarily caused by viscoelastic relaxation (Liu et al. 2017; Yamagiwa et al. 2015).

$$\varepsilon = \frac{|\bar{e}_{obs} - \bar{e}_{cal}|}{|\bar{e}_{obs}|}. \tag{52}$$

Herein, $\bar{e}_{obs}$ and $\bar{e}_{cal}$ represent the GPS-deduced post-seismic strain changes and calculations, respectively. The relative errors are shown in Table 2. We find that the relative error is large in the near field. This is because the near-field deformation is sensitive to the slip model. The details of the distribution using the slip model can significantly affect the near-field deformation (Zhou et al. 2012). As the epicentral distance increases, the relative errors decrease, showing the advantages of our method for research in the far field.

6. Conclusions

We present a set of Green's functions for the post-seismic surface strain changes for four independent point events (strike-slip, dip-slip, and horizontal and vertical tensile sources) in a realistic spherical Earth model. Using these Green's functions, we can calculate the post-seismic strain changes caused by an arbitrary earthquake at an arbitrary time afterwards. The agreement between the co-seismic strain changes obtained using our method and those obtained using the method of Sun et al. (2006) yields an elastic response, and the agreement between the post-seismic strain changes in the near field calculated by our method and the half-space dislocation theory (Wang et al. 2006) yields a viscoelastic response. When the epicentral distances are smaller than 15°, the effects of the curvature are less than 5% in total. However, with an increase in the epicentral distance, the effect of the curvature in both co- and post-seismic strain changes increases rapidly. Finally, we calculate the post-seismic strain changes after the Tohoku-Oki M_w 9.0 earthquake using our strain Green's functions. Our calculated results agree with the GPS-derived strain changes as a whole. As the epicentral distance increases, the relative errors between the GPS-deduced and theoretically calculated strain changes over the period from 3.0 to 4.5 years after the earthquake diminishes rapidly (Table 2), confirming the advantages of our strain Green's functions for research in the far field.

Acknowledgements

We thank two anonymous reviewers for their helpful comments and suggestions. We thank Dr. Yoshiyuki Tanaka at Earthquake Research Institute, the University of Tokyo for providing us with his computer codes, which helped us to establish the Green's functions for post-seismic strain changes. The Green's functions are available in Table 1. This work was financially supported by the National Science Foundation of China (41574071; 41874003; 41331066) and the Basic Research Projects of Institute of Earthquake Science, China Earthquake Administration (2016IES010204).

Table 2

The relative error of the calculated and observed strain changes over the period from 3.0–4.5 years after the Tohoku-Oki M_w 9.0 earthquake

Epicentral distance (km)	Relative error (%)
100–150	1400.51
150–200	147.61
200–250	150.78
250–300	99.72
300–350	67.07
350–400	16.53
> 400	9.98

Appendices

Appendix 1: Explicit Expressions for the Homogeneous Terms

According to Takeuchi and Saito (1972), one can obtain the numerical solution of Eqs. (6) and (7) when excluding excitation terms.

The explicit expressions for the homogeneous terms are specifically written as

$$\frac{d\tilde{y}_1}{dr} = \frac{1}{\lambda+2\mu}\left(\tilde{y}_2 - \frac{\lambda}{r}[2\tilde{y}_1 - n(n+1)\tilde{y}_3]\right), \quad (53)$$

$$\frac{d\tilde{y}_2}{dr} = -s^2\rho\tilde{y}_1 + \frac{2}{r}\left(\lambda\frac{d\tilde{y}_1}{dr} - \tilde{y}_2\right)$$
$$+ \frac{1}{r}\left(\frac{2(\lambda-\mu)}{r} - \rho g\right)[2\tilde{y}_1 - n(n+1)\tilde{y}_3]$$
$$+ \frac{n(n+1)}{r}\tilde{y}_4 - \rho\left(\tilde{y}_6 - \frac{n+1}{r}\tilde{y}_5 + \frac{2g}{r}\tilde{y}_1\right), \tag{54}$$

$$\frac{d\tilde{y}_3}{dr} = \frac{1}{\mu}\tilde{y}_4 + \frac{1}{r}(\tilde{y}_3 - \tilde{y}_1), \tag{55}$$

$$\frac{d\tilde{y}_4}{dr} = -s^2\rho\tilde{y}_3 - \frac{\lambda}{r}\frac{d\tilde{y}_1}{dr} - \frac{\lambda+2\mu}{r_2}[2\tilde{y}_1 - n(n+1)\tilde{y}_3]$$
$$+ \frac{2\mu}{r_2}(\tilde{y}_1 - \tilde{y}_3) - \frac{3}{r}\tilde{y}_4 - \frac{\rho}{r}(\tilde{y}_5 - g\tilde{y}_1), \tag{56}$$

$$\frac{d\tilde{y}_5}{dr} = \tilde{y}_6 + 4\pi G\rho\tilde{y}_1 - \frac{n+1}{r}\tilde{y}_5, \tag{57}$$

$$\frac{d\tilde{y}_6}{dr} = \frac{n-1}{r}(\tilde{y}_6 + 4\pi G\rho\tilde{y}_1)$$
$$+ \frac{4\pi G\rho}{r}[2\tilde{y}_1 - n(n+1)\tilde{y}_3], \tag{58}$$

$$\frac{d\tilde{y}_1^T}{dr} = \frac{1}{r}\tilde{y}_1^T + \frac{1}{\mu}\tilde{y}_2^T, \tag{59}$$

$$\frac{d\tilde{y}_2^T}{dr} = \left[\frac{(n-1)(n+2)\mu}{r^2} + s^2\rho\right]\tilde{y}_1^T - \frac{3}{r}\tilde{y}_2^T, \tag{60}$$

where $\lambda(s) = \frac{\lambda s + \mu K/\eta}{s + \mu/\eta}$, $\mu(s) = \frac{\mu s}{s + \mu/\eta}$, $K = \lambda + \frac{2}{3}\mu$, based on the Laplace transform of Maxwell's constitutive equation (Peltier 1974); and $g = |\mathbf{g}(\mathbf{r})|$ denotes the magnitude of the gravity (Takeuchi and Saito 1972).

Appendix 2: The Solution of the Differential Equation

According to Tanaka et al. (2006, 2007), the expressions for F_v^i ($i = 1,...,4$), F_u^i ($i = 1,...,4$) and F_t^i ($i = 1, 2$) can be expressed as follows:

$$F_u^1(t;n) = -\frac{1}{2\pi i}\frac{G}{g_0 a}$$
$$\oint\left[\frac{3\lambda(r_s,s)+2\mu(r_s,s)}{\lambda(r_s,s)+2\mu(r_s,s)}\frac{1}{r_s}X^{\text{Press}}(r_s,s;n)\right. \quad (61)$$
$$\left.+\frac{\lambda(r_s,s)}{\lambda(r_s,s)+2\mu(r_s,s)}x_2^{\text{Press}}(r_s,s;n)\right]\frac{e^{st}}{s}ds,$$

$$F_u^2(t;n) = -\frac{1}{2\pi i}\frac{G}{g_0 a}\oint x_2^{\text{Press}}(r_s,s;n)\frac{e^{st}}{s}ds, \tag{62}$$

$$F_u^3(t;n) = -\frac{1}{2\pi i}\frac{G}{g_0 a}\oint x_3^{\text{Press}}(r_s,s;n)\frac{e^{st}}{s}ds, \tag{63}$$

$$F_u^4(t;n) = -\frac{1}{2\pi i}\frac{G}{g_0 a}\oint x_4^{\text{Press}}(r_s,s;n)\frac{e^{st}}{s}ds, \tag{64}$$

$$F_v^1(t;n) = -\frac{1}{2\pi i}\frac{G}{g_0 a}$$
$$\oint\left[\frac{3\lambda(r_s,s)+2\mu(r_s,s)}{\lambda(r_s,s)+2\mu(r_s,s)}\frac{1}{r_s}X^{\text{Shear}}(r_s,s;n)\right. \quad (65)$$
$$\left.+\frac{\lambda(r_s,s)}{\lambda(r_s,s)+2\mu(r_s,s)}x_2^{\text{Shear}}(r_s,s;n)\right]\frac{e^{st}}{s}ds,$$

$$F_v^2(t;n) = -\frac{1}{2\pi i}\frac{G}{g_0 a}\int x_2^{\text{Shear}}(r_s,s;n)\frac{e^{st}}{s}ds, \tag{66}$$

$$F_v^3(t;n) = -\frac{1}{2\pi i}\frac{G}{g_0 a}\oint x_3^{\text{Shear}}(r_s,s;n)\frac{e^{st}}{s}ds, \tag{67}$$

$$F_v^4(t;n) = -\frac{1}{2\pi i}\frac{G}{g_0 a}\oint x_4^{\text{Shear}}(r_s,s;n)\frac{e^{st}}{s}ds, \tag{68}$$

$$F_t^1(t;n) = \frac{1}{2\pi i}\frac{G}{g_0 a}\oint x_1^{\text{T}}(r_s,s;n)\frac{e^{st}}{s}ds, \tag{69}$$

$$F_t^2(t;n) = -\frac{1}{2\pi i}\frac{G}{g_0 a}\int x_2^{\text{T}}(r_s,s;n)\frac{e^{st}}{s}ds. \tag{70}$$

In the above equations, the relationship between X, x_1 and x_3 can be found in Okubo (1993) as

$$X^{\text{Press}} = 2x_1^{\text{Press}} - n(n+1)x_3^{\text{Press}}, \tag{71}$$

$$X^{\text{Shear}} = 2x_1^{\text{Shear}} - n(n+1)x_3^{\text{Shear}}. \tag{72}$$

135

Publisher's Note Springer Nature remains neutral with regard to jurisdictional claims in published maps and institutional affiliations.

REFERENCES

Araya, A., Takamori, A., Morii, W., Hayakawa, H., Uchiyama, T., & Ohashi, M. (2010). Analyses of far-field coseismic crustal deformation observed by a new laser distance measurement system. *Geophysical Journal International, 181,* 127–140.

Araya, A., Takamori, A., Morii, W., Miyo, K., Ohashi, M., & Hayama, K. (2017). Design and operation of a 1500-m laser strainmeter installed at an underground site in Kamioka, Japan. *Earth Planets Space, 69,* 77.

Chinnery, M. A. (1961). The deformation of ground around surface faults. *Bulletin of the Seismological Society of America, 51,* 355–372.

Chinnery, M. A. (1963). The stress changes that accompany strike-slip faulting. *Bulletin of the Seismological Society of America, 53,* 921–932.

Diao, F., Xiong, X., Wang, R., Zheng, Y., Walter, T. R., Weng, H., et al. (2014). Overlapping post-seismic deformation processes: Afterslip and viscoelastic relaxation following the 2011 M_w 9.0 Tohoku (Japan) earthquake. *Geophysical Journal International, 196,* 218–229.

Dong, J., Sun, W., Zhou, X., & Wang, R. (2014). Effects of earth's layered structure, gravity and curvature on coseismic deformation. *Geophysical Journal International, 199,* 1442–1451.

Dong, J., Sun, W., Zhou, X., & Wang, R. (2016). An analytical approach to estimate curvature effect of coseismic deformations. *Geophysical Journal International, 206,* 1327–1339.

Dziewonski, A. M., & Anderson, A. (1981). Preliminary reference earth model. *Physics of the Earth and Planetary Interiors, 25,* 297–356.

Fang, M., & Hager, B. H. (1994). A singularity free approach to post glacial rebound calculations. *Geophysical Research Letters, 21,* 2131–2134.

Freed, A. M., Hashima, A., Becker, T. W., Okaya, D. A., Sato, H., & Hatanaka, Y. (2017). Resolving depth-dependent subduction zone viscosity and afterslip from postseismic displacements following the 2011 Tohoku-oki, Japan earthquake. *Earth and Planetary Science Letters, 459,* 279–290.

Fu, G., & Sun, W. (2004). Effects of spatial distribution of fault slip on calculating co-seismic displacement: case studies of the Chi-Chi earthquake (M_w 7.6) and the Kunlun earthquake (M_w 7.8). *Geophysical Research Letters, 31,* 177–178.

Fukahata, Y., & Matsu'ura, M. (2005). General expressions for internal deformation fields due to a dislocation source in a multilayered elastic halfspace. *Geophysical Journal International, 161,* 507–521.

Fukahata, Y., & Matsu'ura, M. (2006). Quasi-static internal deformation due to a dislocation source in a multilayered elastic/viscoelastic half-space and an equivalence theorem. *Geophysical Journal International, 166,* 418–434.

Gao, S., Fu, G., Liu, T., & Zhang, G. (2017). A new code for calculating post-seismic displacements as well as geoid and gravity changes on a layered visco-elastic spherical earth. *Pure and Applied Geophysics, 174,* 1167–1180.

Hashima, A., Fukahata, Y., Hashimoto, C., & Matsu'ura, M. (2014). Quasi-static strain and stress fields due to a moment tensor in elastic–viscoelastic layered half-space. *Pure and Applied Geophysics, 171,* 1669–1693.

Hashima, A., Takada, Y., Fukahata, Y., & Matsu'ura, M. (2008). General expressions for internal deformation due to a moment tensor in an elastic/viscoelastic multilayered half-space. *Geophysical Journal International, 175,* 992–1012.

Lee, E. H. (1955). Stress analysis in visco-elastic bodies. *Quarterly of Applied Mathematics, 13,* 183–190.

Liu, T., Fu, G., Zhou, X., & Su, X. (2017). Mechanism of post-seismic deformations following the 2011 Tohoku-Oki M_w 9.0 earthquake and general structure of lithosphere around the sources. *Chinese Journal of Geophysics, 60,* 3406–3417. **(in Chinese)**.

Maruyama, T. (1964). Statical elastic dislocations in an infinite and semi-infinite medium. *Bulletin of the Earthquake Research Institute, University of Tokyo, 42,* 289–368.

Noda, A., Takahama, T., Kawasato, T., & Matsu'ura, M. (2018). Interpretation of offshore crustal movements following the 2011 Tohoku-oki earthquake by the combined effect of afterslip and viscoelastic stress relaxation. *Pure and Applied Geophysics, 175,* 559–572.

Ohzono, M., Yabe, Y., Iinuma, T., Yusaku, O., Miura, S., & Tachibana, K. (2012). Strain anomalies induced by the 2011 Tohoku Earthquake (M_w, 9.0) as observed by a dense GPS network in northeastern Japan. *Earth Planets Space, 64,* 1231–1238.

Okada, Y. (1985). Surface deformation due to shear and tensile faults in a half-space. *Bulletin of the Seismological Society of America, 75,* 1135–1154.

Okubo, S. (1991). Potential and gravity changes raised by point dislocations. *Geophysical Journal International, 105,* 573–586.

Okubo, S. (1992). Potential and gravity changes due to shear and tensile faults in a half-space. *Journal of Geophysical Research, 97,* 7137–7144.

Okubo, S. (1993). Reciprocity theorem to compute the static deformation due to a point dislocation buried in a spherically symmetric earth. *Geophysical Journal International, 115,* 921–928.

Ozawa, S., Nishimura, T., Munekata, H., Suito, H., Kobayashi, T., Tobita, M., et al. (2012). Preceding, coseismic, and postseismic slips of the 2011 Tohoku earthquake, Japan. *Journal of Geophysical Research, 117,* B07404. https://doi.org/10.1029/2011JB009120.

Ozawa, S., Nishimura, T., Suito, H., Kobayashi, T., Tobita, M., & Imakiire, T. (2011). Coseismic and postseismic slip of the 2011 magnitude-9 Tohoku-oki earthquake. *Nature, 475,* 373–376.

Peltier, W. R. (1974). The impulse response of a Maxwell Earth. *Reviews of Geophysics, 12,* 649–669.

Piersanti, A., Spada, G., Sabadini, R., & Bonafede, M. (1995). Global postseismic deformation. *Geophysical Journal International, 120,* 544–566.

Pollitz, F. F. (1997). Gravitational viscoelastic postseismic relaxation on a layered spherical Earth. *Journal of Geophysical Research, 102,* 17921–17941.

Radok, J. R. M. (1957). Visco-elastic stress analysis. *Quarterly of Applied Mathematics, 15,* 198–202.

Saito, M. (1967). Excitation of free oscillations and surface waves by a point source in a vertically heterogeneous earth. *Journal of Geophysical Research, 72,* 3689–3699.

Savage, J. C., Gan, W., & Svarc, J. L. (2001). Strain accumulation and rotation in the Eastern California Shear Zone. *Journal of Geophysical Research, 106,* 21995–22007.

Savage, J. C., Prescott, W. H., & Gu, G. (1986). Strain accumulation in southern California, 1973–1984. *Journal of Geophysical Research, 91,* 7455–7473.

Steketee, J. A. (1958). On Volterra's dislocations in a semi-infinite elastic medium. *Canadian Journal of Physics, 36,* 192–205.

Sun, W., & Okubo, S. (1998). Surface potential and gravity changes due to internal dislocations in a spherical earth -II. Application to a finite fault. *Geophysical Journal International, 132,* 79–88.

Sun, W., & Okubo, S. (2002). Effects of earth's spherical curvature and radial heterogeneity in dislocation studies—for a point dislocation. *Geophysical Research Letters, 29,* 46-1–46-4.

Sun, W., Okubo, S., & Fu, G. (2006). Green's functions of co-seismic strain changes and investigation of effect of earth's spherical curvature and radial heterogeneity. *Geophysical Journal International, 167,* 1273–1291.

Sun, W., Okubo, S., Fu, G., & Araya, A. (2009). General formulations of global co-seismic deformations caused by an arbitrary dislocation in a spherically symmetric earth model—applicable to deformed earth surface and space-fixed point. *Geophysical Journal International, 177,* 817–833.

Sun, W., Okubo, S., & Vaníček, P. (1996). Global displacements caused by point dislocations in a realistic Earth model. *Journal of Geophysical Research, 101,* 8561–8578.

Sun, T., Wang, K., Iinuma, T., Hino, R., He, J., Fujimoto, H., et al. (2014). Prevalence of viscoelastic relaxation after the 2011 Tohoku-oki earthquake. *Nature, 514*(7520), 84–87.

Takagi, Y., & Okubo, S. (2017). Internal deformation caused by a point dislocation in a uniform elastic sphere. *Geophysical Journal International, 208,* 973–991.

Takahashi, H. (2011). Static strain and stress changes in eastern Japan due to the 2011 off the Pacific coast of Tohoku Earthquake, as derived from GPS data. *Earth Planets Space, 63,* 741–744.

Takeuchi, H., & Saito, M. (1972). Seismic surface waves. *Methods in Computational Physics Advances in Research & Applications, 11,* 217–295.

Tanaka, T., Okuno, J., & Okubo, S. (2006). A new method for the computation of global viscoelastic post-seismic deformation in a realistic earth model (I)—vertical displacement and gravity variation. *Geophysical Journal International, 164,* 273–289.

Tanaka, T., Okuno, J., & Okubo, S. (2007). A new method for the computation of global viscoelastic post-seismic deformation in a realistic earth model (II)—Horizontal displacement. *Geophysical Journal International, 170,* 1031–1052.

Tang, H., & Sun, W. (2017). Asymptotic expressions for changes in the surface co-seismic strain on a homogeneous sphere. *Geophysical Journal International, 209,* 202–225.

Tape, C., Musé, P., Simons, M., Dong, D., & Webb, F. (2009). Multiscale estimation of GPS velocity fields. *Geophysical Journal International, 179,* 945–971.

Vermeersen, L. L. A., & Sabadini, R. (1997). A new class of stratified viscoelastic models by analytical techniques. *Geophysical Journal International, 129,* 531–570.

Wang, H. (1999). Surface vertical displacements, potential perturbations and gravity changes of a viscoelastic earth model induced by internal point dislocations. *Geophysical Journal International, 137,* 429–440.

Wang, R., Lorenzo-Martin, F., & Roth, F. (2003). Computation of deformation induced by earthquakes in a multi-layered elastic crust—FORTRAN programs EDGRN/EDCMP. *Computer & Geosciences, 29,* 195–207.

Wang, R., Lorenzo-Martin, F., & Roth, F. (2006). PSGRN/PSCMP—a new code for calculating co- and post-seismic deformation, geoid and gravity changes based on the viscoelastic-gravitational dislocation theory. *Computer & Geosciences, 32,* 527–541.

Wei, S., Graves, R., Helmberger, D., Avouac, J. P., & Jiang, J. (2012). Sources of shaking and flooding during the Tohoku-Oki earthquake: a mixture of rupture styles. *Earth and Planetary Science Letters, 333–334,* 91–100.

Yamagiwa, S., Miyazaki, S., Hirahara, K., & Fukahata, Y. (2015). Afterslip and viscoelastic relaxation following the 2011 Tohoku-oki earthquake (M_w 9.0) inferred from inland GPS and seafloor GPS/acoustic data. *Geophysical Research Letters, 42,* 66–73.

Zhao, Q., Fu, G., Wu, W., Liu, T., Su, L., Su, X., et al. (2018). Spatial-temporal evolution and corresponding mechanism of the far-field post-seismic displacements following the 2011 M_w 9. 0 Tohoku earthquake. *Geophysical Journal International, 214,* 1774–1782.

Zhou, X., Cambiotti, G., Sun, W., & Sabadini, R. (2014). The coseismic slip distribution of a shallow subduction fault constrained by prior information: the example of 2011 Tohoku (M_w 9.0) megathrust earthquake. *Geophysical Journal International, 199,* 981–995.

Zhou, X., Sun, W., Zhao, B., Fu, G., Dong, J. & Nie, Z. (2012). Geodetic observations detecting coseismic displacements and gravity changes caused by the M_w = 9.0 Tohoku-Oki earthquake. *Journal of Geophysical Research, 117,* https://doi.org/10.1029/2011jb008849.

(Received June 6, 2018, revised November 8, 2018, accepted November 17, 2018, Published online November 27, 2018)

Pure Appl. Geophys. 176 (2019), 3951–3973
© 2019 Springer Nature Switzerland AG
https://doi.org/10.1007/s00024-018-02089-w

Pure and Applied Geophysics

Quasi-Dynamic 3D Modeling of the Generation and Afterslip of a Tohoku-oki Earthquake Considering Thermal Pressurization and Frictional Properties of the Shallow Plate Boundary

BUNICHIRO SHIBAZAKI,[1] (iD) HIROYUKI NODA,[2] (iD) and MATT J. IKARI[3] (iD)

Abstract—The generation of the 2011 Tohoku-oki earthquake has been modeled by many authors by considering a dynamic weakening mechanism such as thermal pressurization (TP). Because the effects of TP on afterslip have not been investigated, this study develops a 3D quasi-dynamic model of the earthquake cycle to investigate afterslip of the Tohoku-oki earthquake, considering TP and the geometry of the plate boundary. We employ several velocity-weakening (VW) patches for M_w 7 class events, and two large shallow VW patches. The frictional properties are set as velocity-strengthening (VS) outside the VW patches. The results show that, during megathrust earthquakes, fast slip propagates to the surrounding VS regions near the VW patches owing to weakening by TP. Following M_w 9 events, large afterslips occur in regions below the northern shallow rupture area in the off-Fukushima region close to the Japan Trench, which is consistent with observations. In the VS region near the VW patches, during the early afterslip period, frictional behavior exhibits less VS with increasing slip velocity due to pore pressure reduction. We also consider the frictional properties of the shallow plate boundary fault off Tohoku, which exhibits a transition from VW to VS from low to high slip velocities. The results show the occurrence of slow slip events (SSEs) at intervals of a few decades at the shallow plate boundary. During megathrust events, the VW property at low slip velocity promotes slip along the shallow SSE region more than the case with VS property throughout the entire velocity range.

Key words: 2011 Tohoku-oki earthquake, 3D earthquake cycle model, afterslip, rate- and state-dependent friction law, thermal pressurization, slow slip events.

1. Introduction

The M_w 9 Off the Pacific Coast of Tohoku Earthquake (the Tohoku-oki earthquake) occurred on 11 March 2011 in the Japan Trench subduction zone, and exhibited large coseismic slip near the trench off Miyagi (Fujii et al. 2011; Ide et al. 2011; Ito et al. 2011; Lay et al. 2011; Yagi and Fukahata 2011; Yokota et al. 2011; Iinuma et al. 2012). As a result of the earthquake, a large fraction of the background deviatoric stress was relieved, as indicated by several studies showing large changes in the principal stress axes near the source region (Hasegawa et al. 2011) and in the prism (Lin et al. 2013; Brodsky et al. 2017). In the region off Tohoku, large thrust earthquakes (M_w 7.0–7.5) have occurred with recurrence intervals of several decades (e.g., Yamanaka and Kikuchi 2004), and megathrust earthquakes have occurred with longer recurrence intervals of around 600 years (Satake et al. 2008; Sawai et al. 2015).

Recently, various earthquake cycle models for the megathrust Tohoku-oki earthquake (M_w 9) have been developed, considering a shallow, strong asperity (Kato and Yoshida 2011), a hierarchical asperity distribution (Hori and Miyazaki 2011; Ohtani et al. 2014; Nakata et al. 2016), a dynamic weakening mechanism caused by thermal pressurization (TP) (e.g., Noda and Lapusta 2013; Cubas et al. 2015; Noda et al. 2017), or a two-state variable friction law (Shibazaki et al. 2011). Recent laboratory experimental studies using fault zone materials have shown that the friction coefficient drops dramatically at slip velocities of ~ 1 m/s (e.g., Di Toro et al. 2011). This dynamic weakening at high slip velocities should be considered in the cycle models of the Tohoku-oki earthquake. Shibazaki et al. (2011)

[1] International Institute of Seismology and Earthquake Engineering, Building Research Institute, Tsukuba, Ibaraki 305-0802, Japan. E-mail: bshiba@kenken.go.jp
[2] Disaster Prevention Research Institute, Kyoto University, Uji 611-0002, Japan.
[3] Marum, Center for Marine Environmental Science and Faculty of Geosciences, University of Bremen, 28359 Bremen, Germany.

performed 3D quasi-dynamic earthquake cycle modeling for the Tohoku-oki earthquake using a rate- and state-dependent friction law, with two state variables to represent the dynamic weakening process at high slip velocities. Based on the experimental results by Tsutsumi et al. (2011), they assumed that velocity weakening (VW) with small critical displacement or velocity strengthening (VS) occurs at low to intermediate slip velocities, whereas strong VW with large critical displacement occurs at high slip velocities. Although their model can reproduce megathrust and M_w 7 class earthquakes, the constitutive law used was empirically derived.

Models of large earthquakes that consider TP have been developed by several authors. Noda and Lapusta (2010) performed 3D earthquake sequence modeling that considered all wave effects with TP through shear heating, in addition to the effect of heterogeneous hydraulic diffusivity. The authors concluded that the region of more efficient TP produced larger slip in the model, resulting in events with long interseismic periods. Noda and Lapusta (2010, 2013) also proposed a model in which stable VS behavior at low slip rates was combined with coseismic weakening owing to TP, allowing unstable and destructive slip to occur. Mitsui et al. (2012) performed quasi-dynamic modeling and showed that very large slip is caused by the TP of pore fluid on a shallow fault plane, and successfully simulated M_w 7 and M_w 9 megathrust earthquakes. Cubas et al. (2015) performed two-dimensional (2D) dynamic simulations of earthquake cycles to investigate fault properties explaining the generation of the Tohoku-oki earthquake.

Ozawa et al. (2012) described the spatial and temporal evolution of the coseismic slip and afterslip of the Tohoku-oki earthquake, based on GPS data. Their study revealed that the afterslip of the Tohoku-oki earthquake occurred in an area of low coseismic slip, complementing the large coseismic slip zone. Iinuma et al. (2016) also investigated the spatial and temporal evolution of afterslip on the plate interface using terrestrial GPS data, in addition to seafloor geodetic data. Inversion analysis of afterslips, after removing the viscoelastic deformation, revealed that large afterslips occurred in the very shallow (< 20 km) part of the plate interface off Ibaraki and

Fukushima and in the region off Sanriku, where large coseismic slip did not occur. Their study also noted that a significant amount of slip was distributed at the deep (> 50 km) plate interface beneath Miyagi Prefecture, where no seismic events occurred, and that afterslip was potentially absent at the asperities in the Miyagi-oki region, which ruptured during the Miyagi-oki earthquake in 1978 and during the M_w 9.0 mainshock in 2011.

Several authors have developed models of earthquake afterslip assuming rate- and state-dependent friction laws (Marone et al. 1991; Perfettini and Avouac 2004; Barbot et al. 2012). Miyazaki et al. (2004) suggested that the stress-velocity paths in the afterslip regions of the 2003 Tokachi-oki earthquake exhibited steady state VS properties. Johnson et al. (2012) investigated a rate- and state-dependent asperity model to explain afterslip following the 2011 Tohoku-oki earthquake. They suggested that either afterslip is required within historical earthquake asperities or that afterslip surrounding asperities must exceed the fully relaxed limit. Fukuda et al. (2013) reported that the evolution of afterslip and postseismic shear stress on the plate interface is consistent with slip-rate-dependent frictional properties that exhibit less VS with increasing slip velocity.

Although many authors have already developed models for Tohoku-oki earthquakes considering TP, there have been no studies on afterslip which consider the realistic distribution of asperities, fault geometry, and TP. Therefore, it is crucial to understand whether the afterslip of the 2011 Tohoku-oki earthquake can be reproduced by models using rate- and state-dependent friction laws and TP.

Recently, the Integrated Ocean Drilling Program (IODP) Expedition 343, Japan Trench Fast Drilling Project (JFAST), revealed the structure of the very shallow plate boundary fault of the 2011 Tohoku-oki earthquake (Chester et al. 2013). Deformation was localized within a pelagic clay unit less than 5 m thick. The apparent friction coefficient during the earthquake was estimated to be 0.08, a very small value, based on a 0.31 °C temperature anomaly at the plate boundary fault (Fulton et al. 2013). High-velocity laboratory friction experiments using fault zone materials retrieved from the plate boundary also showed a small strength drop, with a very low peak and steady-state shear stress (Ujiie et al. 2013; Remitti et al. 2015). Sawai et al. (2017) reported that the friction properties of the shallow plate boundary are dependent on complex temperature conditions and slip velocity. Noda et al. (2017) developed a 2D dynamic earthquake cycle model considering this friction property to reproduce megathrust events and shallow slow earthquakes.

Ikari (2015) and Ikari et al. (2015) also found in laboratory experiments that the frictional properties of the shallow plate boundary show VW at low slip velocity and VS at high slip velocity. These frictional properties are favorable for generating slow slip events (SSEs) (Ikari et al. 2015; Shibazaki and Shimamoto 2007). Ito et al. (2013) detected SSEs in the Japan subduction zone prior to the 2011 Tohoku-oki earthquake. Moreover, shallow very low-frequency earthquakes (Matsuzawa et al. 2015) and SSEs (Uchida et al. 2016) were detected off the Pacific coast of Tohoku. It is unclear how the frictional properties of shallow fault zone material, which cause SSEs, affect the generation of megathrust earthquakes. There are many models of SSEs along subduction plate boundaries which use a rate- and state-dependent friction law (e.g., Shibazaki and Iio 2003; Shibazaki and Shimamoto 2007; Liu and Rice 2007; Liu and Rubin 2010; Segall et al. 2010; Ariyoshi et al. 2012). Here, however, we perform simulations based on the frictional properties obtained by Ikari (2015).

The present study develops a 3D quasi-dynamic earthquake cycle model of the Tohoku-oki earthquake considering TP and the realistic distribution of asperities and fault geometry to investigate afterslip. We use a spectral solver for the one-dimensional (1D) diffusion problem, as developed by Noda and Lapusta (2010), to efficiently calculate the temperature and pore pressure evolution on the fault plane. Moreover, we investigate whether a model accounting for TP is able to reproduce an afterslip pattern similar to that observed after the Tohoku-oki earthquake. We identify the regions of afterslip and the occurrence of afterslip at locations of VS and where weakening occurs at high slip velocities owing to TP. We then consider the effects of the frictional properties of shallow fault zone material at very low to intermediate slip velocities (Ikari 2015; Ikari et al.

2015) and investigate how this behavior affects the generation of SSEs and megathrust earthquakes.

2. Model of Earthquake Cycle and Afterslip with Thermal Pressurization

2.1. Methods

In general, frictional resistance τ can be written as follows:

$$\tau = \mu(\sigma_n - P) = \mu\sigma_n^{eff}, \tag{1}$$

where μ is the sliding friction coefficient and σ_n^{eff} is the effective normal stress, defined as the difference between the normal stress σ_n and the pore fluid pressure P. Here, σ_n is assumed to be lithostatic pressure. We use a rate- and state-dependent friction law in which the friction coefficient μ depends on slip velocity v and a state variable Θ (e.g., Dieterich 1981):

$$\mu(\Theta, v) = \mu_* - a\ln\left(\frac{v_0}{v} + 1\right) + b\ln\left(\frac{v_1\Theta}{D_c} + 1\right), \tag{2}$$

where μ_* is the base friction, a and b are empirical parameters, v_0 is the cut-off velocity for the direct effect, v_1 is the cut-off velocity for the evolution effect, and D_c is a critical displacement. The aging law (Dieterich 1981) is incorporated, in which the evolution law for the state variable can be written as follows:

$$\frac{d\Theta}{dt} = 1 - \frac{\Theta v}{D_c}. \tag{3}$$

Taking the y' axis to be perpendicular to a triangular element located at $y' = 0$, the temperature T and pore fluid pressure P in Eq. (1) evolve following the diffusion equation for temperature with a heat source (e.g., Lachenbruch 1980; Noda and Lapusta 2010):

$$\frac{\partial T}{\partial t} = \alpha_{th}\frac{\partial^2 T}{\partial y'^2} + \frac{\tau v}{\rho c}\omega, \tag{4}$$

$$\frac{\partial P}{\partial t} = \alpha_{hy} \frac{\partial^2 P}{\partial y'^2} + \Lambda \frac{\partial T}{\partial t}, \qquad (5)$$

where α_{th} and α_{hy} are the thermal and hydraulic diffusivity, respectively; ρc is the volume-specific heat capacity; and Λ is the pore pressure change per unit temperature change under undrained conditions. α_{hy} is proportional to the permeability, and ω is the shear heating source caused by fault slip. For example, Noda and Lapusta (2010) assumed that:

$$\omega = \tau v \frac{\exp(-y'^2/2w^2)}{\sqrt{2\pi}w}, \qquad (6)$$

where w is the half-width of the shearing layer.

We perform quasi-dynamic analysis, the details of which are explained in the "Appendix". We assume a curved plate interface in a 3D elastic half-space. The shear stress on an element is accumulated by the delay of the fault slip relative to the long-term average slip $v_{pl}t$, where v_{pl} is the velocity of plate convergence. In this model, as reported by Shibazaki et al. (2010), the direction of fault slip is fixed and the shear stress on each element is considered along the fixed direction. The direction of plate convergence is set to N70°W. The vector has both horizontal (N70°W) and vertical components, and is parallel to the surface of the triangular elements. The magnitude of the vector is fixed at the rate of plate convergence (i.e., $v_{pl} = 8$ cm/year) for all elements.

2.2. Model Parameters and Setting

Figure 1 shows the configuration of the curved plate interface. At a distance of over 400 km from the trench, the depth of the ocean floor becomes almost constant (around 6 km), therefore we set a free surface at this depth. We also set the maximum depth of the plate interface to 100 km because the main afterslip area of the Tohoku-oki earthquake is located at a maximum depth of ca. 80 km (Ozawa et al. 2012). The horizontal extension of the afterslip distribution obtained by Ozawa et al. (2012) is also included in this modeled region.

We assume the rupture areas of previous earthquakes to be VW patches within which VW occurs. Outside the VW patches, VS is assumed. We set the base friction value at 0.52–0.56. Cut-off velocities v_0

and v_1 are set at 1 m/s. In the VW patches, a and b are set at 0.008 and 0.012, respectively. In the VS region, a and b are set at 0.008 and 0.0, respectively. In this case, steady state friction at a velocity of 10^{-6} m/s is 0.58–0.62 in the VW region. This is consistent with experimental friction measurements on siliceous mudstones at the JFAST borehole (Ikari 2015) and incoming sediments at the Japan Trench (Sawai et al. 2014), which may act as asperities at seismogenic depths. VS regions would be expected in a weaker, more phyllosilicate-rich lithology. In the present parameter setting, the steady state friction in the VS region is always smaller than that in the VW region. For example, the steady state friction at velocity of 10^{-6} m/s is 0.41–0.45 in the VS region.

We set nine different VW patches (Fig. 1), based on studies of the rupture areas of past large earthquakes by Yamanaka and Kikuchi (2003, 2004), Murotani (2003) and the Central Disaster Prevention Council (http://www.bousai.go.jp/jishin/nihonkaikou/houkoku/sankou1.pdf). The assumed distribution of VW patches is similar to that reported by Shibazaki et al. (2011). We set the values of the constitutive law parameters as shown in Table 1.

Among the VW patches are two large VW patches, ST and MT, located near the trench in the offshore Sanriku and Miyagi regions, respectively. Following previous studies (e.g., Fujii et al. 2011; Yokota et al. 2011), we set the MT VW patch to represent the largest slip during the 2011 Tohoku-oki earthquake. In Fig. 1b, we assume a small size for the MT VW patch. We set the ST asperity to correspond to the source region of the 1896 M_w 8.6 Meiji–Sanriku tsunamigenic earthquake. Even during the 2011 Tohoku-oki earthquake, some slip was estimated to have occurred in the Sanriku region near the trench (e.g., Yokota et al. 2011; Satake et al. 2013). The D_c of the ST and MT VW patches is as assumed to have a relatively large value of 0.1 m (Table 1) to reproduce slip that rise relatively slowly, compared with those in the deeper VW patches. In our model, a region of VS exists near the trench.

The absolute shear stress of the source region of the Tohoku-oki earthquake was estimated to be around 21–22 MPa (Hasegawa et al. 2011) and 0–20 MPa for depths from the trench to 30 km

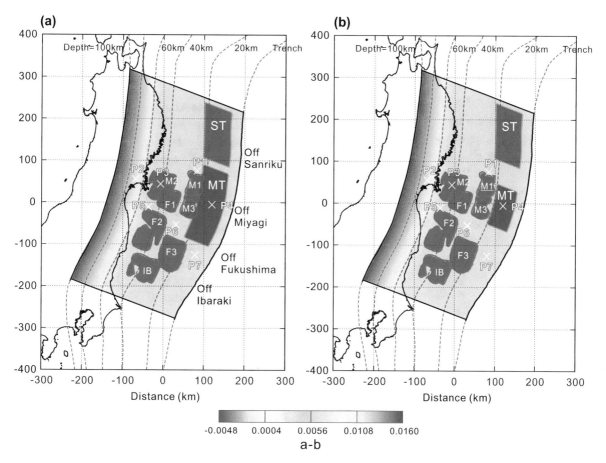

Figure 1

Distribution of constitutive law parameter $a - b$ on the configuration of the curved plate interface between the Pacific and Okhotsk plates. Dashed blue lines represent isodepth contour lines on the plate interface. The acronyms used for VW patches, corresponding events and constitutive law parameters for the VW patches and other regions are given in Table 1. Regions with solid red lines indicate regions of VW patches determined by observational analysis. The MT and ST VW patches are located in the offshore regions of Sanriku and Miyagi near the trench. **a** A shallow large MT patch is assumed. **b** A shallow small MT patch is assumed

Table 1

Constitutive law parameters for each region

Region	Event	Region	a	b	D_c (m)
ST	1611 M8.1, 1896 M8.5	Off Sanriku	0.008	0.012	0.1
MT		Off Miyagi	0.0096	0.0144	0.1
M1	1981 M7.1	Off Miyagi	0.008	0.012	0.04
M2	1978 M7.4	Off Miyagi	0.008	0.012	0.04
M3	1936 M7.4	Off Miyagi	0.008	0.012	0.04
F1	2003 M6.8	Off Fukushima	0.008	0.012	0.04
F2	1938 M7.3	Off Fukushima	0.008	0.012	0.04
F3	1938 M7.5	Off Fukushima	0.008	0.012	0.04
IB	1938 M7.0	Off Ibaraki	0.008	0.012	0.04
Outside VW patches	–	–	0.008 (depth $\leq$ 50 km)	–	–
			Linearly increases from 0.008 to 0.016 (50 km < depth $\leq$ 100 km)		

(Gao and Wang 2014). In this study, we assume a depth distribution of the initial effective stress $\sigma_{n,ini}^{eff}$, shown in Fig. 2. To model the observed low shear stress, $\sigma_{n,ini}^{eff}$ is set to be 28 MPa at depths greater than approximately 15 km. Although the absolute shear stress may vary depending on the timing of the last earthquake, we assume a loaded fault so that the pre-Tohoku stress levels are appropriate.

In our model, for simplicity, we assume that the TP is effective over the entire area (both VW and VS regions) and assume the uniform values of hydraulic and thermodynamic parameters as shown in Table 2. Based on the experimental results of Tanikawa and Shimamoto (2009), Noda and Lapusta (2013) estimated the value of hydraulic diffusivity of the Chelungpu fault to be $7.2–7.5 \times 10^{-5}$ m^2/s in the northern region and $2.7 \times 10^{-2}–7.3 \times 10^{-4}$ m^2/s in the southern region. Tanikawa and Shimamoto (2009) showed that the frictional properties in the northern and southern regions are VS and VW, respectively. In this model, the value of hydraulic

Figure 3 ▶

Figure 3
Slip distributions of the largest event for case **a** 1, **b** 2, **c** 3, **d** 4 and **e** 5. The number shown in each panel represents the elapsed time in years when coseismic slip occurs

diffusivity is assumed to be 0.01 m^2/s, corresponding to the value of the southern VW region. This is also the value of the patch where rupture nucleates in the study by Noda and Lapusta (2013). Even if the value of hydraulic diffusivity is large, TP can occur depending on the size of the VW patch and the base friction level. The width of the slip zone is reported to be < 1 to 20 cm for several subduction zones (Rowe et al. 2013), consistent with the 1–5 cm in the exhumed fault of the inland region (Wibberley and Shimamoto 2005) and less than 1–20 cm in exhumed accretionary complexes (Ujiie and Kimura 2014). Therefore, we assume the width of the slip zone to be 1 cm or 5 cm.

3. Numerical Results

3.1. Large and Megathrust Events

We investigate several cases in which the friction law parameters and their distributions are changed, as shown in Table 3. Figure 3 shows the slip distributions for the largest event, obtained by numerical simulations in each case. In cases 1 (Fig. 3a) and 2 (Fig. 3b), the base friction is set to be 0.56; the VW patch distribution is shown in Fig. 1a. For cases 1 (Fig. 3a) and 2 (Fig. 3b), w is taken to be 5 cm and 1 cm, respectively. The amounts of fault slip for the largest events in the two cases are similar. By using a

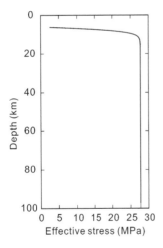

Figure 2
Depth distribution of initial effective normal stress $\sigma_{n,ini}^{eff}$ assumed in the simulation

Table 2

Thermal and hydraulic parameters used in the simulation

Parameter	Value
Specific heat ρC	2.7×10^6 Pa/K
Thermal diffusivity α_{th}	10^{-6} m^2/s
Hydraulic diffusivity α_{hy}	0.01 m^2/s
Undrained $\Delta P/\Delta T$ Λ	0.1×10^6 Pa/K

Table 3

Cases with differing values of base friction, half width of shear zone and size of MT VW patch

Case	Base friction μ_*	Half width of shear zone w (m)	Remarks
1	0.56	0.05	Large MT VW patch (Fig. 1a)
2	0.56	0.01	Large MT VW patch (Fig. 1a)
3	0.54	0.01	Large MT VW patch (Fig. 1a)
4	0.52	0.01	Large MT VW patch (Fig. 1a)
5	0.56	0.01	Small MT VW patch (Fig. 1b)

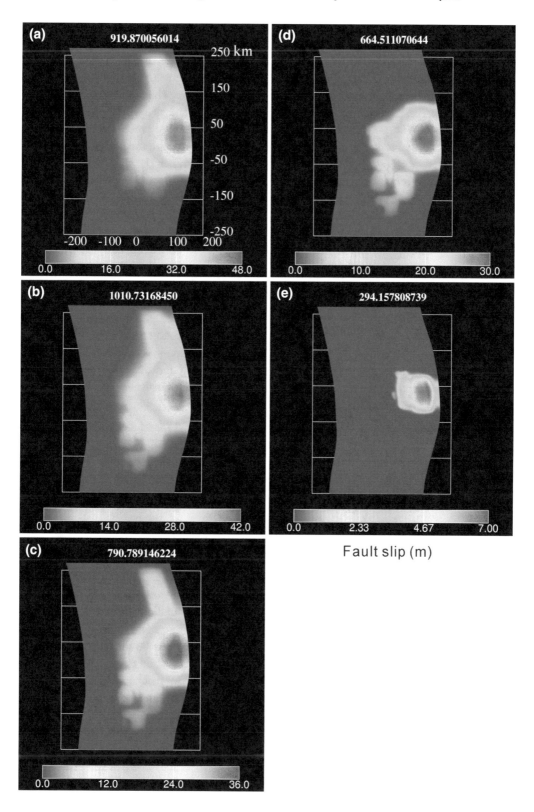

Fault slip (m)

base friction of 0.56, we can obtain megathrust events which are similar to the Tohoku-oki earthquake.

We investigate several cases in which the base friction values are changed with a fixed w of 1 cm (cases 2–4 in Table 3). The slip distribution of the largest event for each case is shown in Fig. 3b–d (cases 2–4, respectively). With an increase in base friction, TP occurs more effectively because the absolute stress increased, thereby generating more frictional heat. In case 5 (Fig. 3e), we assume a smaller, shallow VW patch (MT) than that used in case 2 (Fig. 3b). In this case, the size of the VW patch is not large enough for significant TP to occur.

Figure 4 shows snapshots of the slip velocity and temperature at different elapsed times for case 1. In this case, megathrust events cause a maximum temperature increase of approximately 1500 °C in the largest slip region. Figure 4a, b shows the distribution of slip velocity for the nucleation and the rupture of the F3 VW patch in Fig. 1 (a Fukushima-oki earthquake). Figure 4c and d shows the distribution of slip velocity for the ruptures of the F2 VW patch in Fig. 1 (a Fukushima-oki earthquake) and the M2 + F1 VW patch in Fig. 1 (a Miyagi-oki earthquake). Figure 5a–c, e–g shows the stress drop and maximum fault slip of these earthquakes to be approximately 4 MPa and 4 m, respectively. The recurrence interval of these events is several decades. During these ruptures, TP does not work effectively; therefore, the rupture does not propagate into the surrounding region.

Figure 4e–h shows the rupture propagation of a megathrust earthquake. The rupture initiates at the lower part of the largest VW patch and propagates to the surrounding VS region and to the VW patches of the M_w 7.5 earthquakes. Figure 4m–p shows the temperature increase during the rupture propagation of the megathrust earthquake. The maximum slip reaches ca. 44 m (Fig. 5h). In this event, coseismic fault slip occurs even in the VS regions, which surround the VW region because dynamic weakening occurred owing to TP. The stress drop of the megathrust event is approximately 8 MPa in the largest VW patch (Fig. 5d). The recurrence interval of the megathrust event is approximately 600 years.

Figure 6 shows the relationship between slip and shear stress or pore pressure at certain locations

during the megathrust event. At point P4, within the large VW patch, stress increases rapidly, then decreases rapidly, finally decreasing gradually with large displacement. The total stress drop is approximately 7 MPa. The first weakening process is caused by the rate- and state-friction properties and TP. The second weakening process is caused by TP, because during this process, the pore fluid pressure continues to increase. When the slip velocity becomes smaller, the pore pressure decreases significantly and stress increase occurs. At point P3, within the Miyagi VW patch, stress increases rapidly and then decreases rapidly owing to the rate- and state-friction properties and TP. At points P6 and P7, in the stable zone between or near the VW patches, stress increases rapidly and then decreases gradually with displacement during the megathrust event. This is caused by TP, because pore pressure increases during dynamic rupture. The final stress level, when the megathrust event terminates, is greater than the initial stress level. During the afterslip period, stress decreases gradually with displacement.

3.2. Afterslip

Figure 7 shows the distributions of slip, slip velocity and stress change during afterslip for case 1. To calculate the stress change, the stress state prior to a megathrust event is assumed as the reference stress state. The time $t = 0$ corresponds to 30 h after the earthquake because the starting time of the afterslip analysis by Ozawa et al. (2012) was around 30 h after the earthquake. Figure 7h shows that within VW patches, no afterslip takes place. In the regions surrounding the VW patches, the stress increased after a megathrust event and afterslip occurred. Even 1.88 months after $t = 0$, the slip velocities of afterslip exceed 10^{-7} m/s in the regions outside the VW patches. The slip velocities decrease gradually with time, and the afterslip extends to marginally greater depths.

As shown in Fig. 7a–h, the regions of high slip velocity and large slip occur at the deep extension of the ST VW patch, where the coseismic slip is small or zero. In this region, the maximum slip reaches 5 m about 3 years after the mainshock. Off the Fukushima and Ibaraki regions, the regions between VW patches

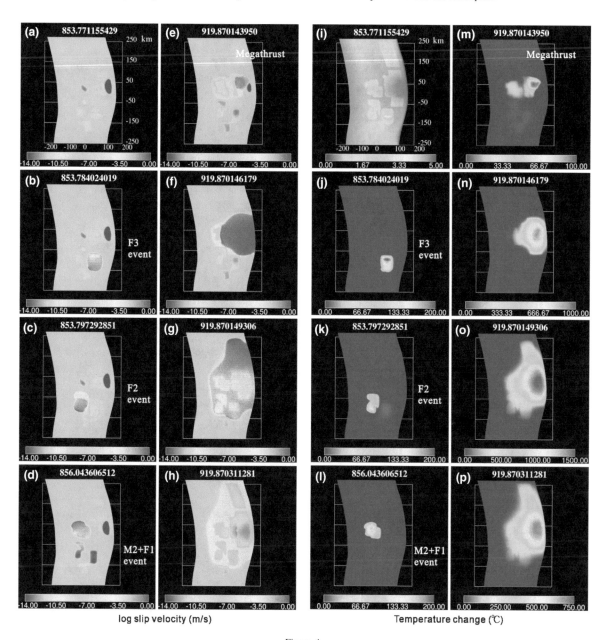

Figure 4

a–h Slip velocity distributions and **i–p** temperature distributions at certain time steps for case 1. **b** Slip velocity and **i** temperature distributions for the M_w 7.5 Ibaraki event. **c** Slip velocity and **k** temperature distribution for the first M_w 7.5 Fukushima event. **d** Slip velocity and (l) temperature distribution for the second M_w 7.5 Fukushima event. **e–h** Slip velocity and **m–p** temperature distributions for the megathrust event. The number shown in each panel represents the elapsed time in years

are characterized by low levels of afterslip. In the stable region near the trench off Fukushima, the slip exceeds 6 m approximately 3 years after the mainshock. Such a large afterslip occurs in this region because it is surrounded by coseismic slip regions.

Figure 8 shows the changes in the moment ($M_0 = G \int u \, ds$) of coseismic slip and afterslip with time. G is assumed to be 30 and 40 GPa for the results in Fig. 8. If $G = 40$ GPa, all values of σ_n^{eff}, τ, T and P increase by a factor of 40/30 (Appendix). In case 1

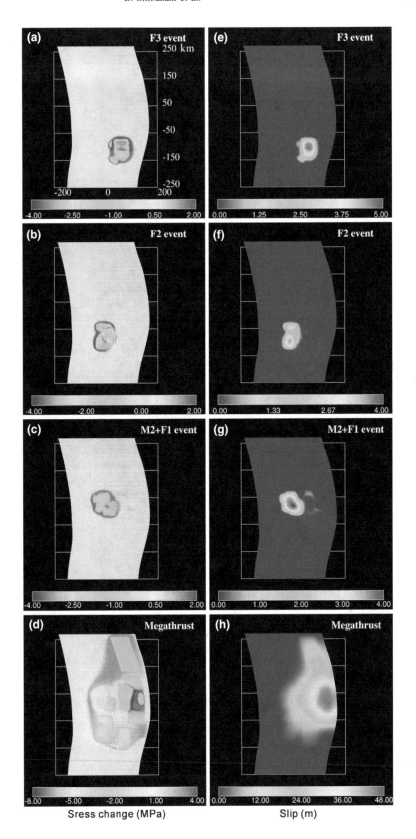

Sress change (MPa)   Slip (m)

◄Figure 5
a, d Slip and stress drop distribution for the M_w 7.5 Fukushima event; **b, e** the M_w 7.5 Fukushima event; **c, f** the M_w 7.5 Miyagi event; and **d, h** the M_w 9.1 event for case 1

(Fig. 8a), the magnitudes of the megathrust events were M_w 9.09 and M_w 9.15 for rigidities of 30 and 40 GPa, respectively. The moments of aftershocks for these rigidities reaches approximately 5×10^{21} and 6.7×10^{21} Nm 1 year after the main event, respectively (Fig. 8). The moment of afterslip reproduced in our model is slightly smaller than the observed value of 8.5×10^{21} Nm at 1 year after the main event (Ozawa et al. 2012).

Figure 9 shows the relationships between stress and slip velocity during the coseismic and postseismic periods at several points. P4 is located in the VW region and is characterized by a significant stress drop during the earthquake owing to the weakening

processes of rate- and state-dependent friction and TP. Therefore, after the megathrust slip, the slip velocity becomes low, and no afterslip occurs.

P1 and P6 are located in the VS region near the VW patches. Both coseismic slip and afterslip occur at these points. During a megathrust event, a stress drop occurs, associated with the coseismic slip; the stress then gradually decreases, coupled with a decrease in the slip velocity with afterslip. Afterslip can occur as a result of the VS property at low slip velocities. P5 is located in the VS region near the Miyagi and Fukushima asperities. At this location, stress increases with an increase in slip velocity. A small stress drop then occurs during the megathrust event, followed by a stress decrease with a decrease in slip velocity; afterslip is able to occur in this region (Fig. 7h). The behavior of P7 is similar to that of P5. The overlap of coseismic rupture and afterslip in the VS region near the VW patch due to dynamic

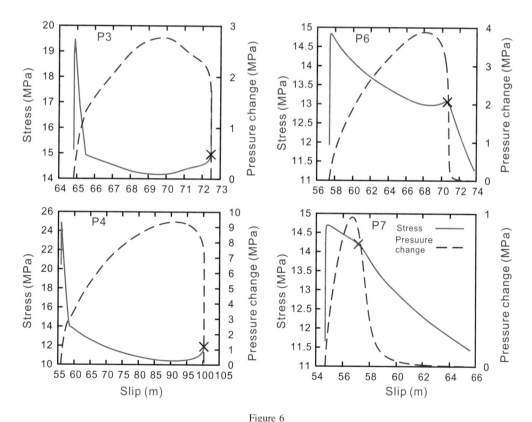

Figure 6
Relationships between stress and slip with time at points P3 within off Miyagi patch (M2), P4 within off Miyagi large patch (MT), P6 within off Fukushima stable zone, and P7 within off Fukushima stable zone close to the trench for the megathrust event in case 1. Dashed lines indicate the relationships between pore pressure and slip. The cross indicates the stress and slip when a megathrust event is almost stopped. Figure 1 shows the locations of the points

Pure Appl. Geophys.

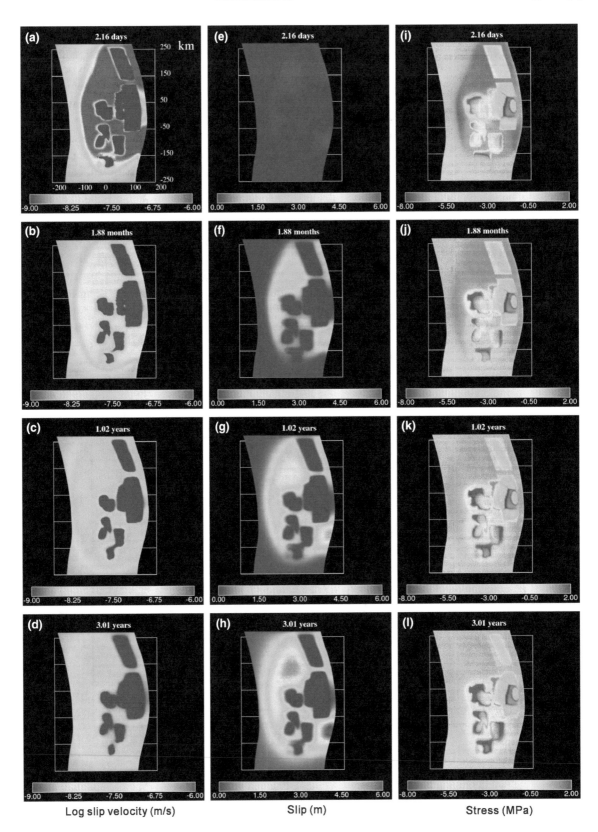

◀Figure 7
Snapshots of slip velocity (left column), slip (center column) and stress (right column) during afterslip following the megathrust event for case 1. The number (with units) shown in each panel represents the elapsed time. Shear stress was calculated as the difference between absolute stress and stress just prior to the megathrust event. $t = 0$ corresponds to 30 h after the megathrust event

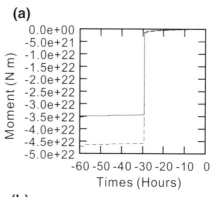

(a)

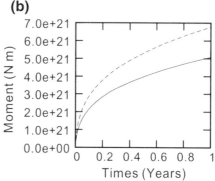

(b)

weakening was also suggested by Salman et al. (2017).

P2 is located in the VS region more than 30 km from the asperity and experienced no coseismic slip during the megathrust event. At this location, stress increases with an increase in slip velocity during the megathrust event, and then decreases with a decrease in slip velocity. Large afterslip is noted in this region (Fig. 7h).

4. Model of Earthquake Cycle Considering the Friction Properties of Shallow Fault Zones

Ikari and Kopf (2017) examined frictional properties at low (plate tectonic rates) to intermediate slip velocities from several fault zone materials, including those from the shallow plate boundary off Tohoku. Their results and those from Ikari et al. (2015) for the Tohoku fault sample showed VW or VS for velocity-step tests in the range 0.0027–0.081 μm/s, and VS at slip velocities of 0.3–100 μm/s (Fig. 10). Here, we investigate the effects of the frictional properties of shallow fault zone material on the generation of SSEs and megathrust events.

We use a rate- and state-dependent friction law with a cut-off velocity to an evolution effect (Shibazaki and Shimamoto 2007) to represent this frictional behavior. Using Eq. 2, $\mathrm{d}\mu_{ss}/\mathrm{d}\ln v$ is

$$\frac{\mathrm{d}\mu_{ss}}{\mathrm{d}\ln v} = \frac{a}{1 + v/v_0} - \frac{b}{1 + v/v_1}. \quad (7)$$

Based on experimental results of Ikari (2015) and Ikari et al. (2015) (reported in the supplementary material of Ikari and Kopf 2017), we set the friction parameters a, b, and the cut-off velocity to 0.002,

(c)

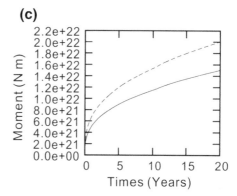

Figure 8
Moment functions of **a** megathrust events and **b, c** afterslip at two different time scales with assumed rigidities of 30 GPa (solid line) and 40 GPa (dashed line) for case 1. The value of the moment was set to zero at $t = 0$, which corresponds to 30 h after the megathrust event

0.007 and 10^{-7} m/s, respectively. In this case, as shown by the solid line in Fig. 10, $\mathrm{d}\mu_{ss}/\mathrm{d}\ln v$ roughly matches experimental results. We consider the shallow region where VW occurs at low slip velocities and VS occurs at high slip velocities. Because this frictional property is favorable for generating SSEs, we refer to this region as the SSE region. The value

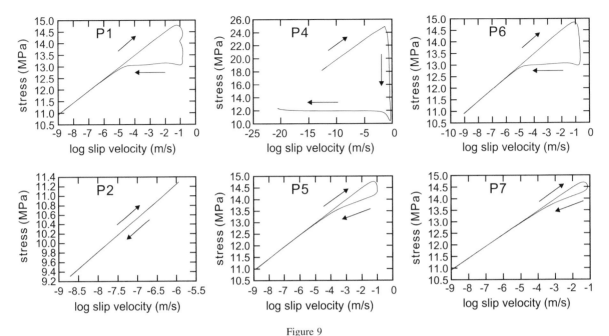

Figure 9

Relationships between stress and slip velocity during coseismic and postseismic periods at points P1 in the stable zone just north of M1 patch, P2 and P5 in stable zone in the deep extension of M2 patch, P4 within the off Miyagi large patch (MT), P6 within off Fukushima stable zone and P7 within off Fukushima stable zone close to the trench for case 1. Arrows indicate the directions of movement. Figure 1 shows the locations of the points

of the base friction μ_* is set to be 0.2, based on the experimental results of Ikari (2015).

As with Fig. 1, we set the rupture areas of previous earthquakes to be patches in which VW occurs and outside which VS occurs, as shown in Fig. 11a. We set the values of the constitutive law parameters as shown in Table 1 for VW patches. We do not consider an ST VW patch. The value of a increases more rapidly at the deeper part than the case shown in Fig. 1. Tsuru et al. (2002) found a thick sedimentary layer along the shallow plate boundary, which is narrow in the off-Miyagi region but wider in the off-Sanriku region. We assume that this thick sedimentary layer contains the pelagic clay that exists in the JFAST fault zone material and that VW occurs at low slip velocity and VS occurs at high slip velocity in the zone of the thick sediment layer. Considering the results reported by Tsuru et al. (2002), we assume the SSE zone to be narrower in the off-Miyagi region and wider in the off-Sanriku region. We also set low effective stress for the SSE zone (Fig. 11b) to reproduce the temperature change consistent with the observed value. The effective stress in the SSE zone

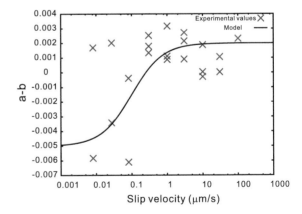

Figure 10

Friction properties of fault zone material for low to intermediate slip velocities at the shallow plate boundary off Tohoku (Ikari 2015). The solid line indicates the friction property $d\mu_{ss}/d\ln v$, where parameters a, b, and the cut-off velocity are set to be 0.002, 0.007 and 10^{-7} m/s

decreases from around 14 MPa at a depth of 10 km to around 1.27 MPa at a depth of 0.2 km.

We perform 3D quasi-dynamic modeling of the slip process. In this model, dynamic weakening owing to TP occurs during coseismic slip. The half-width of the shearing layer w is taken to be 1 cm. The

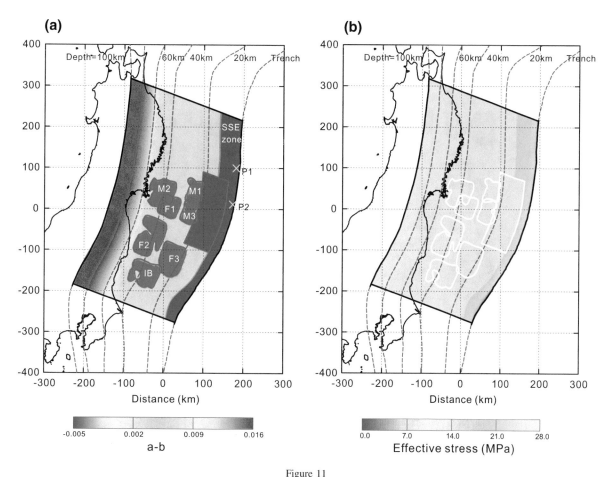

Figure 11
a Distribution of constitutive law parameter $a - b$ on the configuration of the plate interface between the Pacific and Okhotsk plates. The MT patch is located offshore Miyagi near the trench. The shallow slow slip event (SSE) region where velocity weakening (strengthening) occurs at low (high) velocity is considered. **b** Distribution of effective stress

shallow plate boundary fault material in the Japan Trench has higher clay content and a very low hydraulic diffusivity (Tanikawa et al. 2013). Therefore, the value of hydraulic diffusivity is assumed to be 0.0001 m²/s and 0.01 m²/s for the shallow SSE zone and the deeper zone, respectively. For other parameters we use the values shown in Table 2.

Figure 12 shows the distributions of slip velocity for an SSE (a–c), a foreshock (d), and an M_w 9 class earthquake (e–h). The slip velocity of the SSE is approximately 10^{-5}–10^{-7} m/s, consistent with velocity estimation for the SSE immediately preceding the Tohoku-oki earthquake (Ito et al. 2013), although the location of the observed SSEs is different from that of simulated SSEs. Figure 13 indicates changes in slip velocity over time at the

point P1 on the northern plate boundary. This result shows the occurrence of SSEs at intervals of around 13 years at the shallow plate boundary. When an earthquake begins at the largest VW patch, the rupture grows to the megathrust events. In this case, the rupture extends into the SSE region more easily than the case with VS property throughout the entire velocity range.

Figure 14a shows the distribution of the temperature increase 1.983 years after the main earthquake. The temporal evolution of temperature at the shallow fault zone near the trench is shown in Fig. 14b, c. The depth and effective stress of this point are around 0.98 km and 5.7 MPa, respectively. Two years after the megathrust event, the temperature increase is

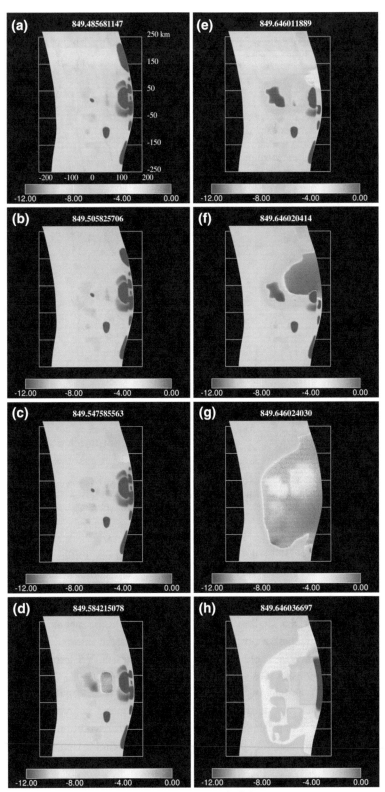

Log slip velocity (m/s)

◀Figure 12
a–h Slip velocity distributions at certain time steps. **a–c** Slip velocity distributions for a slow slip event. **e–h** Slip velocity distributions for the megathrust event. The number shown in each panel represents the elapsed time in years

0.3 °C, which is close to the observed value (Fulton et al. 2013).

5. Discussion

5.1. Model Assumptions on VW Patch Distribution and Shallow Friction Property

We set several VW patches in the regions off Miyagi, Fukushima and Ibaraki. Following previous studies (Yokota et al. 2011), our model assumes a large VW patch (off Miyagi) near the shallow subduction zone to reproduce the largest slips during the 2011 Tohoku-oki earthquake. However, no direct evidence indicates strong coupling in this area. Therefore, there is another possibility that a shallow VS region with low hydraulic diffusivity allows unstable and destructive slip to occur via strong coseismic weakening by TP (e.g., Noda and Lapusta 2013).

In our model, megathrust events cause a maximum temperature increase of approximately 1500 °C at the largest slip region. Although frictional melting should be considered, incorporating this effect is not straightforward in the present study. It is expected that strength will rise and then weakening will occur dramatically when frictional melting occurs (Tsutsumi and Shimamoto 1997). Considering the effect of frictional melting should be investigated in future work.

Laboratory friction experiments on samples retrieved from the very shallow subduction plate boundary showed the value of the base friction μ_* is 0.2 and that VW should be considered at slip velocities of < 0.1 μm/s and VS at slip velocities > 0.1 μm/s (Ikari 2015). The occurrences of the megathrust events and afterslips are mainly controlled by the asperities. However, during the dynamic rupture, the VW property at low slip velocity promotes slip along the shallow SSE region more than the case with VS property at all velocity range. There are many types of slow earthquakes in the subduction zone of NE Japan (Ito et al. 2013; Matsuzawa et al. 2015; Uchida et al. 2016). Investigation of mechanisms on various slow earthquakes should be the focus of future studies.

5.2. Comparison of Observed Afterslip and Numerical Results

Following the M_w 9 events in our simulations, large amounts of afterslip occurred in an area in which coseismic slip was limited. Large afterslip occurred in the region off Sanriku and below the northern, shallow rupture area of the simulated Tohoku-oki earthquake. Afterslip also occurred in the deep off-Miyagi region, where coseismic slip was small. The afterslip characteristics obtained by our simulations explain some features of the distribution of afterslip obtained by Ozawa et al. (2012). Ozawa et al. (2012) noted that small afterslips occurred between the VW patch regions off Fukushima and Ibaraki. It is expected that as the VW patches become smaller, the amount of afterslip increases in the regions off Fukushima and Ibaraki.

Johnson et al. (2012) reported that a model that allows afterslip in the source regions of the 1978 and 1936 earthquakes can fit the observations well. However, Iinuma et al. (2016) pointed out that postseismic slip was absent in the region of the 1978 asperity where a large coseismic slip occurred. In

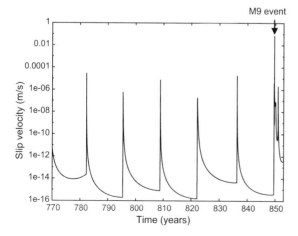

Figure 13
History of fault slip at the shallow slow slip event (SSE) zone (P1) indicated in Fig. 11

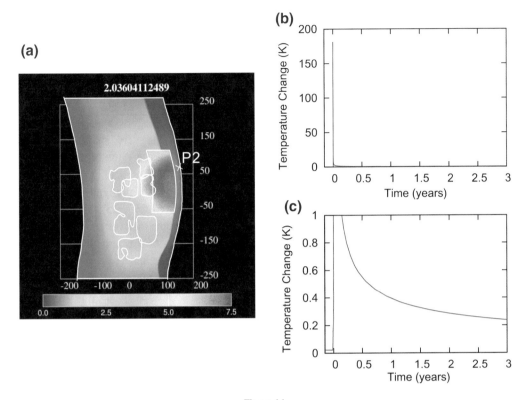

Figure 14
a Temperature distribution 1.983 years after the M_w 9.0 earthquake. **b**, **c** Changes in temperature over time with different axis scales at the point (P2)

their results, postseismic slip was present in the region of the 1936 earthquake although location of asperity was not well determined. In our results, afterslip does not occur in VW patches. However, in the regions surrounding the VW patches, both coseismic slip and afterslip can occur. Whether there was afterslip of the Tohoku-oki earthquake in the asperities of the previous earthquakes is still unclear. Therefore, observational results cannot sufficiently constrain the model for verification.

In our model, we set the VS region offshore Fukushima close to the Japan Trench. Therefore, a large amount of afterslip occurred in this region close to the trench. If the frictional property in this region is VS at a low slip velocity, and if coseismic slip did not occur during the 2011 Tohoku-oki earthquake, significant amounts of afterslip could continue for several years, which would be detectable through the observation of ocean-bottom crustal deformation. Sun and Wang (2015) identified large shallow afterslip at

the southern extension of the largest slip region using data from seafloor GPS sites (Watanabe et al. 2014). The inversion analysis by Iinuma et al. (2016) also indicates the occurrence of large afterslip at the shallow fault zone off Fukushima and Ibaraki.

Iinuma et al. (2016) performed inversion analysis of afterslip after removing the viscoelastic deformation. In their results, the amount of afterslip becomes smaller compared with the results without considering viscoelastic relaxation. Large afterslips in the deep part of off-Sanriku and shallow part of off Fukushima and off Ibaraki are obtained, but the afterslip in the deep extension of off Miyagi earthquake area diminishes. Our results reproduced large afterslips in the deep part of off-Sanriku and shallow part of off Fukushima and off Ibaraki. However, our model did not consider the effect of viscoelastic relaxation. Modeling of the interactions between viscoelastic deformation and afterslips should be considered in future work.

In the VS region near the VW patches, coseismic slip with a small stress drop occurred during simulated megathrust earthquakes owing to TP, causing subsequent afterslips (e.g., P1 and P6 in Fig. 9). The relationship between stress and slip velocity during slip deceleration in the early afterslip stage indicates less VS at slip velocities greater than 10^{-5} m/s. Fukuda et al. (2013) investigated the relationship between shear stress change and slip velocity and found slip-velocity-dependent frictional behavior with less VS at high slip velocities. They pointed out the possibility that this velocity dependence was caused by the reduction of elevated pore pressure in the afterslip region. Our results are qualitatively consistent with those reported by Fukuda et al. (2013).

Nakata et al. (2016) simulated the Tohoku-oki earthquake and its afterslips assuming that the M_w 9 source area is a large-fracture-energy area and that the M_w 7–8 asperities are smaller-fracture-energy areas. In their model, 3.5 years after the main events, afterslips occur in the deeper extension of the seismogenic zone. However, shallow afterslips in the off Fukushima and Ibaraki regions were not reproduced. On the other hand, in our model we could reproduce afterslips similar to the observed results just by assuming the distribution of VW patches and geometry of plate boundary.

Finally, comparing our results with those of the previous model shows that our model reproduced results similar to those obtained by Shibazaki et al. (2011). As shown in Fig. 6, inside the VW patches two weakening processes occurred: the rapid weakening with small displacement caused by the rate-and state-friction properties and TP, and the gradual weakening with large displacement was caused by TP. In the VS region near the VW region, gradual weakening occurred by TP. Shibazaki et al. (2011) considered a rate- and state-dependent friction law with two state variables to represent dynamic weakening at high slip velocities. In their model, at high slip velocities, rapid weakening with small displacement and gradual weakening with large displacement occur. Therefore, the frictional behaviors in their model are similar to those in the present model with TP. Shibazaki et al. (2011) derived their constitutive law empirically. However,

this study reproduced a Tohoku-oki earthquake and its afterslip by considering physical processes of TP. Although there have been several models of the Tohoku-oki earthquake considering TP (e.g., Noda and Lapusta 2013; Cubas et al. 2015), this is the first study to investigate afterslips considering the realistic distribution of asperities and fault geometry and TP.

6. Conclusions

We developed a 3D quasi-dynamic earthquake cycle model of the Tohoku-oki earthquake considering TP. We set several VW patches in the regions off Miyagi, Fukushima and Ibaraki, and we set a large strong VW patch near the trench. We set the frictional properties to VS outside the VW patches. When rupture occurred near the large asperity, significant TP occurred which resulted in large and fast slip. This rupture propagated to the surrounding regions and to the asperities of M_w 7.5 earthquakes owing to weakening caused by TP. When the size of the shallow VW patch became smaller, the effect of TP also became small so that the rupture did not propagate to the surrounding regions and did not grow to a megathrust event. When the base friction level was larger, more frictional heat was generated, and TP worked more effectively to generate larger events.

In our model, following the M_w 9 class events, large afterslips occurred in areas where the coseismic slip was not large, complementing the large coseismic slip zone. For example, in the region below the northern shallow rupture area of the simulated earthquake, large afterslips occurred. In the off-Miyagi region, afterslips occurred in the deep area where the coseismic slip was small. In the off-Fukushima region close to the Japan Trench, a large amount of afterslip occurred. The distribution of afterslips obtained by our simulation is roughly consistent with the observed distribution. In the VS region near the VW patches, large coseismic slip can occur due to dynamic weakening by TP, followed by afterslips. During the early afterslip period, friction behavior exhibit less VS with increasing slip velocity due to the effect of TP. The results are qualitatively consistent with those reported by Fukuda et al. (2013).

We also considered the frictional properties of fault zone material of low to intermediate slip velocity at the shallow plate boundary off Tohoku, which exhibits VW at low slip velocity and VS at higher slip velocity. We investigated the effect of this property on the generation of SSEs and megathrust events. The numerical results showed that SSEs occurred at intervals of decades at the shallow plate boundary, with stress accumulation during the periods between SSEs. During a megathrust event, the VW property at low slip velocity promotes slip along the shallow SSE region more than the case with VS property at all velocity ranges.

In the present study, we considered the simple friction property that exhibits VW at a low slip velocity and VS at high slip velocity. The friction properties of the shallow plate boundary depend on complex temperature conditions and slip velocity (Sawai et al. 2017). Noda et al. (2017) developed a 2D dynamic model of megathrust earthquakes and shallow slow earthquakes that considers these friction properties. Development of a 3D model that considers the complex friction properties will be necessary to understand the generation of SSEs and megathrust earthquakes. Furthermore, more realistic modeling of SSEs, earthquake cycles, and afterslips considering viscoelastic effects (e.g., Barbot 2018) will be developed for the subduction zone off Tohoku.

Acknowledgements

We are grateful to the guest editor, Sylvain Barbot, and the anonymous reviewers for valuable comments that helped to improve the manuscript. This study was supported by MEXT KAKENHI (26109007, 24340107). This study was also supported by the Ministry of Education, Culture, Sports, Science and Technology (MEXT) of Japan, under its Earthquake and Volcano Hazards Observation and Research Program. For this study, we used the computer systems of the Earthquake Information Center of the Earthquake Research Institute, University of Tokyo.

Appendix

We perform a quasi-dynamic analysis using the following equation (e.g. Rice 1993):

$$\tau_i = \sum_{i_s} k_{i-i_s}\left(v_{\mathrm{pl}}t - u_{i_s}\right) - \frac{G}{2\beta}\frac{\mathrm{d}u_i}{\mathrm{d}t}, \qquad (8)$$

where k_{i-i_s} is the stiffness which is the stress at the center of gravity of triangular element i caused by uniform slip over triangular element i_s, G is the rigidity and β is the shear wave velocity. To calculate the stiffness for the triangular elements, we use the program developed by Stuart et al. (1997). In the quasi-dynamic modeling, we add seismic radiation damping to the second term introduced by Rice (1993) to approximate the effects of inertia during earthquakes. The centers of gravity of the triangular elements are taken as the nodes (i.e., co-location points).

From Eqs. (1) and (8), we obtain the following equation:

$$\mu(v, \Theta)\frac{\sigma_{\mathrm{n},i}^{\mathrm{eff}}}{G} = \sum k'_{i-i_s}\left(v_{\mathrm{pl}}t - u_{i_s}\right) - \frac{1}{2\beta}\frac{\mathrm{d}u_i}{\mathrm{d}t}, \qquad (9)$$

where $k'_{i-i_s} = k_{i-i_s}/G$. From this equation, it is clear that we get the same solution for the same $\sigma_{\mathrm{n},i}^{\mathrm{eff}}/G$. In this calculation, G is assumed to be 30 GPa. However, we get the same result when $G = 40$ GPa and $\sigma_{\mathrm{n}}^{\mathrm{eff}}$ increases by a factor of 40/30. In this case, τ, T and P also increase by a factor of 40/30.

We solve Eqs. (1)–(3) and (8) using a fifth-order Runge–Kutta method with adaptive step-size control (Press et al. 1992), to obtain v^{n+1} and τ^{n+1} at time step $n + 1$ using P^n and T^n at step n. We calculate the average slip rate $\underline{v}^{n+1} = (v^{n+1} + v^n)/2$ and shear stress $\underline{\tau}^{n+1} = (\tau^{n+1} + \tau^n)/2$ between steps n and $n + 1$. Using the values of $\underline{v}^{n+1}$ and $\underline{\tau}^{n+1}$ to calculate the heat source (ω), we obtain P^{n+1} and T^{n+1} by solving Eqs. (4) and (5) using a spectral method (Noda and Lapusta 2010). We then update P^{n+1} in Eq. (1). A few iterations are performed with a fixed step size to obtain the converged solutions.

Publisher's Note Springer Nature remains neutral with regard to jurisdictional claims in published maps and institutional affiliations.

REFERENCES

Ariyoshi, K., Matsuzawa, T., Ampuero, J. P., Nakata, R., Hori, T., Kaneda, Y., et al. (2012). Migration process of very low-frequency events based on a chain-reaction model and its application to the detection of preseismic slip for megathrust earthquakes. *Earth, Planets and Space, 64*, 693–702. https://doi.org/10.5047/eps.2010.09.003.

Barbot, S. (2018). Asthenosphere flow modulated by megathrust earthquake cycles. *Geophysical Research Letters, 45*, 6018–6031. https://doi.org/10.1029/2018GL078197.

Barbot, S., Lapusta, N., & Avouac, J. P. (2012). Under the hood of the earthquake machine: Toward predictive modeling of the seismic cycle. *Science, 336*, 707–710. https://doi.org/10.1126/science.1218796.

Brodsky, E. E., Saffer, D., Fulton, P., Chester, F., Conin, M., Huffman, K., et al. (2017). The postearthquake stress state on the Tohoku megathrust as constrained by reanalysis of the JFAST breakout data. *Geophysical Research Letters, 44*, 22. https://doi.org/10.1002/2017GL074027.

Chester, F. M., Rowe, C., Ujiie, K., Kirkpatrick, J., Regalla, C., Remitti, F., et al. (2013). Structure and composition of the plate-boundary slip zone for the 2011 Tohoku-Oki earthquake. *Science, 342*, 1208–1211. https://doi.org/10.1126/science.1243719.

Cubas, N., Lapusta, N., Avouac, J.-P., & Perfettini, H. (2015). Numerical modeling of long-term earthquake sequences on the NE Japan megathrust: Comparison with observations and implications for fault friction. *Earth and Planetary Science Letters, 419*, 187–198. https://doi.org/10.1016/j.epsl.2015.03.002.

Di Toro, G., Han, R., Hirose, T., De Paola, N., Nielsen, S., Mizoguchi, K., et al. (2011). Fault lubrication during earthquakes. *Nature, 471*, 494–498. https://doi.org/10.1038/nature09838.

Dieterich, J. H. (1981). Constitutive properties of rock with simulated gouge, chap. 8. In N. L. Carter, et al. (Eds.), *Mechanical behavior of crustal rocks: the handing volume* (pp. 103–120). Washington, D. C.: AGU. https://doi.org/10.1029/GM024p0103.

Fujii, Y., Satake, K., Sakai, S., Shinohara, M., & Kanazawa, T. (2011). Tsunami source of the 2011 off the Pacific coast of Tohoku earthquake. *Earth Planets Space, 63*, 815–820. https://doi.org/10.5047/eps.2011.06.010.

Fukuda, J., Kato, A., Kato, N., & Aoki, Y. (2013). Are the frictional properties of creeping faults persistent? Evidence from rapid afterslip following the 2011 Tohoku-oki earthquake. *Geophysical Research Letters, 40*, 3613–3617. https://doi.org/10.1002/grl.50713.

Fulton, P. M., Brodsky, E. E., Kano, Y., Mori, J., Chester, F., Ishikawa, T., et al. (2013). Low coseismic friction on the Tohoku-Oki fault determined from temperature measurements. *Science, 342*, 1214–1217. https://doi.org/10.1126/science.1243641.

Gao, X., & Wang, K. (2014). Strength of stick-slip and creeping subduction megathrusts from heat flow observations. *Science, 345*, 1038–1041. https://doi.org/10.1126/science.1255487.

Hasegawa, A., Yoshida, K., & Okada, T. (2011). Nearly complete stress drop in the 2011 M_w 9.0 off the Pacific coast of Tohoku Earthquake. *Earth Planets Space, 63*, 703–707. https://doi.org/10.5047/eps.2011.06.007.

Hori, T., & Miyazaki, S. (2011). A possible(mechanism of M 9 earthquake generation cycles in the area of repeating M 7 and 8

earthquakes surrounded by aseismic sliding. *Earth Planets Space, 63*, 773–777. https://doi.org/10.5047/eps.2011.06.022.

Ide, S., Baltay, A., & Beroza, G. C. (2011). Shallow dynamic overshoot and energetic deep rupture in the 2011 M_w 9.0 Tohoku-Oki earthquake. *Science, 332*, 1426–1429. https://doi.org/10.1126/science.1207020.

Iinuma, T., Hino, R., Kido, M., Inazu, D., Osada, Y., Ito, Y., et al. (2012). Coseismic slip distribution of the 2011 off the Pacific Coast of Tohoku Earthquake (M 9.0) refined by means of seafloor geodetic data. *Journal of Geophysical Research: Solid Earth, 117*, B07409. https://doi.org/10.1029/2012JB009186.

Iinuma, T., Hino, R., Uchida, N., Nakamura, W., Kido, M., Osada, Y., et al. (2016). Seafloor observations indicate spatial separation of coseismic and postseismic slips in the 2011 Tohoku earthquake. *Nature Communications, 7*, 13506. https://doi.org/10.1038/ncomms13506.

Ikari, M.J. (2015). Data report: rate- and state-dependent friction parameters of core samples from Site C0019, IODP Expedition 343 (JFAST). In: Chester, F.M., Mori, J., Eguchi, N., Toczko, S., & the Expedition 343/343T Scientists (Eds.) *Proc. IODP, 343/343T: Tokyo (Integrated Ocean Drilling Program Management International, Inc.).* https://doi.org/10.2204/iodp.proc.343343t.203.2015.

Ikari, M. J., Ito, Y., Ujiie, K., & Kopf, A. J. (2015). Spectrum of slip behaviour in Tohoku fault zone samples at plate tectonic slip rates. *Nature Geoscience, 8*, 870. https://doi.org/10.1038/ngeo2547.

Ikari, M. J., & Kopf, A. J. (2017). Seismic potential of weak, near-surface faults revealed at plate tectonic slip rates. *Science Advances, 3*, e1701269. https://doi.org/10.1126/sciadv.1701269.

Ito, Y., Hino, R., Kido, M., Fujimoto, H., Osada, Y., Inazu, D., et al. (2013). Episodic slow slip events in the Japan subduction zone before the 2011 Tohoku-Oki earthquake. *Tectonophysics, 600*, 14–26. https://doi.org/10.1016/j.tecto.2012.08.022.

Ito, Y., Tsuji, T., Osada, Y., Kido, M., Inazu, D., Hayashi, Y., et al. (2011). Frontal wedge deformation near the source region of the 2011 Tohoku-Oki earthquake. *Geophysical Research Letters, 38*, L00G05. https://doi.org/10.1029/2011GL048355.

Johnson, K. M., Fukuda, J., & Segall, P. (2012). Challenging the rate-state asperity model: Afterslip following the 2011 M9 Tohoku-oki, Japan, earthquake. *Geophysical Research Letters, 39*, L20302. https://doi.org/10.1029/2012GL052901.

Kato, N., & Yoshida, S. (2011). A shallow strong patch model for the 2011 great Tohoku-oki earthquake: A numerical simulation. *Geophysical Research Letters, 38*, L00G04. https://doi.org/10.1029/2011GL048565.

Lachenbruch, A. H. (1980). Frictional heating, fluid pressure, and the resistance to fault motion. *Journal of Geophysical Research: Solid Earth, 85*(B11), 6097–6112.

Lay, T., Ammon, C. J., Kanamori, H., Xue, L., & Kim, M. J. (2011). Possible large near trench slip during the 2011 M_w 9.0 the Pacific coast of Tohoku Earthquake. *Earth Planets Space, 63*, 687–692. https://doi.org/10.5047/eps.2011.05.006.

Lin, W., Conin, M., Moore, J. C., Chester, F. M., Nakamura, Y., Mori, J. J., et al. (2013). Stress state in the largest displacement area of the 2011 Tohoku-Oki earthquake. *Science, 339*, 687–690. https://doi.org/10.1126/science.1229379.

Liu, Y., & Rice, J. R. (2007). Spontaneous and triggered aseismic deformation transients in a subduction fault model. *Journal of Geophysical Research: Solid Earth, 112*, B09404. https://doi.org/10.1029/2007JB004930.

Liu, Y., & Rubin, A. M. (2010). Role of fault gouge dilatancy on aseismic deformation transients. *Journal of Geophysical Research: Solid Earth, 115*, B10414. https://doi.org/10.1029/2010JB007522.

Marone, C. J., Scholz, C. H., & Bilham, R. (1991). On the mechanics of earthquake afterslip. *Journal of Geophysical Research: Solid Earth, 96*(B5), 8441–8452. https://doi.org/10.1029/91JB00275.

Matsuzawa, T., Asano, Y., & Obara, K. (2015). Very low frequency earthquakes off the Pacific coast of Tohoku, Japan. *Geophysical Research Letters, 42*, 4318–4325. https://doi.org/10.1002/2015GL063959.

Mitsui, Y., Kato, N., Fukahata, Y., & Hirahara, K. (2012). Megaquake cycle at the Tohoku subduction zone with thermal fluid pressurization near the surface. *Earth and Planetary Science Letters, 325*, 21–26. https://doi.org/10.1016/j.epsl.2012.01.026.

Miyazaki, S., Segall, P., Fukuda, J., & Kato, T. (2004). Space time distribution of afterslip following the 2003 Tokachi-oki earthquake: Implications for variations in fault zone frictional properties. *Geophysical Research Letters, 31*, L06623. https://doi.org/10.1029/2003GL019410.

Murotani, S., (2003). Rupture processes of large Fukushima-oki Earthquakes in 1938. Master's thesis, University of Tokyo, Tokyo, Japan.

Nakata, R., Hori, T., Hyodo, M., & Ariyoshi, K. (2016). Possible scenarios for occurrence of M ∼ 7 interplate earthquakes prior to and following the 2011 Tohoku-Oki earthquake based on numerical simulation. *Scientific Reports, 6*, 25704.

Noda, H., & Lapusta, N. (2010). Three-dimensional earthquake sequence simulations with evolving temperature and pore pressure due to shear heating: Effect of heterogeneous hydraulic diffusivity. *Journal of Geophysical Research: Solid Earth, 115*, B12314. https://doi.org/10.1029/2010JB007780.

Noda, H., & Lapusta, N. (2013). Stable creeping fault segments can become destructive as a result of dynamic weakening. *Nature, 493*, 518–521. https://doi.org/10.1038/nature11703.

Noda, H., Sawai, M., & Shibazaki, B. (2017). Earthquake sequence simulations with measured properties for JFAST core samples. *Philosophical Transactions of the Royal Society A, 375*(2103), 20160003.

Ohtani, M., Hirahara, K., Hori, T., & Hyodo, M. (2014). Observed change in plate coupling close to the rupture initiation area before the occurrence of the 2011 Tohoku earthquake: Implications from an earthquake cycle model. *Geophysical Research Letters, 41*, 1899–1906. https://doi.org/10.1002/2013GL058751.

Ozawa, S., Nishimura, T., Munekane, H., Suito, H., Kobayashi, T., Tobita, M., et al. (2012). Preceding, coseismic, and postseismic slips of the 2011 Tohoku earthquake. *Japan, Journal of Geophysical Research: Solid Earth, 117*, B07404. https://doi.org/10.1029/2011JB009120.

Perfettini, H., & Avouac, J.-P. (2004). Postseismic relaxation driven by brittle creep: A possible mechanism to reconcile geodetic measurements and the decay rate of aftershocks, application to the Chi-Chi earthquake, Taiwan. *Journal of Geophysical Research: Solid Earth, 109*, B02304. https://doi.org/10.1029/2003JB002488.

Press, W. H., Teukolsky, S. A., Vetterling, W. T., & Flannery, B. P. (1992). *Numerical recipes in Fortran: The art of scientific computing* (2nd ed.). Cambridge: Cambridge University Press.

Remitti, F., Smith, S. A. F., Mittempergher, S., Gualtieri, A. F., & Di Toro, G. (2015). Frictional properties of fault zone gouges from teh J-FAST drilling project (M_w 9.0 Tohoku-Oki earthquake). *Geophysical Research Letters, 42*, 2691–2699. https://doi.org/10.1002/205GL063507.

Rice, J. R. (1993). Spatio-temporal complexity of slip on a fault. *Journal of Geophysical Research: Solid Earth, 98*(B6), 9885–9907. https://doi.org/10.1029/93JB00191.

Rowe, C. D., Moore, J. C., Remitti, F., & IODP Expedition 343/343T Scientists. (2013). Thickness of subduction plate boundary faults from the seafloor into the seismogenic zone. *Geology, 419*, 991–994. https://doi.org/10.1130/G34556.1.

Salman, R., Hill, E. M., Feng, L., Lindsey, E. O., Mele Veedu, D., Barbot, S., et al. (2017). Piecemeal rupture of the Mentawai patch, Sumatra: The 2008 M_w 7.2 North Pagai earthquake sequence. *Journal of Geophysical Research: Solid Earth, 122*, 9404–9419. https://doi.org/10.1002/2017JB014341.

Satake, K., Fujii, Y., Harada, T., & Namegaya, Y. (2013). Time and space distribution of coseismic slip of the 2011 Tohoku earthquake as inferred from tsunami waveform data. *Bulletin of the Seismological Society of America, 103*, 1473–1492. https://doi.org/10.1785/0120120122.

Satake, K., Namegaya, Y., & Yamamoto, S. (2008). Numerical simulation of the AD 869 Jogan tsunami in Ishinomaki and Sendai plains. *Annual Report of Active Fault and Paleoearthquake Research, 8*, 71–89.

Sawai, M., Hirose, T., & Kameda, J. (2014). Frictional properties of incoming pelagic sediment at the Japan Trench: implications for large slip at a shallow plate boundary during the 2011 Tohoku earthquake. *Earth Planets Space, 66*, 65.

Sawai, Y., Namegaya, Y., Tamura, T., Nakashima, R., & Tanigawa, K. (2015). Shorter intervals between great earthquakes near Sendai: Scour ponds and a sand layer attributable to A.D. 1454 overwash. *Geophysical Research Letters, 42*, 4795–4800. https://doi.org/10.1002/2015GL064167.

Sawai, M., Niemeijer, A. R., Hirose, T., & Spiers, C. J. (2017). Frictional properties of JFAST core samples and implications for slow earthquakes at the Tohoku subduction zone. *Geophysical Research Letters, 44*, 8822–8831. https://doi.org/10.1002/2017GL073460.

Segall, P., Rubin, A. M., Bradley, A. M., & Rice, J. R. (2010). Dilatant strengthening as a mechanism for slow slip events. *Journal of Geophysical Research: Solid Earth, 115*, 12305. https://doi.org/10.1029/2010JB007449.

Shibazaki, B., Bu, S., Matsuzawa, T., & Hirose, H. (2010). Modeling the activity of short-term slow slip events along deep subduction interfaces beneath Shikoku, southwest Japan. *Journal of Geophysical Research: Solid Earth, 115*, 12301. https://doi.org/10.1029/2010JB007566.

Shibazaki, B., & Iio, Y. (2003). On the physical mechanism of silent slip events along the deeper part of the seismogenic zone. *Geophysical Research Letters, 30*, 1489. https://doi.org/10.1029/2003GL017047.

Shibazaki, B., Matsuzawa, T., Tsutsumi, A., Ujiie, K., Hasegawa, A., & Ito, Y. (2011). 3D modeling of the cycle of a great Tohoku-oki earthquake, considering frictional behavior at low to high slip velocities. *Geophysical Research Letters, 38*, L21305. https://doi.org/10.1029/2011GL049308.

Shibazaki, B., & Shimamoto, T. (2007). Modelling of short-interval silent slip events in deeper subduction interfaces considering the frictional properties at the unstable–stable transition regime. *Geophysical Journal International, 171*, 191–205. https://doi.org/10.1111/j.1365-246X.2007.03434.x.

Stuart, W. D., Hildenbrand, T., & Simpson, R. (1997). Stressing of the New Madrid Seismic Zone by a lower crust detachment fault. *Journal of Geophysical Research: Solid Earth, 102,* 27623–27633. https://doi.org/10.1029/97JB02716.

Sun, T., & Wang, K. (2015). Viscoelastic relaxation following subduction earthquakes and its effects on afterslip determination. *Journal of Geophysical Research: Solid Earth, 120,* 1329–1344. https://doi.org/10.1002/2014JB011707.

Tanikawa, W., Hirose, T., Mukoyoshi, H., Tadai, O., & Lin, W. (2013). Fluid transport properties in sediments and their role in large slip near the surface of the plate boundary fault in the Japan Trench. *Earth and Planetary Science Letters, 382,* 150–160. https://doi.org/10.1016/j.epsl.2013.08.052.

Tanikawa, W., & Shimamoto, T. (2009). Frictional and transport properties of the Chelungpu fault from shallow borehole data and their correlation with seismic behavior during the 1999 Chi-Chi earthquake. *Journal of Geophysical Research: Solid Earth, 114,* B01402. https://doi.org/10.1029/2008JB005750.

Tsuru, T., Park, J. O., Miura, S., Kodaira, S., Kido, Y., & Hayashi, T. (2002). Along arc structural variation of the plate boundary at the Japan Trench margin: Implication of interplate coupling. *Journal of Geophysical Research: Solid Earth, 107,* 2357. https://doi.org/10.1029/2001JB001664.

Tsutsumi, A., Fabbri, O., Karpoff, A. M., Ujiie, K., & Tsujimoto, A. (2011). Friction velocity dependence of clay-rich fault material along a megasplay fault in the Nankai subduction zone at intermediate to high velocities. *Geophysical Research Letters, 38,* L19301. https://doi.org/10.1029/2011GL049314.

Tsutsumi, A., & Shimamoto, T. (1997). High-velocity frictional properties of gabbro. *Geophysical Research Letters, 24,* 699–702. https://doi.org/10.1029/97GL00503.

Uchida, N., Iinuma, T., Nadeau, R. M., Bürgmann, R., & Hino, R. (2016). Periodic slow slip triggers megathrust zone earthquakes in northeastern Japan. *Science, 351*(6272), 488–492. https://doi.org/10.1126/science.aad3108.

Ujiie, K., & Kimura, G. (2014). Earthquake faulting in subduction zones: insights from fault rocks in accretionary prisms. *Progress in Earth and Planetary Science, 1,* 7. https://doi.org/10.1186/2197-4284-1-7.

Ujiie, K., Tanaka, H., Saito, T., Tsutsumi, A., Mori, J. J., Kameda, J., et al. (2013). Low coseismic shear stress on the Tohoku-oki megathrust determined from laboratory experiments. *Science, 342,* 1211–1214. https://doi.org/10.1126/science.1243485.

Watanabe, S., Sato, M., Fujita, M., Ishikawa, T., Yokota, Y., Ujihara, N., et al. (2014). Evidence of viscoelastic deformation following the 2011 Tohoku-Oki earthquake revealed from seafloor geodetic observation. *Geophysical Research Letters, 41,* 5789–5796. https://doi.org/10.1002/2014GL061134.

Wibberley, C. A., & Shimamoto, T. (2005). Earthquake slip weakening and asperities explained by thermal pressurization. *Nature, 436,* 689–692. https://doi.org/10.1038/nature03901.

Yagi, Y., & Fukahata, Y. (2011). Rupture process of the 2011 Tohoku-oki earthquake and absolute elastic strain release. *Geophysical Research Letters, 38,* L19307. https://doi.org/10.1029/2011GL048701.

Yamanaka, Y. & Kikuchi, M. (2003). Oct. 31, Fukushima-oki earthquake (Mj = 6.8). *EIC Seismological Notes,* 141. http://wwweic.eri.u-tokyo.ac.jp/sanchu/Seismo_Note/EIC_News/031031.html **(in Japanese)**. Accessed 18 Jan 2019.

Yamanaka, Y., & Kikuchi, M. (2004). Asperity map along the subduction zone in northeastern Japan inferred from regional seismic data. *Journal of Geophysical Research: Solid Earth, 109,* B07307. https://doi.org/10.1029/2003JB002683.

Yokota, Y., Koketsu, K., Fujii, Y., Satake, K., Sakai, S., Shinohara, M., et al. (2011). Joint inversion of strong motion, teleseismic, geodetic, and tsunami datasets for the rupture process of the 2011 Tohoku earthquake. *Geophysical Research Letters, 38,* L00G21. https://doi.org/10.1029/2011GL050098.

(Received April 12, 2018, revised November 10, 2018, accepted December 29, 2018, Published online February 5, 2019)

Pure Appl. Geophys. 176 (2019), 3975–3992
© 2019 The Author(s)
https://doi.org/10.1007/s00024-019-02094-7

Effect of Slip-Weakening Distance on Seismic–Aseismic Slip Patterns

PIOTR SENATORSKI[1] (iD)

Abstract—In order to explain the seismic–aseismic slip patterns observed on megathrust faults, numerical simulations were carried out using the quasi-dynamic asperity fault model with the slip-dependent friction and stress-dependent healing. Two friction law parameters, strength and slip-weakening distance, are interpreted in the subduction channel context. The parameters are treated as random fields with specified characteristics. Their distributions define heterogeneities of the interplate frictional coupling. The higher strength regions accumulate stresses, whereas the slip-weakening distance lengths control the stress release rates. The simulation results indicate that the slip-dependent asperity model reproduces key features of real megathrust fault behavior. First, stable and unstable slip movements can occur at the same locations, even if the friction parameters are fixed. Second, two rupture styles, single asperity breaks and wide, smooth, propagating rupture fronts, can be distinguished. The latter style is responsible for large slips near the free surface, where lower fault strengths are expected. The reason for these effects is that slip instabilities depend both on local friction and on the system stiffness, which is related to the slipping area size and distribution of slips. It is also shown that the high-strength interplate patches, such as subducted seamounts, can both promote and restrain large earthquakes, depending on the slip-weakening distance lengths.

Key words: Earthquake dynamics, slip-dependent friction, slow and fast slip, numerical modeling, subduction zone processes, megathrust earthquakes.

1. Introduction

Slip movements on the subduction interface can be considered in terms of the slip budget. At any point, the long-term slip is controlled by the plate convergence rate, V_P. The slip deficit is accumulated over a period of time in a given region, if the slip rate is lower than the convergence rate. The slip deficit is released by both seismic and aseismic slips, which complement each other on the plate interface. Understanding of the interplay between fast and slow slips, such as regular earthquakes and slow slip events (Peng and Gomberg 2010), respectively, is crucial for earthquake hazard estimations.

The interplate coupling coefficient is defined as the ratio of the slip deficit rate over the plate convergence rate. Regions of high interseismic coupling roughly correspond with high-coseismic-slip regions in most cases (Moreno et al. 2010, 2012; Hashimoto et al. 2012; Loveless and Meade 2011).

A general approach towards the seismic and aseismic slip interplay is given by asperity fault models. Such models assume that the fault plane consists of strong patches of episodic, unstable slips, which are surrounded by weaker, creeping regions (Lay et al. 1982). The slow slips in the non-asperity regions concentrate stresses in the asperity regions during interseismic periods. Asperities accumulate stresses until they break during earthquakes. More sophisticated models assume hierarchical asperity structures (Uchida and Matsuzawa 2011; Ohtani et al. 2014).

Fast and slow slips are commonly interpreted in terms of frictional fault characteristics, with asperities and non-asperity regions treated as fixed, spatially separated fault features. That view is inconsistent with observations of the Tohoku–Oki 2011 earthquake. First, the largest coseismic slip occurred in the shallow fault area considered to be a velocity-strengthening and slowly creeping region (Ide et al. 2011). Second, aseismic afterslip occurred within the areas of previous smaller earthquakes considered to be velocity-weakening asperities (Johnson et al. 2012). Explanations of these inconsistencies, or the slow and fast slip paradox, can refer to the applied friction law modifications (Noda and Lapusta 2013), non-planar fault geometry (Fukuyama and Hok

---
[1] Institute of Geophysics, Polish Academy of Sciences, ul. Księcia Janusza 64, 01-452 Warsaw, Poland. E-mail: psenat@igf.edu.pl

2013), or rupture dynamics and free surface effects (Huang et al. 2013).

The present paper refers to fault heterogeneities and the system stiffness to solve the problem. Stability of the slip movement at a given location depends both on the local friction characteristics and the rate at which the resisting stress decreases with the ongoing slip, i.e., on the critical stiffness, and on the rate at which the driving stress changes with the ongoing slip, i.e., on the system stiffness (Senatorski 2002; Mansinha and Smylie 1971). The latter factor changes with time since distribution of slips, which contribute to the driving stress, evolves in time. The proposed model enables us to explain the observed fault behavior without introducing any extra, velocity-dependent, weakening mechanism to the friction law.

Spatial variations of the interplate frictional coupling are considered in terms of the asperity fault model. Structural and material heterogeneities along the fault plane, due to subducted seamounts, oceanic ridges, sediments, and released fluids, are thought to be responsible for the coupling variations. The specific role of subducted seamounts or sediments have been debated (Wang and Bilek 2011; Scholz and Campos 2012; Heuret et al. 2012). It is not clear, whether the seamounts start or stop large earthquakes (Wang and Bilek 2011).

In the present paper, topographic features are modeled by slip-dependent cohesive stresses acting along the fault plane. The cohesive stress is defined at each point of the fault plane as a function of slip by two parameters: the peak stress or fault strength, and the slip-weakening distance at which friction or cohesive stress decreases to its residual level. The strength can be related to the normal stress variations due to the hilly topography of the subducting plate surface (Scholz and Small 1997). The slip-weakening distance can be related to the material fault characteristics, such as subducted sediment layers or fluids (Heuret et al. 2012; Vannucchi et al. 2017). Thus, variations of the interplate frictional coupling are defined as distributions of these two parameters. The first aim is to relate these distributions to the stable and unstable slip patterns. The second aim is to find out whether such distributions promote small or large earthquakes.

A subduction zone plate interface is a subduction channel hundreds of meters to several kilometers thick, lubricated by a layer of shearing unconsolidated sediment (Cloos and Shreve 1988; Vannucchi et al. 2012). Consequently, assumed friction or cohesive stress variations applied to the model fault plane approximate all processes taking place in the channel: not only sliding along a well-defined thin rupture plane, but also sediment viscous flow dragged by the descending plate, the sediment dewatering, consolidation and underplating to the hanging wall, the upper plate erosion and seamounts being jammed against the roof of the thinning channel (Cloos and Shreve 1996). Therefore, fault strength and slip-weakening distance parameters, as well as the shear stress critical value for slip healing described below, even if well-supported by laboratory experiments with the sliding friction (Ohnaka and Shen 1999; Brantut 2015), here are also interpreted within the subduction channel context.

To study seismic–aseismic slip patterns, long-term fault dynamics and a large range of spatial and temporal scales should be modeled. To that end, quasi-dynamic fault models have been used for computer simulations by many authors as the optimal solution (Rice 1993; Senatorski 1995, 2002; Ziv and Cochard 2006; Hillers et al. 2006; Ohtani et al. 2014; Ohtani and Hirahara 2015). Such models enable us to generate realistic, both slow and fast, slip movements. It should be borne in mind, however, that details of coseismic slips obtained from quasi-dynamics can differ from those obtained from fully dynamic models. The difference depends on the friction law applied and related fault rupture style (Thomas et al. 2014).

In its general form, the present model is similar to other models based on the quasi-dynamic approximation. Other authors (Hillers et al. 2006) investigated the spatially variable critical slip distance and the resultant complexity in seismicity patterns, using the model based on rate-and-state friction law. The main difference between their approach and the present one concerns details of the interplate frictional coupling formulation, including the applied friction law and distribution of its parameters, which defines asperity and non-asperity regions on the megathrust fault.

The model used in the present work is similar to the model described in the previous work (Senatorski 2002). The main objective of this paper concerns the relation between the megathrust fault structural and material characteristics and the slow and fast slip interplay, so the model is adapted to megathrust seismicity context. Its formulations of slip-dependent friction or cohesive stress, healing and asperity distributions, interpreted in the context of subduction channel processes, enable us to find an explanation of the slow and fast slip paradox mentioned above, as well as the role of subducted seamounts as both promoting and restraining large earthquakes.

2. Methods

2.1. Theory

Geometry An earthquake source is modeled as a planar cut in an elastic half-space subjected to the background tectonic shear stress, s^T. An earthquake rupture process is represented by evolution of stresses and slip displacements, $s(\mathbf{r}, t)$ and $q(\mathbf{r}, t)$, respectively, defined over the rupture area, A, and the earthquake slip duration, T. For simplicity, only one component of slips and shear stresses defined on a planar fault, the subducting direction component, is considered. Distributions of slips and stresses change from $q(\mathbf{r}, t = 0)$ to $q(\mathbf{r}, t = T)$ and from $s(\mathbf{r}, t = 0)$ to $s(\mathbf{r}, t = T)$, respectively, during the earthquake. The slip movement gradually reduces the stress to the residual value of the cohesive or frictional resistance, s^F.

Quasi-dynamics. The quasi-dynamic evolution equations for slips can be derived in two steps. First, the stress pulse, $\Delta s = s^T - s^F$, is applied over the $z = 0$ fault plane instantaneously at $t = 0$ (Brune 1970). Second, the resulting equation is generalized to the heterogeneous rupture case by introducing the interaction stress term, s^I. Assume that the uniform stress pulse propagates along the z-axis perpendicular to the fault plane at the shear wave speed, v_S.

$$\frac{\mu}{2v_S} \dot{q}(\mathbf{r}, t) = \Delta s(\mathbf{r}, t) = s^T + s^I(\mathbf{r}, t) - s^F(\mathbf{r}, t) , \quad (1)$$

where μ is the shear modulus, $s = s^T + s^I$ is the driving stress, s^F is friction or the cohesive stress, and

Δs represents the net stress. In that approximation, the radiated energy can be illustrated as the area between the driving and frictional or cohesive stress lines in the stress vs. slip plots (Senatorski 2002). Equation (1) represents the instantaneous radiation of a plane wave from the fault when the shear stress is released. Its left-hand side can be interpreted as the radiational stress, with the mechanical impedance, $\kappa = \mu/v_S$, i.e., ratio of the stress and the particle velocity, $\dot{q}/2$ (Achenbach 1975). The same equation has been derived rigorously as the overdamped dynamics (Senatorski 1994, 1995) from the energy functional (Rundle 1989), from the stress balance (Rice 1993), or from the full dynamics by assuming that stress transfer is instantaneous (Cochard and Madariaga 1994) in the two-dimensional (2D) case. For the assumed one component of slip and shear stress, its general form is the same in the three-dimensional (3D) case.

Interactions. The interaction stress describes the long-range elastic interactions along the fault plane. It is due to heterogeneities of the slip field. It is expressed as

$$s^I(\mathbf{r}, t) \approx \int_A d\mathbf{r}' G(\mathbf{r}, \mathbf{r}') q(\mathbf{r}', t) , \quad (2)$$

where A is the rupture area, and function $G(\mathbf{r}, \mathbf{r}')$ represents the static stress at point $\mathbf{r}$ due to unit slip at point $\mathbf{r}'$. It is assumed that (1) the wavefront at $|\mathbf{r} - \mathbf{r}'| = t \cdot v_S$ simply leaves the static stress field behind it, so variations of the transient stress between the longitudinal and transverse wavefronts are neglected (Hirth and Lothe 1967); and (2) that effects of finite wave speed, or the stress correction due to the stress signal delay, are neglected. Such an approximation enables us to express stresses in terms of slips at the same instead of at earlier instants, which defines the overdamped or quasi-dynamics. The meaning and validity of the approximation are discussed in Senatorski (2014).

Function G depends on the system geometry. For simplicity, the vertical fault in a half-space solution, with vertical slip direction, is used in the present work. Consequently, the attractive force between slips and a free surface is taken into account, but effects of a fault dip angle value are ignored. In particular, the normal stress variations, which are responsible for the fault strength (Scholz and Campos 2012), are not considered. Instead, the fault strength is introduced directly by defining cohesive stresses, s^F, acting along the fault plane. Such an approach enables us to study effects of the cohesive stress parameters, such as strength or slip-weakening distance, for megathrust fault seismicity. The Green's function G used in previous works (Senatorski 2002, 2004) has been adapted to the simplified megathrust geometry by changing fault boundary conditions. It is consistent with the solutions presented by other authors (Chinnery 1963; Mansinha and Smylie 1971; Okada 1992).

Cohesive stress. The cohesive stress field, s^F, is defined at each point of the fault interface as a Gaussian function of the slip displacement from the previous event, or the last healing, $\tilde{q}(\mathbf{r}, t)$,

$$s^F(\tilde{q}) = \frac{b}{d}(\tilde{q} + \alpha d) \exp\left\{0.5 - \frac{(\tilde{q} + \alpha d)^2}{2d^2}\right\}, \quad (3)$$

where b represents the cohesive stress peak, d represents the slip-weakening distance, and α is related to the initial stress: for $\alpha = 0$, the initial stress is zero and it attains the peak value b for $\alpha = 1$ (Fig. 1). For $\alpha < 1$, the cohesive stress increases with increasing slip until its peak value is attained, then it decreases with the ongoing slip to the residual value assumed as zero. Here, $\alpha = 0.6$ is assumed in simulations. The peak stress, $b(\mathbf{r})$, defines the fault strength at a given point $\mathbf{r}$. The higher the peak is, the higher stress is needed to break the site. The slip-weakening distance, $d(\mathbf{r})$, is responsible for frictional fault behavior at $\mathbf{r}$. The fracture energy is represented by the area under the cohesive stress line. Therefore, for a given strength, larger d implies that more fracture energy is consumed when the site moves. Small d implies sudden stress drop and higher slip acceleration, whereas its larger value implies that the stress is gradually released. These two cases can be related to intact rock failure and frictional sliding, respectively (Ohnaka 2013). The choice of the Gaussian function is motivated by the fact that it exhibits basic characteristics of the cohesive stress, with the stress peak separating its slip-strengthening and slip-weakening parts.

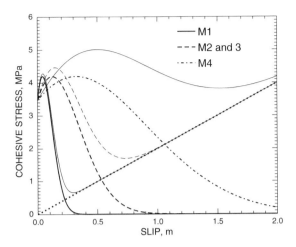

Figure 1
Illustration of a slip-dependent friction law and its stability regimes for models M1–4. Cohesive stress vs. slip displacement relations for four models are characterized by peak stresses and the slip displacements at which the peak stresses are attained (thick lines). The residual friction stress is zero. The area under the curve gives density of the fracture energy consumed by the cohesive forces. The back stress, kq, due to the elastic medium response for the case of uniform slip along a square patch (the rest of the fault unmoved), is determined by the system stiffness, $k \approx 8/L$ [MPa/m], where L is the patch size in kilometers (dotted line; $L = 4$ km is assumed). The resisting stress is defined as a sum of the cohesive stress and the back stress (thin lines)

Healing mechanism. After the driving stress drops below the healing stress level, $s < s_H$, the slip movement stops and the cohesive stress is instantaneously rebuilt, i.e., $\tilde{q} \to 0$ and $s^F \to s^F(\tilde{q} = 0)$. The healing stress is assumed to be proportional to the peak stress, $s_H = \beta b$, with $\beta = 0.1$. Then, when the driving stress attains the cohesive stress initial value, $s > s^F(0)$, s^F starts to increase and, after attaining its peak value, $s^F = b$, to decrease with the increasing slip, according to Eq. (3), till the next healing. Note that the healing mechanism activation occurs at low driving stress level, when the slip velocity is low too. Therefore, the healing is not activated shortly after the slip begins, when the slip velocity is low, but the driving stress is high. The model behavior depends on the healing parameter β: Its lower value smooths out heterogeneities in slips on the fault after consecutive earthquakes, which leads to more regular earthquake cycles. In fact, transition from regular behavior ($\beta = 0.04$) through period doubling ($\beta = 0.06$) to chaos ($\beta = 0.1$) has been observed in simulations with constant slip-weakening distances (Senatorski

2002). On the other hand, a larger β would lock slipping at too high rates, so $\beta = 0.1$ is here assumed as generating realistic slip patterns.

The abrupt arrest of seismic slip at a given point can be interpreted as the effect of the short-scale variations of the dynamic friction above its residual value, due to, for instance, encountered small asperities. For low enough driving stress, such variating resistance locks the movement locally at the final value of sliding friction higher than its residual value (Brune 1976). The resistance variation level is assumed to be related to the local fault strength. The healing stress, s_H, can be regarded as the upper limit of the size of the variations. The chosen $\beta = 0.1$ value means that the upper limit of the variations is 10 percent of the local strength, b. Here it is assumed that the slip movement stops with probability 1 if the driving stress drops below the s_H level.

The proposed healing mechanism, despite its simplifications, leads to realistic fault behavior: local stress and slip velocity variations, rupture propagation, earthquake parameters, and seismic cycle complexity, are similar, at least qualitatively, to those estimated for real seismicity or laboratory experiments (Senatorski 2002, 2006). Note that (1) the mechanism operates when the local slip velocity decreases; (2) there is no healing if high driving stresses overcome high cohesive stresses, i.e., restrengthening occurs after the stresses at a point had been mostly released. On the other hand, the mechanism causes an instantaneous increase of the cohesive stress to its initial value after healing. Note, however, that (3) the healing rule is local. The collective behavior of larger fault patches is more complicated: strengths of the locked patches depend on their size and distribution of strengths in their surroundings, so it changes gradually with the rupture extension.

Subduction channel Slips along a thin fault plane actually represent deformations within a subduction channel, or a subduction interface that is hundreds of meters to several kilometers thick (Cloos and Shreve 1988). Natural heterogeneities, such as the seafloor topography and subduction channel material characteristics, are mapped as distributions of the cohesive stress parameters. Thus, the cohesive stress concept can be thought of as representing all subduction channel processes and characteristics, including variations of normal stresses due to the plate topography, pore pressure and buoyancy of the subduction channel material, sediment consolidation, its viscosity and density, as well as the channel thickness. Seamounts and other topographic features within the channel, which is filled with sediments and fluids, create a complicated structure. Its detailed and realistic description in terms of a hilly surface moving across the non-elastic medium requires an adequate theory. Using the cohesive stresses as representing basic, averaged characteristics of the subduction channel, including both slip along rupture planes and volume deformations within the channel, is chosen as an alternative approach. Note that horizontal dimensions of the topographic features, such as seamounts, are of an order of kilometers, whereas earthquake slips are measured in meters, so the related strength heterogeneities can be treated as fixed characteristics.

Also, threshold stress for healing can be interpreted in the subduction channel context. Sediment dewatering and underplating can not occur where the shear stress acting on the material near the channel roof exceeds a critical value. On the other hand, strong asperities, such as seamounts, can still slowly move with the descending plate where the shear stress is high enough to rasp the overriding plate, causing its erosion (Cloos and Shreve 1988). The proposed model suggests that two processes should be distinguished: frictional sliding, and underplating or erosion processes. The latter is related to enabling or inhibiting the healing mechanism. That is why the slip-dependent cohesive stress and the healing mechanism are treated separately in this approach.

Tectonic stress The tectonic stress, s^T, is due to the tectonic loading,

$$s^T = s^T(t_0) + \frac{\mu}{w} \left\{ V_P(t - t_0) - \frac{1}{A} \int_A d\mathbf{r} \left[q(t, \mathbf{r}) - q(t_0, \mathbf{r}) \right] \right\},$$
(4)

where V_P is the relative plate velocity; t_0 means the simulation starting time; w denotes the width of the elastic medium layer surrounding the fault, as measured perpendicularly to the fault plane (Senatorski 2002). The tectonic stress increases as within the elastic layer of width w, which is deformed by a steady movement of its sides at the relative speed V_P. Here $w = 30$km is assumed in the simulations. The tectonic stress decreases with the mean slip along the planar cut within the layer. The stress s^T can be thought of as the uniform component of driving stresses.

Stable and unstable slips The unstable slip movements start if the driving stress, $s = s^T + s^I$, decreases slower than friction, s^F, during sliding. Consider the uniform slip q on the $L \times L$ square fault patch. In this case, the interaction stress is the back stress due to the elastic medium response, which resists the slip movement; it depends on the slipping area size. The driving stress is

$$s = s^T - k_R q , \qquad (5)$$

where s^T is assumed to be constant. The driving stress decrease with increasing slip, due to the medium response or the back stress, is represented by the interaction stress term, $s^I = -k_R q$, where the system stiffness is (Senatorski 2002; Okada 1992)

$$k_R = (\mu/4\pi)[(2 - v)/(1 - v)](\sqrt{2}/L) . \qquad (6)$$

This gives for the back stress $k_R q \approx 8q/L$ for $\mu = 3 \times 10^4$ MPa and the Poisson's ratio, $v = 0.25$ (L in kilometers, q in meters; Fig. 1).

The cohesive stress decrease can be approximated as

$$s^F \approx b - k_C q, \qquad (7)$$

where b denotes the cohesive stress peak value. The slip movement starts when the tectonic stress attains the cohesive stress peak stress or strength value, $s^T = b$. The critical stiffness, k_C, is proportional to the rate of the cohesive shear stress decrease,

$$k_C \approx b/d . \qquad (8)$$

Both parameters, b and d, represent material fault characteristics.

The stable to unstable slip transition occurs after the system stiffness exceeds its critical value, $k_R < k_C$. The stiffness k_R depends on distribution of slips, or the area of possible slip (it is easier to move larger rupture area), whereas the critical stiffness is related to the plate frictional coupling or cohesive stress dependence on slip displacement (it increases with

the increasing slope of the stress vs. slip-weakening distance line). The critical stiffness can be exceeded in two ways: (1) by decrease of the system stiffness k_R (dynamical factor), or (2) by increase of the critical stiffness k_C (material or structural factor).

The unstable rupture condition, $k_R < k_C$, can be written as

$$L > (\mu/4\pi)[(2 - v)/(1 - v)](\sqrt{2}(d/b) . \qquad (9)$$

Thus, for given strength, larger d requires larger slipping area size for an unstable rupture. In general, stable or unstable slip movements depend on distribution of slips and on the concurrently slipping area size, as well as frictional or cohesive stress local characteristics.

Figure 1 illustrates the condition for the unstable slip. For the same strength, $b = 4.2$ MPa, three cases of small (M1, $d \approx 0.2$ m), medium (M2 and M3, $d \approx 0.6$ m) and large (M4, $d \approx 1.7$ m) slip-weakening distances are considered. The critical stiffness values are, respectively $k_C = b/d = 21$ MPa/m, 7 MPa/m, and 2.5 MPa/m. The system stiffness due to the medium back stress response, represented by the $k_R q = -s^I$ dotted line, is $k_R = 2$ MPa/m ($L = 4$km), so $k_R < k_C$ in all cases. For each model, M1–M4, two lines are shown: the cohesive stress, s^F (thick line), and the fault resistance line, $s^R = s^F + k_R q$ (thin line). Assuming constant tectonic stress, s^T, the difference $\Delta s = s^T - s^R = s^T + s^I - s^F$ is the net stress given by Eq. (1). Therefore, the unstable slip is possible, if the s^R line has a minimum value lower than the tectonic stress. Note that the s^R stress peak value is higher and is attained at a larger q than the s^F peak value. This effect is due to the initial slip-strengthening part of the cohesive stress function. Only stable slip movements (i.e., slips under increasing tectonic stress) would be possible for: (1) lower strength, b, (2) larger slip-weakening distance, d, or (3) smaller rupture area size, L.

Radiated energy Intensity of the fault slip movements is measured by using the radiated energy rate signal. The radiated seismic energy, E_S, can be defined as the portion of the released potential energy, W, that is not consumed by the fault resistance to slip. For quasi-dynamics, such definition leads to the seismic energy rate expressed as (Senatorski 1994, 2014)

$$\dot{E}_S(t) = \frac{\mu}{2v_S} \int_A d\mathbf{r}\ \dot{q}(\mathbf{r}, t)^2. \qquad (10)$$

Equation (10) can be obtained directly from Eqs. (1), (2), and (3) by noting that they represent a gradient dynamical system, or the overdamped dynamics (Lichtenberg and Lieberman 1992; Senatorski 1994, 1995). The quasi-dynamic solution for the radiated seismic energy, E_S, when applied to real slip velocity fields, $\dot{q}(\mathbf{r})$, approximates the correct, fully dynamic solution for the radiated seismic energy. The approximation can be explained as follows: Term $(\mu/2v_S)\dot{q}$ in Eq. (10) stands for the net stress, Δs, as defined by Eq. (1), whereas the correct solution depends on the static part of the net stress only. This is because the released potential energy, W, depends on the long-range elastic interactions (or concurrent slips), as they occur in Eq. (2) (Senatorski 2014).

2.2. Specific Models

Distributions of parameters $b(\mathbf{r})$ and $d(\mathbf{r})$ are generated as random fields defined by 2D stochastic processes. Two different processes are combined. The fractional Gaussian noise enables us to model a rough surface, where the Fourier coefficients of the surface elevation, A_N, are related to the corresponding wavenumbers k_N by the relation $A_N \propto k_N^{-\beta/2}$. It results from filtering a Gaussian white noise ($\beta = 0$; Turcotte 1997). The Neymann–Scott process enables us to represent larger structures. It results from clustering applied to a Poisson process, where the daughter points are scattered around the parent points (Stoyan et al. 1995). The modeled objects, or asperities, have their internal structure, as suggested by hierarchical asperity models (Uchida and Matsuzawa 2011).

Four models, M1–M4, with the same distribution of b and different distributions of d are considered (Fig. 2; Table 1). The fault surface of 100 km length and 60 km width is discretized into 1 km × 1 km segments. The fault plane is divided into three parts, 25, 10, and 25 km of width, respectively. The fractional Gaussian noise with $\beta = 2$, $\beta = 2.8$, and

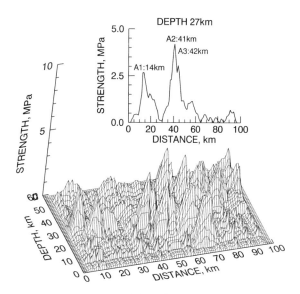

Figure 2

Distribution of peak cohesive stresses on the 100×60 km^2 fault (with 1×1 km^2 discretization cells). The strength random field is generated by using the combined Neumann–Scott (with maximum value of 4 MPa) and fractional Gaussian (with mean value of 0.4 MPa) stochastic processes. The 27 km at depth section with three measurement points, A1–3, is shown in the subplot. The peak stress $b = 0$ at the bottom, leftmost and rightmost boundary cells

$\beta = 0$ is applied to obtain the fault strengths, $b(\mathbf{r})$, on the deep, medium, and shallow part, respectively. Then, the Neymann–Scott process is used to create larger structures with higher strengths. The noise is rescaled, so that its mean value $b_{\mathrm{noise}}^{\mathrm{mean}} = 0.4$ MPa. The larger structures are rescaled, so that the maximum strength $b^{\mathrm{max}} = 4$ MPa.

Slip-weakening distances, $d(\mathbf{r})$, are assumed proportional to the strengths, with their maximum values $d^{\mathrm{max}} = 0.1$m and 0.8 m for M1 and M4, respectively. If $d < d^{\mathrm{min}}$, d is replaced by the minimum values $d^{\mathrm{min}} = 0.02$m for both models. For M2 and M3, $d^{\mathrm{max}} = 0.3$ m, and the minimum values are $d^{\mathrm{min}} = 0.01$ m. A larger value of $d^{\mathrm{min}} = 0.4$ m is assumed in the upper 20 km layer in M3 to investigate possible effects of the higher slip-weakening distance values at smaller depth on the fault dynamics. The initial stress parameter, $\alpha(\mathbf{r})$, is fixed as $\alpha = 0.6$. Fracture energies can be obtained by integration of the cohesive stress, s^{F}, over slip, q. Therefore, the role of different fracture energy distributions (strengths are the same) can be analyzed by using M1–M4. Both

Table 1

Model parameters

Model	d^{max}, m	d^{min}, m	α	b^{max}, MPa	b^{min}_{noise}, MPa
M1	0.1	0.02	0.6	4	0.4
M2	0.3	0.01	0.6	4	0.4
M3	0.3	0.01, 0.4	0.6	4	0.4
M4	0.8	0.02	0.6	4	0.4

Each of the models, M1–M4, has the same distribution of strengths, b, and initial stress parameters, α. Distributions of slip-weakening distances, d, are proportional to the strengths, with scale coefficients that are different for each model. Maximum and minimum slip-weakening distances, d^{max} and d^{min}, maximum strength, b^{max}, and its mean noise value, b^{mean}_{noise}, are given. Higher minimum critical slip (0.4 m) at the 20-km upper layer is assumed for M3

long-term seismicity and individual earthquake rupture patterns generated by the models are compared.

For a slip on a single discretization cell, $L = 1$ km, the system stiffness $k_R = 8$ MPa, so $k_R < k_C$ for M1, and $k_R > k_C$ for M2, M3, and M4. This means that spontaneous slip is possible (since the net stress increases above zero at constant tectonic stress) at a single discretization cell in the case of M1. The slip distance and slip velocity are limited during such an event: the asperity breaks, and the slip velocity pulse occurs, but it does not propagate by breaking its neighboring cells. In the case of M2, M3, and M4, the spontaneous slip is only possible if more discretization cells are involved. Small slip-weakening distances of M1 enable single, localized asperity breaking without rupture propagation. Thus, the 1 km $\times$ 1 km discretization cell determines the smallest asperity size that can break singly, as the smallest earthquakes, or initiate larger events by rupture propagation. The assumed discretization cell size enables us to model both single asperity breaks and wide, smooth, propagating rupture fronts.

Note that estimation of the cohesive zone size for the slip-weakening friction models (Day et al. 2005) gives about 0.7 km in the case of M1, which is less than the computational grid size. This is consistent with the stiffness analysis: Single grid failures should be considered as the smallest asperity breaks. In the case of M2–M4, the estimation gives about 2 and 5 km, respectively, so well above the computational grid size.

Variations of strengths, b, correspond to topographic or structural fault characteristics, such as

seamounts and ridges. Variations of slip-weakening distances, d, correspond to frictional fault characteristics, which may be controlled by sediments or fluids. A role played by those geological features, whether they increase or decrease the interplate coupling coefficient, have been debated (Wang and Bilek 2011; Scholz and Small 1997). In the present paper, effects of related strength and friction variations are studied.

The relative plate velocity $V_P = 0.1$ m/year is assumed in Eq. (4). This value defines the long-term time scale of tectonic loading. For computational reasons, to enable simulations of both earthquake cycles (the long-term time scale in years) and single events (the short-term time scale in seconds) by using the same model, a higher impedance value of $\kappa^y = 0.001\mu$ MPa year/km $= 3 \cdot 10^4$ MPa s/km is used. Then, when details of single events or multiple events are analyzed, the rescaling procedure is applied, with the scale factor $\kappa^s/\kappa^y = 0.006$ s/year, to recover the correct order of magnitude of slip rates during earthquakes (Senatorski 2002). Interpretation of the simulation results needs caution since the tectonic plate rate-to-slip rate ratio (so earthquake recurrence time to earthquake duration time) is higher in simulations than in the real world. To distinguish between multiple events and separated earthquakes, one should verify whether the event is triggered by slips at neighboring sites or by the tectonic loading (Senatorski 2002). The approach enables us, however, to analyze the seismic vs. aseismic slip interplay effectively.

Because of the model fault size, 100 km $\times$ 60 km, up to ≈ 4 MPa strengths, earthquakes with up to

$\approx 1m$ of slip are generated at an average rate determined by a 0.1m/year slip budget.

3. Results

3.1. Long-Term Scale

The radiated energy rate variations (Fig. 3a), computed by using Eq. (10), exhibit irregular earthquake sequences. The largest events are separated by smaller ones. Time between the largest earthquakes varies, but subsequent seismic cycles can be distinguished by looking at the tectonic stress variations: New cycles start after the tectonic stress is mostly released (Fig. 3b). Thus, the tectonic stress signal provides extra information about system behavior.

Larger energy rates and largest tectonic stress drops occur for lower d^{max} (M1). Larger tectonic stresses are attained for higher d^{max} (M3), but they decrease more gradually. Larger d, or larger fracture energies, stabilize the system, causing its evolution to be gentler, even if high peak stresses are attained. Note that the higher stresses do not imply larger stress drops.

Large earthquakes can follow sequences of smaller events. Such patterns illustrate the problem of earthquake forecasting based on a short-term seismicity. In fact, the regularly repeating moderate earthquakes within the Tohoku–Oki region before the great 2011 event might have suggested that they are characteristic for the region. On the other hand, old tsunami traces suggest that much larger earthquakes

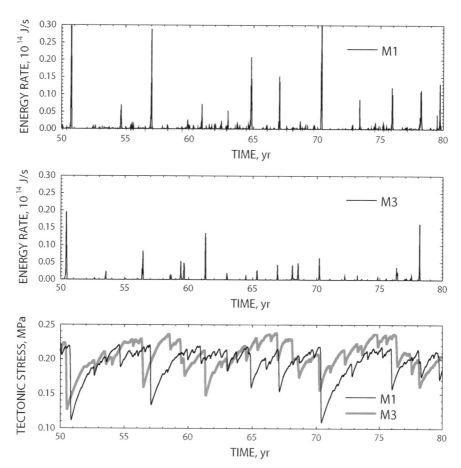

Figure 3
Long-term fault behavior. Seismic energy release rates for models M1 and M3, and time variations of the tectonic stress for the same models

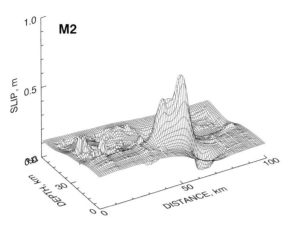

Figure 4

Distributions of slips on the fault during 1-year-long interval for the model with larger slip-weakening distances (M2): aseismic slips and coseismic slips at broken asperities

occur with the repeat time of about a thousand years (Hashimoto et al. 2012).

Both seismic and aseismic slips can be recognized in Fig. 4 (M2), where distribution of slip distances

covered during a 1-year-long interval is shown. The elevated, curved surface represents the aseismic slip of about 10–20 cm. The hills represent coseismic slips and the valleys represent coupled, slip-deficit regions. Note that the slips are measured relative to their reference distribution at a given time; the coseismic slips do not exceed the maximum aseismic slip value as measured from the beginning, zero slip distribution. In the long-term perspective, the same slip distance is covered at each point. For shorter time intervals, however, changing areas of slip deficits and excess occur around asperity regions. Although the hill and valley patterns change with time to some extent, their general outlines remain permanent fault features.

Despite slight difference between slip-weakening distance values at smaller depth for M2 and M3, general characteristics of their dynamics are similar, as determined mainly by the large structures. The difference may be essential for the resulting ground

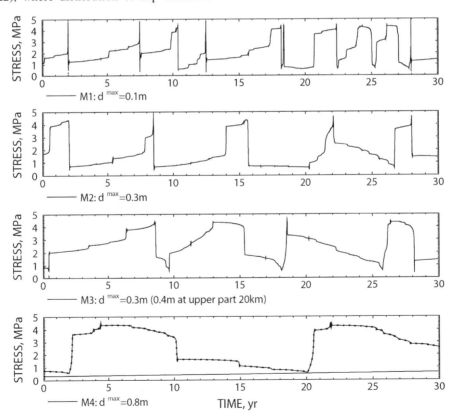

Figure 5

Driving stresses at the same asperity (41 km along strike, 27 km of depth) for M1–M4. The stresses exhibit slow increase and sudden drops for M1. Sudden stress drops related to large earthquakes are followed by slow slips for M2 and M3

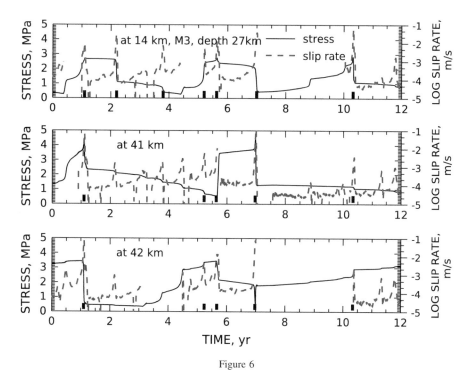

Figure 6
Driving stresses and slip rates (in log scale, dashed lined) at different asperities (A1: 14 km; A2: 41 km; and A3: 42 km along the strike, 27 km of depth) for M3. Different stress patterns between large events are shown. Fast slips during earthquakes are indicated by vertical bars. Zero slip rates mean locked asperity

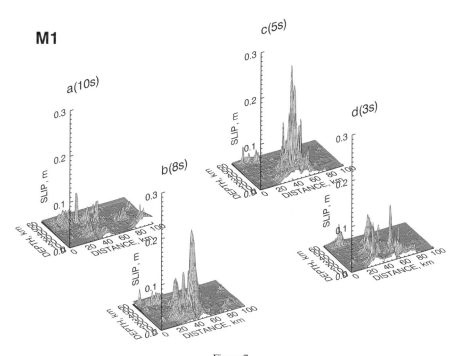

Figure 7
Distributions of slips on the fault for consecutive 3 − 10s intervals: a 0–10 s, b 11–18 s, c 19–24 s, and d 25–28 s, for the model with shortest slip-weakening distances (M1); a breaking of individual asperities; b–d propagating rupture front that activates individual asperities

motions, which is outside the scope of the present work.

3.2. Short-Term Scale

Driving stress variations at the same asperity (A2: 41 km along strike, 27 km of depth; see Fig. 2) are compared for M1–M4 (Fig. 5). The stresses increase slowly until they suddenly drop during large earthquakes for M1. For M2 and M3, steep stress drops are followed by periods of gentle stress decrease, which reflect slow slips. This means that locked asperities can exhibit both unstable, coseismic rupture and stable afterslip. Slips are more stable in the case of M4: The slip movement is too slow at the asperity to propagate into other asperities, so the area of simultaneous slip is too small for unstable movements. It can be concluded that slow and fast slip occurrences at the same place requires relatively large d, so that the unstable movement can occur only if the slipping area is adequately large.

The driving stress variations and the slip rates at different locations (A1–A3: 14, 41, and 42 km along the strike) for M2 are shown in Fig. 6. Different patterns of stress at chosen sites occur between two large events. The first event starts when A3 (42 km) breaks, inducing unstable slip at A2 (41 km). The strong asperity at 41 km exhibits fast slip during the earthquakes and slow slip between them. This slow slip increases stresses at adjacent sites, leading the next fast slip at A3, and, consequently, the sudden stress growth at A2 without its breaking. The A2 breaks later, during the second great earthquake that starts close to A1 and involves all the illustrated sites. Again, the unstable slip at 41 km occurs after the neighboring asperities become ready for being broken and, consequently, the slipping area becomes large enough. It can be concluded that both fast and slow slips are possible at the same asperity, depending on distribution of slips and stresses in its surrounding.

Distributions of slips during earthquake ruptures are illustrated in Figs. 7 and 8. Slips during consecutive several seconds-long intervals are shown. Two types of the rupture patterns can be distinguished. First, stronger patches break separately when their strengths and sizes allow unstable slip. This is

possible only if the slip-weakening distance, d, is short enough. Isolated, sharp hills representing slips that emerge from the relatively flat surface illustrate the case (Fig. 7a; M1). Second, when a larger area starts to slip, sharp peaks emerge from an extended, smooth elevation that propagates along the fault (Fig. 7b–d; M1). The strongest asperities break or re-break when the spreading slip wave sweeps them away. The unstable slip movement extends on weaker places that can slip slowly otherwise. For larger slip-weakening distance, the propagating rupture front extends, sweeping increasingly stronger asperities that cannot break separately (Fig. 8a–e; M3). Larger fault area is involved by the rupture process.

Two earthquakes simulated by using M1 ($d = 0.1$ m) and M2 ($d = 0.3$ m) are illustrated in Figs. 9 and 10, respectively. Temporal variations of the radiated energy rate are more irregular in the small d case of M1 (Fig. 9). The event consists of two subevents, which can be also treated as an earthquake doublet since slower movements occur between these subevents. The two largest asperities (A2 and A3) break almost separately during these subevents (Fig. 9c, e), with only small slips induced on the neighboring asperity. Consequently, characteristic stress switching occurs at these two asperities (Fig. 9d, f). In contrast, more regular variations of the energy rate occur in the larger d case of M2 (Fig. 10). Both asperities break during a single event (Fig. 10c), so the stress is released at both locations (Fig. 10d).

4. Discussion

The slow versus fast slip interplay is more than just about the friction law problem. This is because slip instability depends both (1) on the critical stiffness, i.e., on the local friction variations with slip, and (2) on the system stiffness, i.e., on the rate at which the driving stress decreases with slip at a given location due to the medium response. The system stiffness is related to the slipping area size and distribution of slips, so it changes from one event to another, even if the local friction parameters remain unchanged.

Computer simulations of earthquake sequences illustrate fault activity in different scales. The model

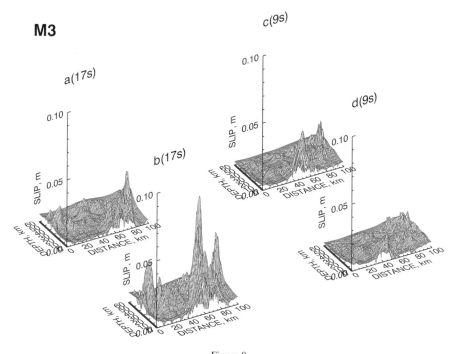

Figure 8
Distributions of slips on the fault for consecutive 9–17-s intervals: a 0–17 s, b 18–34 s, c 35–44 s, and d 45–54 s, for the model with longer slip-weakening distances (M3): smooth rupture front

is simplified, but it enables us to analyze patterns of slow and fast slips and other rupture characteristics, and to reveal relations between these patterns and underlying fault heterogeneities. Therefore, the model can be qualitatively interpreted in terms of a subduction megathrust with heterogeneous frictional coupling distribution.

Strong, coupled interface patches (asperities) and their weaker surroundings resulting from a variety of material and structural fault features, such as subducted seamounts or ridges, sediments, and released fluids, are modeled as respective distributions of frictional peak stresses and slip-weakening distances (Scholz and Campos 2012; Heuret et al. 2012; Vannucchi et al. 2017). The patches of high peak stresses are interpreted as asperities. The asperities have their internal structures: More or less densely distributed stronger places are surrounded by weaker ones (Uchida and Matsuzawa 2011). The asperities are characterized by different slip-weakening distances. Due to such asperity structure, the coseismic slip area depends on which of the strong sites break. Locked asperities limit the possible slip area size of other

asperities located within their areas. The largest earthquakes occur after the strongest asperity in a given region breaks. Then, weaker asperities start to define smaller rupture areas surrounding them (Senatorski 2017).

Healing processes are enabled if the driving stress drops below a threshold value. The fault does not heal under high driving stress, even if the net stress and the slip rate are low. Such an approach is motivated by subduction channel characteristics, where high driving stress inhibits underplating and enables erosion at the channel roof and the overriding plate. Thus, the healing mechanism is distinguished from the laboratory-derived friction law, which is adequate for a well-defined rupture plane. On the other hand, the stress-dependent healing can be also supported by laboratory experiments (Brantut 2015), where the proposed mechanism of the observed damage recovery is related to the microcrack closure. Further studies on the healing mechanism in the subduction channel context are needed.

The tectonic stress is understood as the uniform stress component in a major fault region. Cyclic

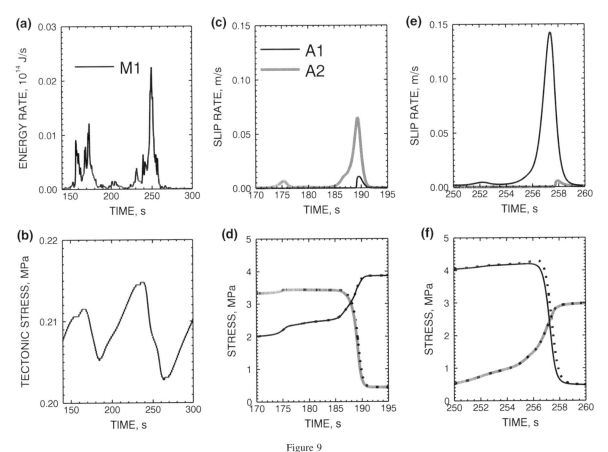

Figure 9
Simulated earthquake (M1). Temporal variations of the radiated energy rate (**a**), tectonic stress (**b**), slip velocities (**c**), and stresses (**d**) at A1 and A2 for the first and second (**e** and **f**, respectively) subevent. Dotted lines represent driving stresses and solid lines represent cohesive stresses (see Eq. 1)

stress accumulations and reliefs are reflected by vertical motions observed on the surface above megathrust seismogenic fault segments. Therefore, saw blade patterns of continuous GPS and sea level time series can be compared with the simulated tectonic stress variations (Natawidjaja et al. 2004; Sieh et al. 2008; Konca et al. 2008).

The present paper is focused on the role played by the slip-weakening distance values, d, in the stable vs. unstable slip interplay and large earthquake occurrences. Unlike in other papers (Hillers et al. 2006), distributions of d are not defined separately; instead, values of d are proportional to strengths, b, which define asperity distributions. Higher or lower d in models M1–M4 are related to their assumed maximum value.

Larger d values lead, on average, to higher tectonic stresses, but smaller stress drops, since stresses

decrease gradually to minimum stress levels. Thus, large d stabilizes the system. The model limits seismic activity to the fault plane only; in the real world, higher tectonic stresses may be responsible for higher seismicity levels in the surrounding fault.

Sequences of small or moderate earthquakes do not exclude large events. A great earthquake can occur after the toughest asperity breaks and a large-scale slip becomes possible. Analyses of the slip budget and rupture style of smaller earthquakes, assumed to be related to distribution of d, could be used to estimate the potential for large events. In the real world, sequences of moderate earthquakes occur within the Tohoku–Oki 2011 earthquake region (Hashimoto et al. 2012).

Large d values can both promote and restrain moderate or large earthquakes. Large events are promoted for two reasons, related to (1) the

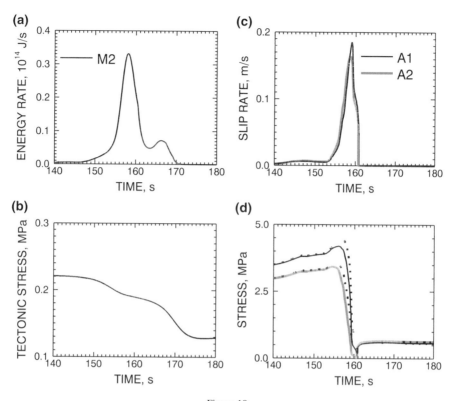

Figure 10
Simulated earthquake (M2). Temporal variations of the radiated energy rate (**a**), tectonic stress (**b**), slip velocities (**c**) and stresses (**d**) at A1 and A2. Dotted lines represent driving stresses and solid lines represent cohesive stresses (see Eq. 1)

system stiffness and (2) interaction strength. First, the unstable movements at the strongest asperities become possible only when the slipping area is large enough (so, the system stiffness becomes small enough), say larger than the asperity size. Second, a larger fault area is involved by the rupture process, because longer slips at instability imply higher stresses acting on neighboring asperities (so, interactions among asperities are stronger).

If, due to a large d value, individual asperities cannot release all the accumulated stresses by unstable slip, two scenarios are possible: (1) They break within a larger slipping area, or (2) they slip gradually by slow movements at high driving stresses. In the first case, the strong patches promote large events, as described above, accumulating stress for a long time. In the second case, however, they restrain large earthquakes. Depending on both strength and slip-weakening distance distributions, large events or aseismic creeping occur. The system can switch

between these two scenarios, if the distributions change due to some physical processes.

Small d values allow the same asperities to break singly. In this case, small and moderate earthquakes are promoted, depending on asperity sizes and the fault structure. This leads to the effect of stress switching at neighboring asperities that break one after another (Fig. 9). Large earthquakes, involving more asperities breaking in groups, are possible, but as a random effect: The asperities break together if they just happen to be in the same phase. For large d, asperities tend to break in groups, embracing large regions with both high and low strengths.

Possibility of different roles played by asperities, depending on the fault structure and its material characteristics, shed some light on the controversy about whether subducted seamounts start or stop large earthquakes (Mochizuki et al. 2008; Wang and Bilek 2011; Scholz and Campos 2012). Both roles can be played in the real world, and parameter d seems to be an important factor. The seamounts are thought of as the regions of higher normal stresses, which enable unstable slips (Scholz and Small 1997). However, the slip-weakening distance, d, is another factor deciding the slip style besides strength. Thus, the subducted seamounts can play both roles, moving by unstable slips or creeping, depending on the material characteristics on the plate interface.

Both slow and fast slip movements are possible at the sites with large d values, depending on the slipping area size. Only slow slip is possible, if the size is limited by strong, locked asperities. Fast slip occurs after consecutive asperities break and the area of possible slip increases.

Large and smooth waves of the slip velocity field propagate into weaker segments that slip slowly otherwise. This effect explains the unstable slip at the shallow fault segment close to the trench during Tohoku–Oki 2011 earthquake.

The stiffness dependence on the slipping area size leads to a rupture hierarchy. Locked asperities cause the back stress in their surroundings and limit the possible slip area size of other asperities. Both stable and unstable slips are possible within the locked asperity region, but such slips become larger and faster after the locking site breaks. Such a view is consistent with the hierarchical asperity model outlined by Uchida and Matsuzawa (2011) and Lay (2015). According to that model, the plate frictional coupling is defined by hierarchical asperity structures. The present paper shows that the hierarchical megathrust nature may result both from heterogeneous frictional coupling, as well as the system dynamics dependent on changing stability conditions along the fault.

By using the quasi-dynamics approximation in simulations, inertial effects of the rupture process are neglected, as described in Sect. 2.1. Such effects can be important during large earthquakes, when the area of simultaneous slip is large and the rupture front propagates fast (Senatorski 2014). Their significance for the interplay between slow and fast fault movements, and for the fault stability conditions that lead to seismic or aseismic character of slip, is debatable. The present work is an attempt to find a possible explanation of the unexpected seismic–aseismic slip patterns without referring to the inertial effects.

5. Conclusions

The heterogeneous fault model with slip-dependent friction enables us to explain some unexpected seismicity patterns in subduction zones. The debated observations concern seismic and aseismic slips that occur at the same sites, hierarchical nature of the megathrust dynamics, and different roles played by subducted seamounts. Two related effects, the fault stiffness and slip-weakening distance, provide a key to understand these findings. The role of both effects have been often neglected. Although continuum fault models have inherently incorporated the fault stiffness, most works are focused on local friction law and its modifications only to explain the observed seismicity patterns.

The main findings can be summarized as follows.

- Complex processes within a subduction channel can be modeled by using the slip-dependent friction law with strength and slip-weakening distance as two model parameters.
- Slow or fast slip is more than the friction law problem only; it depends also on the changing system stiffness, which is related to the rupture area size.
- Relatively large slip-weakening distance value leads to a variety of slip patterns with both small and large earthquakes possible.

Future works should reveal specific relations between model parameters, such as strengths, slip-weakening distance and healing stress values, and subduction channel characteristics in the real world.

Acknowledgements

Constructive comments by the editor Sylvain Barbot and two anonymous reviewers helped me to improve the manuscript. This work was partially supported within statutory activities no. 3841/E-41/S/2018 of the Ministry of Science and Higher Education of Poland.

REFERENCES

Achenbach, J. D. (1975). *Wave propagation in elastic solids.* Amsterdam: Elseviere Science Publishers B.V.

Brantut, N. (2015). Time-dependent recovery of microcrack damage and seismic wave speeds in deformed limestone. *Journal of Geophysical Research: Solid Earth, 120,* 80888109. https://doi.org/10.1002/2015JB012324.

Brune, J. N. (1970). Tectonic stress and the spectra of seismic shear waves from earthquakes. *Journal of Geophysical Research, 75,* 4997–5009.

Brune, J.N. (1976). The physics of earthquake strong motion. In: Rosenbluthe, E., Lomnitz, C. (Eds.) Seismic Risk And Engineering Decisions. *Develop. in Geotech. Eng.,* 15.

Chinnery, M. (1963). The stress changes that accompany strike-slip faulting. *Bulletin of the Seismological Society of America, 53,* 921–932.

Cloos, M., & Shreve, R. L. (1988). Subduction-channel model of prism accretion, melange formation, sediment subduction, and subduction erosion at convergent plate margins: 1.Background and description. *Pure and Applied Geophysics, 128,* 455–499.

Cloos, M., & Shreve, R. L. (1996). Shear zone thickness and the seismicity of Chilean- and Marianas-type subduction zones. *Geology, 24,* 107–110.

Cochard, A., & Madariaga, R. (1994). Dynamic faulting under rate-dependent friction. *Pure and Applied Geophysics, 142,* 420–445.

Day, S. M., Dalguer, L. A., Lapusta, N., & Liu, Y. (2005). Comparison of finite difference and boundary integral solutions to three-dimensional spontaneous rupture. *Journal of Geophysical Research, 110,* B12307. https://doi.org/10.1029/2005JB003813.

Fukuyama, E., & Hok, S. (2013). Dynamic overshoot near trench caused by large asperity break at depth. *Pure and Applied Geophysics.* https://doi.org/10.1007/s00024-013-0745-z.

Hashimoto, C., Noda, A., & Matsu'ura, M. (2012). The Mw 9.0 northeast Japan earthquake: total rupture of a basement asperity. *Geophysical Journal International, 189*(1), 1–5. https://doi.org/10.1111/j.1365-246X.2011.05368.x.

Heuret, A., Conrad, P., Funiciello, F., Lallemand, S., & Sandri, L. (2012). Relation between subduction megathrust earthquakes, trench sediment thickness and upper plate strain. *Geophysical Research Letters, 39,* L05304. https://doi.org/10.1029/2011GL050712.

Hillers, L., Ben-Zion, Y., & Mai, P. M. (2006). Seismicity on a fault controlled by rate- and state-dependent friction with spatial variations of the critical slip distance. *Journal of Geophysical Research, 111,* B01403. https://doi.org/10.1029/2005JB003859.

Hirth, J. P., & Lothe, J. (1967). *Theory of Dislocations.* New York: McGraw-Hill.

Huang, Y., Ampuero, J.-P., & Kanamori, H. (2013). Slip-weakening Models of the Tohoku-Oki earthquake and constraints on stress drop and fracture energy. *Pure and Applied Geophysics, 171,* 2555–25668.

Ide, S., Baltay, A., & Beroza, G. C. (2011). Shallow Dynamic Overshoot and Energetic Deep Rupture in the 2011 Mw 9.0 Tohoku-Oki Earthquake. *Science, 332,* 1426. https://doi.org/10.1126/science.1207020.

Johnson, K. M., Fukuda, J., & Segall, P. (2012). Challenging the rate-state asperity model: Afterslip following the 2011 M9 Tohoku-oki, Japan, earthquake. *Geophysical Research Letters, 39,* L20302. https://doi.org/10.1029/2012GL052901.

Konca, A. O., Jean-Philippe Avouac, J.-P., Sladen, A., Meltzner, A. J., Sieh, K., Fang, P., et al. (2008). Partial rupture of a locked patch of the Sumatra megathrust during the 2007 earthquake sequence. *Nature, 456,* 631–635.

Lay, T. (2015). The surge of great earthquakes from 2004 to 2014. *Earth and Planetary Science Letters, 409,* 133–146.

Lay, T., Kanamori, H., & Ruff, L. (1982). The asperity model and the nature of large subduction zone earthquake occurrence. *Earthquake Prediction Research, 1,* 3–71.

Lichtenberg, A.J., & Lieberman, M.A. (1992). *Regular and chaotic dynamics,* 2nd edn., Applied Mathematical Sciences, series vol. 38, Springer, New York.

Loveless, J. P., & Meade, B. J. (2011). Spatial correlation of interseismic coupling and coseismic rupture extent of the 2011 MW = 9.0 Tohoku-oki earthquake. *Geophysical Research Letters, 38,* L17306. https://doi.org/10.1029/2011GL048561.

Mansinha, L., & Smylie, D. E. (1971). The displacement field of inclined faults. *Bulletin of the Seismological Society of America, 61,* 1433–1440.

Mochizuki, K., Yamada, T., Shinohara, M., Yamanaka, Y., & Kanazawa, T. (2008). Weak Interplate Coupling by Seamounts and Repeating M 7 Earthquakes. *Science, 321,* 1194–1197.

Moreno, M., Rosenau, M., & Oncken, O. (2010). 2010 Maule earthquake slip correlates with pre-seismic locking of Andean subduction zone. *Nature, 467,* 198–202.

Moreno, M., Melnick, D., Rosenau, M., Baez, J., Klotz, J., Oncken, O., et al. (2012). Toward understanding tectonic control on the Mw 8.8 2010 Maule Chile earthquake. *Earth and Planetary Science Letters, 321–322,* 152–165.

Natawidjaja, D. H., Sieh, K., Ward, S. N., Cheng, H., Edwards, R. L., Suwargadi, B. W., et al. (2004). Paleogeodetic records of seismic and aseismic subduction from central Sumatran microatolls, Indonesia. *Geophysical Research Letters, 109,* B04306. https://doi.org/10.1029/2003JB002398.

Noda, H., & Lapusta, N. (2013). Stable creeping fault segments can become destructive as a result of dynamic weakening. *Nature, 493,* 518–521.

Ohnaka, M. (2013). *The Physics of Rock Failure and Earthquakes.* New York: Cambridge University Press.

Ohnaka, M., & Shen, L. (1999). Scaling of the shear rupture process from nucleation to dynamic propagation: implications of geometric irregularity of the rupturing surfaces. *Journal of Geophysical Research, 104,* 817–844.

Ohtani, M., Hirahara, K., Hori, T., & Hyodo, M. (2014). Observed change in plate coupling close to the rupture initiation area before the occurrence of the 2011 Tohoku earthquake: Implications from an earthquake cycle model. *Geophysical Research Letters, 41,* 1899–1906. https://doi.org/10.1002/2013GL058751.

Ohtani, M., & Hirahara, K. (2015). Effect of the Earths surface topography on quasi-dynamic earthquake cycles. *Geophysical Journal International, 203,* 384–398.

Okada, Y. (1992). Internal deformation due to shear and tensile faults in a half-space. *Bulletin of the Seismological Society of America, 82,* 1018–1040.

Peng, Z., & Gomberg, J. (2010). An integrated perspective of the continuum between earthquakes and slow-slip phenomena. *Nature Geoscience, 3*, 599–607.

Rice, J. R. (1993). Spatio-temporal complexity of slip on a fault. *Journal of Geophysical Research, 98*, 9885–9907.

Rundle, J. B. (1989). A physical model for earthquakes: 3. Thermodynamical approach and its relation to nonclassical theories of nucleation. *Journal of Geophysical Research, 94*, 2839–2855.

Scholz, C. H., & Small, C. (1997). The effect of seamount subduction on seismic coupling. *Geology, 25*, 487–490. https://doi.org/10.1130/0091-7613.

Scholz, C. H., & Campos, J. (2012). The seismic coupling of subduction zones revisited. *Journal of Geophysical Research, 117*, B05310. https://doi.org/10.1029/2011JB009003.

Senatorski, P. (1994). Spatio-temporal evolution of faults: deterministic model. *Physica D, 76*, 420–435.

Senatorski, P. (1995). Dynamics of a zone of four parallel faults: A deterministic model. *Journal of Geophysical Research, 100*, 24111–24120.

Senatorski, P. (2002). Slip-weakening and interactive dynamics of an heterogeneous seismic source. *Tectonophysics, 344*, 37–60.

Senatorski, P. (2004). Interactive dynamics of an heterogeneous seismic source. A model with the slip dependent friction. *Pub. Inst. Geophys. PAS, A-27(354)*, Monographic Vol., pp.151.

Senatorski, P. (2006). Fluctuations, trends and scaling of the energy radiated by heterogeneous seismic sources. *Geophysical Journal International, 166*, 267–276.

Senatorski, P. (2008). Apparent stress scaling for tectonic and induced seismicity: Model and observations. *Physics of the Earth and Planetary Interiors, 167*, 98–109.

Senatorski P. (2014). Radiated energy estimations from finite-fault earthquake slip models. *Geophysical Research Letters*. https://doi.org/10.1002/2014GL060013.

Senatorski, P. (2017). Effect of slip-area scaling on the earthquake frequency-magnitude relationship. *Physics of the Earth and Planetary Interiors, 267*, 41–52.

Sieh, K., Natawidjaja, D. H., Meltzner, A. J., Shen, C.-C., Cheng, H., Li, K.-S., et al. (2008). Earthquake supercycles inferred from sea-level changes recorded in the Corals of West Sumatra. *Science, 322*, 1674–1678.

Stoyan, D., Kendall, W. S., & Mecke, J. (1995). *Stochastic Geometry and its Applications* (2nd ed.). Chichester: Wiley.

Thomas, M. Y., Lapusta, N., & Avouac, J.-P. (2014). Quasi-dynamic versus fully dynamic simulations of earthquakes and aseismic slip with and without enhanced coseismic weakening. *Journal of Geophysical Research: Solid Earth, 119*, 1986–2004. https://doi.org/10.1002/2013JB010615.

Turcotte, D. L. (1997). *Fractals and Chaos in Geology and Geophysics* (2nd ed.). New York: Cambridge University Press.

Uchida, N., & Matsuzawa, T. (2011). Coupling coefficient, hierarchical structure, and earthquake cycle for the source area of the 2011 off the Pacific coast of Tohoku earthquake inferred from small repeating earthquake data. *Earth Planets Space, 63*, 675–679. https://doi.org/10.5047/eps.2011.07.006.

Vannucchi, P., Sage, F., Phipps Morgan, J., Remitti, F., & Collot, J.-Y. (2012). Toward a dynamic concept of the subduction channel at erosive convergent margin with implications for interplate material transfer. *Geochemistry, Geophysics, Geosystems, 13*, Q02003. https://doi.org/10.1029/2011GC003846.

Vannucchi, P., Spagnuolo, E., Aretusini, S., Di Toro, G., Ujiie, K., Tsutsumi, A., et al. (2017). Past seismic slip-to-the-trench recorded in Central America megathrust. *Nature Geoscience, 10*, 935–940. https://doi.org/10.1038/s41561-017-0013-4.

Wang, K., & Bilek, S. L. (2011). Do subducting seamounts generate or stop large earthquakes? *Geology, 39*, 819–822. https://doi.org/10.1130/G31856.1.

Ziv, A., & Cochard, A. (2006). Quasi-dynamic modeling of seismicity on a fault with depth-variable rate- and state-dependent friction. *Journal of Geophysical Research, 111*, B08310. https://doi.org/10.1029/2005JB004189.

(Received January 29, 2018, revised October 1, 2018, accepted January 8, 2019, Published online January 22, 2019)

Pure Appl. Geophys. 176 (2019), 3993–4007
© 2019 The Author(s)
https://doi.org/10.1007/s00024-019-02111-9

Physics-Based Scenario of Earthquake Cycles on the Ventura Thrust System, California: The Effect of Variable Friction and Fault Geometry

Su Qing Miranda Ong,[1] (iD) Sylvain Barbot,[1,2,3] and Judith Hubbard[1,2]

Abstract—The Ventura Thrust system in California is capable of producing large magnitude earthquakes. Geological studies suggest that the fault geometry is complex, composed of multiple segments at different dips: thrust ramps dipping 30°–50° linked with bed-parallel décollements dipping < 10°. These latter types of gently dipping faults form due to preexisting weaknesses in the crust, and therefore have different frictional parameters from thrust ramps; the faults also experience different stresses because of how stresses are resolved onto the fault planes. Here, we use a two-dimensional fault model to assess how geometry and frictional properties of the ramp/décollement system should affect the seismic cycle. We test velocity-strengthening, velocity-weakening, and conditionally stable décollements, and in addition explore how the dip angle of the décollement changes the earthquake behavior. A velocity-strengthening décollement cannot replicate the through-going earthquake ruptures that have been inferred for the Ventura fault system. We therefore suggest that this and other décollements may be better represented using a velocity-weakening or conditionally stable response. Our results show that minor variations in fault geometry produce slip amounts and recurrence intervals that differ only by 10–20%, but do not fundamentally alter the types of earthquakes and interseismic slip. We conclude that geological constraints on fault geometry are typically sufficient to produce modeled earthquake sequences that are statistically consistent with paleoseismic records. However, both frictional parameters along the fault and effective normal stress influence earthquake rupture patterns significantly. More research is needed to adequately constrain these quantities in order for earthquake rupture models to work as effective predictors of fault behavior.

1. Introduction

Continental thrust fault systems are capable of producing large magnitude earthquakes, often near large population centers (e.g., 2008 $M_w = 7.9$ Wenchuan earthquake, China; 2015 $M_w = 7.8$ Gorkha earthquake, Nepal). In California, although the majority of the plate boundary is defined by the right-lateral strike-slip San Andreas system, there is also significant hazard from thrust systems due to the complexity of the fault network (Davis and Namson 1994; Hauksson et al. 1995; Rubin et al. 1998; Shaw and Shearer 1999; Dolan and Rockwell 2001; Anderson et al. 2003; Lavé and Burbank 2004; Daout et al. 2016; McPhillips and Scharer 2018; Rollins et al. 2018). The Western Transverse Ranges, 10–20 km northwest of the city of Los Angeles, is one such region. In this area, multiple fault strands may rupture together in large events along the Ventura-Pitas Point fault system (Hubbard et al. 2014; McAuliffe et al. 2015; Rockwell et al. 2016). Uplifted Holocene marine terraces suggest that such earthquakes occur roughly every 400–2400 years with up to 10 m of uplift per event (Rockwell et al. 2016) (Fig. 1).

The geometry of the Ventura fault system is complex, with multiple segments at various angles linking together to form an irregular, kinked fault at depth. Similar geometric variations have been interpreted to have limited the shape and size of the 2015 $M_w = 7.8$ Gorkha earthquake in Nepal (Wang and Fialko 2015; Hubbard et al. 2016). Thus, constraining the subsurface fault geometry may allow us to better forecast the sizes and locations of future earthquakes (Wesnousky 2006; Feng et al. 2015; Elliott et al. 2016; Hubbard et al. 2016; Qiu et al. 2016; Duan

Electronic supplementary material The online version of this article (https://doi.org/10.1007/s00024-019-02111-9) contains supplementary material, which is available to authorized users.

[1] Asian School of Environment, Singapore, Singapore. E-mail: mirandaosq@gmail.com

[2] Earth Observatory of Singapore, Singapore, Singapore.

[3] *Present Address*: Department of Earth Sciences, University of Southern California, Los Angeles, USA.

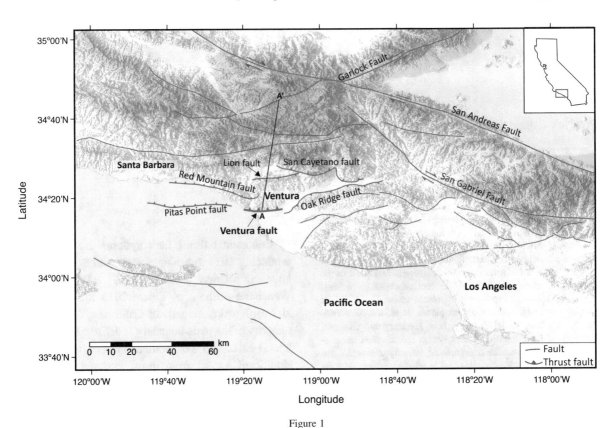

Figure 1
Map of the Ventura region and Ventura Fault in Southern California; red lines show regional faults, with teeth on the hanging wall side. AA′ shows cross section location of fault model. Inset shows location of this map with relationship to the borders of California

et al. 2017). However, given the sparsity of the earthquake record and the long recurrence interval of large earthquakes, such forecasts may rely in part on dynamic rupture models, to assess how geometry affects rupture (e.g., Li and Liu 2016; Qiu et al. 2016; Yu et al. 2018). However, these models rely on detailed descriptions of fault geometry, which can be difficult to constrain, especially in the absence of instrumented earthquakes.

The Ventura Fault system is a good target for investigating the role of fault geometry, as its geometry is constrained by seismic reflection profiles, petroleum well data, and surface geology. Although there is no instrumental record of earthquakes on this system, some information about past earthquakes has been interpreted from Holocene marine terraces. Hubbard et al. (2014) proposed that the fault can be defined by three segments: an upper ramp extending from the surface to ∼ 8 km depth, dipping ∼ 45° to the north; this flattens onto a horizontal or sub-

horizontal décollement for about 10 km, and then steepens onto a deeper, blind ramp that reaches to ∼ 15 to 20 km depth. Ryan et al. (2015) modelled dynamic rupture on a 3D Ventura fault, using a ramp–flat–ramp model modified to have a dip of 8° on the décollement and wide curved hinges between the different fault segments. However, according to structural geology theory (Davis et al. 1983; Dahlen et al. 1984; Zhao et al. 1986; Dahlen 1990; Suppe 2007; Hubbard et al. 2014), the intersections between faults of different dips are typically not smooth, large-scale bends, but rather relatively narrow intersections of two kinds of faults: thrust faults dipping ∼ 20° to 40°, and bedding-parallel décollements (usually horizontal or sub-horizontal), resulting in abrupt, angular transitions between fault segments.

Moreover, the Ventura Fault system is also a good target for investigating the role of frictional properties. A secondary aspect of the structural geology theory is that near-horizontal décollements should be weak compared to more steeply dipping thrust faults, due to the fact that they occur at unfavorable angles for slip (Suppe 2007; Hubbard et al. 2015). Weak fault gouges are often associated with velocity-strengthening or reduced velocity-weakening behavior (Ikari et al. 2011, 2016). Hence, we experiment with different ways of parameterizing friction, allowing for contrasting friction parameters between the décollement and thrust ramps for a suite of kinked fault–bend–fold models, inspired by the Ventura fault system (Hubbard et al. 2014). The dynamic rupture models used in this study can help to explore the frictional and stressing conditions along the fault necessary to produce realistic scenarios, including both single-segment and through-going ruptures (e.g., Biemiller and Lavier 2017).

As complex seismic cycles can result from both nonlinear frictional dynamics (Lapusta and Rice 2003; Kaneko et al. 2010; Wu and Chen 2014; Michel et al. 2017; Biemiller and Lavier 2017) and morphological gradients (Qiu et al. 2016; Romanet et al. 2018), we evaluate how varying the décollement dip and the fault curvature, as well as frictional properties along the décollement, can influence patterns of coseismic and interseismic slip. We also evaluate the effect of smoothing fault intersections to replicate the curved fault geometry studied by Ryan et al. (2015) to evaluate the sensitivity of the models to such geometric variations.

2. Modeling Assumptions: Fault Geometry and Friction Parameters

We simulate a time record of fault slip using the boundary integral method with the radiation damping approximation (Ben-Zion and Rice 1997; Liu and Rice 2005; Shibazaki et al. 2011; Hori and Miyazaki 2011; Qiu et al. 2016; Lambert and Barbot 2016; Goswami and Barbot 2018; Barbot 2018) in two-dimensional in-plane strain conditions. The method allows us to capture all phases of the seismic cycle, except for the radiation of seismic waves, and to include a realistic fault geometry. We properly include the effect of the free surface but we ignore the dynamic evolution of normal stress on the fault plane.

The geometry of the Ventura fault system has been studied using surface geology, seismic reflection profiles, and borehole data (Sarna-Wojcicki et al. 1976; Sarna-Wojcicki and Yerkes 1982; Hubbard et al. 2014). Following these results, we define the system as a series of three fault segments: two fault ramps dipping 40° connected by a gently dipping décollement (Fig. 2a). Although Hubbard et al. (2014) suggested that the décollement is horizontal, following the dip of horizontal bedding planes, Ryan et al. (2015) modeled fault rupture using a dip of 8° to represent the décollement surface. The décollement could have a slight dip, if the bedding planes are not horizontal; this geometry is common in other fault systems (e.g., the Main Himalayan Thrust in Nepal). Realistically, geological fault models will have uncertainties on décollement dip of several degrees in most cases. In order to evaluate the sensitivity of the modelling to the geometry, we tested several décollement dips (0°, 2°, 4°, 6°, 8°) to evaluate the effect on the seismic cycle (Fig. 2a). To evaluate the effect of fault smoothing, we build a series of 2D fault geometries with 40° ramp dips and an 8° décollement dip, with different amounts of curvature at the fault intersections (Fig. 2b).

To calculate fault behavior on our different fault segments, we use the framework of rate-and-state friction with the aging law, which involves a state variable that represents the effect of healing and weakening during the period of quiescence and slip (Dieterich 1978; Ruina 1983; Dieterich 2007). Rate-and-state laws are capable of producing realistic models of earthquake sequences that capture the time dependence of friction, simulating a fault from its nucleation to dynamic rupture propagation, postseismic slip, interseismic period and the strengthening of the fault (Tse and Rice 1986; Lapusta et al. 2000; Daub and Carlson 2008; Barbot et al. 2012; Cubas et al. 2015; Mele Veedu and Barbot 2016; Thomas et al. 2017).

As dynamic frictional properties remain largely unknown for décollements, we explore the frictional parameters of the décollement and determine their effect on the style, magnitude, and recurrence interval of earthquake ruptures. Below 20 km, we assume a velocity-strengthening zone (i.e., $a–b$ is positive, see Table 1), as this is approximately the depth of the

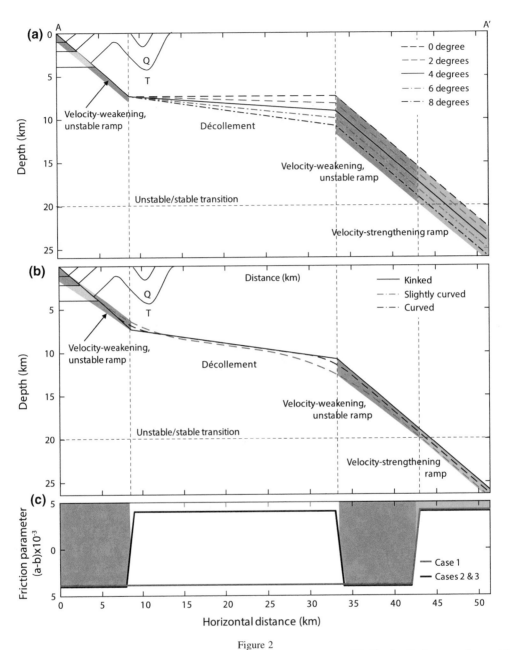

Figure 2

Simplified, two-dimensional geometry for the Ventura Thrust system along cross-section A–A′ in Fig. 1. **a** Fault geometries considered, with the décollement dipping 0°, 2°, 4°, 6° and 8°. **b** Kinked versus curved fault geometry models; these use a décollement dip of 8°. The "Slightly Curved" model is transitional between the "kinked" and "curved" fault geometries. **c** Dynamic friction parameter (a-b) along downdip distance of the fault for Case 1 (velocity-strengthening décollement), Case 2 (velocity-weakening, unstable décollement), and Case 3 (velocity-weakening, conditionally stable décollement). Colors along the fault geometry are used to indicate the frictional properties along the ramp or décollement, with green indicating the velocity-weakening, unstable ramp, blue indicating the velocity-strengthening region beneath the unstable to stable transition and yellow for the décollement where we vary the frictional properties between Cases 1, 2 and 3. Q: Quaternary stream terrace and alluvial fan deposits close to point A, T: Tertiary sandstone and siltstone (Rockwell et al. 1984). Ventura anticline (see folded Q unit) formed due to crustal shortening above the Ventura fault (Hubbard et al. 2014)

unstable to stable transition, and below this depth the fault is expected to creep (Marone et al. 1991; Scholz 1998). In contrast, we set the ramps to have velocity-

weakening behavior (negative a–b values, see Table 1), so that they generate frictional instabilities when the characteristic weakening distance is small

Table 1

Model parameters for rate-and-state friction dynamics. Plate rate is from Marshall et al. (2013)

Frictional parameters	Symbol	Case 1 Velocity-strengthening décollement	Case 2 Velocity-weakening, unstable décollement	Case 3 Velocity-weakening, conditionally stable décollement
Direct effect parameter	a	1×10^{-2}	1×10^{-2}	1×10^{-2}
Evolution effect parameter	b	6×10^{-3} along décollement (velocity-strengthening) 1.4×10^{-2} above 20 km (velocity-weakening) 6×10^{-3} below 20 km (velocity strengthening)	1.4×10^{-2} along décollement (velocity-weakening) 1.4×10^{-2} above 20 km (velocity-weakening) 6×10^{-3} below 20 km (velocity strengthening)	1.4×10^{-2} along décollement (velocity-weakening) 1.4×10^{-2} above 20 km (velocity-weakening) 6×10^{-3} below 20 km (velocity strengthening)
Characteristic weakening distance	L	1 cm	1 cm	10 cm along décollement 1 cm along ramps
Static friction coefficient	μ_0	0.2	0.2	0.2
Reference slip velocity	V_0	10^{-6} m/s	10^{-6} m/s	10^{-6} m/s
Shear modulus	G	30 GPa	30 GPa	30 GPa
Plate rate	V_{pl}	7 mm/year	7 mm/year	7 mm/year

enough and exhibit 'stick–slip' behavior (Ampuero and Rubin 2008; Ariyoshi et al. 2009). We test three frictional behaviors for the décollement: velocity-strengthening (Case 1), unstable velocity-weakening (Case 2), and conditionally stable velocity-weakening (Case 3). In Case 1, the décollement should impede earthquake-style slip, while in Case 2, it should encourage it. In Case 3, the décollement has velocity-weakening properties but a large characteristic weakening distance, which is expected to result in a mixture of stick–slip and creep behavior. We change the characteristic weakening distance by a factor of 10 to place the décollement firmly into the stable-weakening regime.

Numerical studies indicate that complex seismic cycles with full and partial ruptures of the seismogenic zone along a uniform fault occur for large values of the ratio of R/h^*, where R is the length of the fault and h^* is the nucleation size (Kato 2003, 2014; Lapusta and Rice 2003; Gabriel et al. 2012; Wu and Chen 2014; Michel et al. 2017). A smaller L represents a smaller critical nucleation size and hence the ability to produce smaller earthquakes, resulting in a wider range of earthquake magnitudes with increasing complexity. Here, we use a R/h^* ratio as high as approximately 50 in the unstable weakening case (Case 2).

For all models, we evaluate the dynamics of fault slip numerically for a period of 10,000 years to

determine the resulting fault slip patterns. Outputs include slip amounts, recurrence intervals, earthquake magnitudes, and stress drop, which refers to the difference between the initial and final shear stress and is approximately 1–10 MPa for natural earthquakes (Brune 1970). The amount of fault slip in a given earthquake and the recurrence interval between seismic events depend on the frictional parameters $(a–b)$, the effective normal stress $\bar{\sigma}$, characteristic weakening distance, L, and the rupture size (e.g., Tse and Rice 1986; Lapusta and Rice 2003, Liu and Rice 2005, 2007, 2009; Ikari et al. 2011; Barbot et al. 2012; Lapusta and Barbot 2012; Ikari et al. 2016). Earthquake magnitude is empirically related to the length of the fault patch and the average amount of slip on the patch, as follows (Biasi and Weldon 2006):

$$M_w = 6.92 + 1.14 \log_{10}(s),　(1)$$

where s is the average fault slip during a seismic event.

Table 1 summarizes the parameters for the three frictional cases described above. To ensure that our model produces plausible slip amounts, we varied the effective normal stress, $\bar{\sigma}$, for each fault geometry and frictional case, testing 100 and 200 MPa. This range of effective normal stress is typical for models of this type (e.g., Tse and Rice 1986; Segall 2012; Qiu et al. 2016). In all models, we load the fault at a constant rate of $V_l = 7$ mm/year west-northwest, compatible with the rate of contraction across the Ventura region from GPS data (Marshall et al. 2013).

3. Impact of Friction and Geometry on the Seismic Cycle

We compare the predictions resulting from our models to the paleoseismic interpretations of the Ventura Thrust system (Hubbard et al. 2014; Rockwell et al. 2016), which suggest that ruptures should be capable of breaking through the décollement and causing slip on all three segments in a single event. However, when we explore décollement behavior in other parts of the world, we note that these faults display variable behavior. Thus, when we evaluate the results of our modeling, we are looking for not just behavior that can match the interpreted multi-segment ruptures of the Ventura fault but also behavior consistent with earthquakes on other systems. Recorded earthquakes in which we can observe both the geometry and slip patterns on décollements are rare, but they do exist. In the 1999 $M_w = 7.3$ Chi–Chi, Taiwan earthquake, the Chelungpu–Sanyi thrust system produced 3–10 m of slip at the surface, activating both the ramp and the frontal part of the décollement at 5–8 km depth (Lee and Ma 2000; Yu et al. 2001; Johnson et al. 2001; Yue et al. 2005; Rousset et al. 2012). In the 2015 $M_w = 7.8$ Gorkha earthquake, Nepal, most of the slip was limited to a patch of décollement, decreasing to zero along the surrounding ramps (Diao et al. 2015; Bai et al. 2016; Hubbard et al. 2016; Sreejith et al. 2016; Qiu et al. 2016). This event is only one example of how coseismic slip can occur on this fault system. Paleoseismological evidence suggests that previous ruptures have also propagated to the surface, resulting in larger magnitude earthquakes (Bilham et al. 2001; Lavé et al. 2005; Sapkota et al. 2013; Bollinger et al. 2016). We therefore expect that if dynamic rupture models are properly representing realistic physical parameters, they should exhibit multiple styles of earthquakes: décollement-only, ramp and décollement together, and through-going events.

The three sets of frictional parameters that we test produce different earthquake rupture behaviors (Fig. 3). In Case 1 (velocity-strengthening décollement), slip along the décollement occurs primarily as creep, with partial ruptures that either nucleate at the bottom and propagate updip to the décollement, or nucleate at the top of the décollement and rupture in both directions, either partially or to the surface (Fig. 3, Case 1). In Case 2 (velocity-weakening, unstable décollement), partial ruptures occur with occasional larger through-going ruptures (Fig. 3, Case 2). All of the earthquakes nucleate either at the top or bottom of the décollement, where there is a change in orientation of the fault. Patches can undergo full locking, creep and coseismic slip during different periods of the earthquake cycle, similar to models of fault slip evolution by Noda and Lapusta (2013) and Noda and Hori (2014), suggesting that the same segment may experience creep, slow slip, and seismic ruptures during different periods over many earthquake cycles. Lastly, in Case 3, the velocity-weakening, conditionally stable décollement creeps during the interseismic period; earthquakes that occur at depth mostly stop at the décollement, as stresses on the décollement have already been partially released through creeping behavior (Fig. 3, Case 3). This continues until a larger through-going rupture occurs.

We compare the simulated earthquakes that we generate with the characteristics of the paleoseismic catalogue for the Ventura fault (Rockwell et al. 2016). We discard models that produce a recurrence time of large, through-going earthquake greater than 2.4 kyr or lower than 400 years, and those that produce seismic slip amplitudes near the surface greater that 11 m or lower than 4 m (Fig. 4). Although our goal is to observe variations due to fault geometry and frictional properties, the effective normal stress exerts significant control on slip amount and recurrence intervals. An increase from 100 to 200 MPa along the fault results in an approximate doubling of slip amount and recurrence interval for all frictional parameters and dip angles (Fig. 4). We therefore recognize that each physical parameter cannot be effectively constrained simply by comparing with the paleoseismic record, and we focus our analysis on three end-member models with large $R/h*$ ratios that broadly reproduce the variability of recurrence times

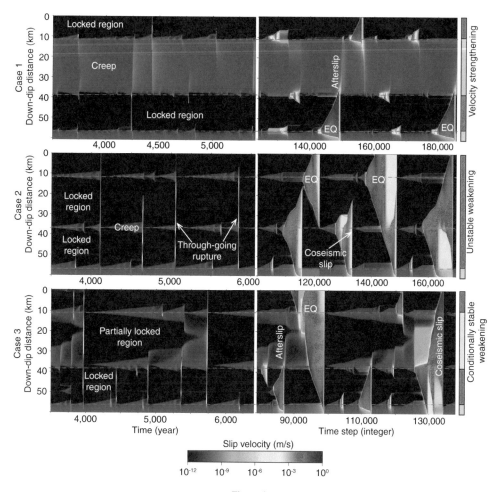

Figure 3
Earthquake cycles for the three sets of frictional parameters on the décollement (décollement dip 0°); the right column with colored boxes represent the dynamic frictional parameters along the fault. **a** Case 1 (velocity-strengthening décollement): the décollement only undergoes creep and afterslip. **b** Case 2 (velocity-weakening, unstable décollement): the décollement exhibits full locking and creep (dark blue) at different times in the earthquake cycle and participates in partial and through-going ruptures. **c** Case 3 (velocity-weakening, conditionally stable décollement); the décollement exhibits full locking and creep, and participates in through-going but not partial ruptures; through-going ruptures are less frequent than in Case 2

and slip amplitude near the fault trace, with complex (non-characteristic) earthquake sequences.

3.1. Sensitivity of Models to Variations in Décollement Dip

Figure 5 compares the seismic cycles obtained for décollement dips of 0° and 8° for Case 3 (velocity-weakening, conditionally stable décollement). The details of the earthquake sequence produced in the two cases are different, including the number of small earthquakes on the deep ramp and the earthquake size. However, the overall pattern of alternating partial and full thrust ruptures is similar, and the average amplitude of slip and recurrence times for the earthquakes that break the surface are comparable (Table 2). In other words, varying the dip from 0° to 8° leads to only mild variations in average earthquake slip amounts and recurrence intervals, despite more significant changes in the pattern of small earthquakes at depth. These results indicate that surface slip measured through paleoseismic techniques near the fault tip have its limitations and may have missed small earthquakes that are confined to the deeper ramp and did not produce slip in the shallow trench.

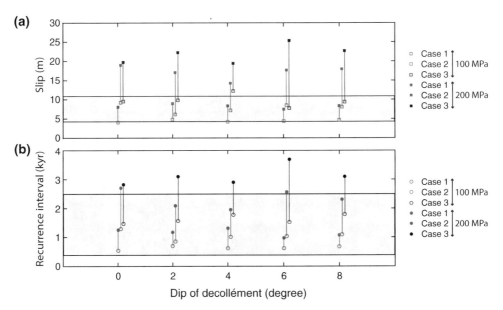

Figure 4

Plots of slip and recurrence interval for all décollement dip angles at effective normal stress levels of 100 MPa and 200 MPa (represented by the unfilled and filled markers respectively). Yellow shaded area represent range of slip amounts and recurrence intervals for the Ventura fault from paleoseismic records (Hubbard et al. 2014; Rockwell et al. 2016). The slip and recurrence interval values vary by approximately 20% across the 0°–8° dips for any given frictional case, but there is no clear pattern to this variation

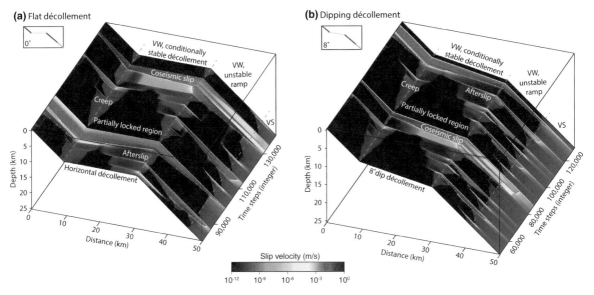

Figure 5

Seismic cycle over 10,000 years for Case 3 (velocity weakening (VW), conditionally stable décollement) for a décollement with dip of **a** 0° and **b** 8°. The x-axis indicates the horizontal distance from the surface trace. The y-axis indicates the computational time steps in the simulation, which emphasizes periods when the slip velocity is high. The vertical axis is depth below the surface trace of the fault. The colors describe the slip velocity: dark colors indicate fault locking, warm colors indicate seismic slip, and dark blue represents fault creep. Both models show a complex seismic cycle with full and partial ruptures, small earthquakes clustering near the fault bends and near the bottom of the seismogenic zone, and occasional creep near the fault bends. While the seismic cycles are different, the average slip and recurrence intervals of earthquakes that rupture to the surface are comparable (Table 2)

Table 2

Slip and recurrence intervals for five fault geometries under Case 3 (velocity weakening, conditionally stable décollement), for $\bar{\sigma} = 100$ MPa

Fault geometry	Range of fault slip (m)	Range of recurrence intervals (year)	Average fault slip (m)	Recurrence interval (year)
0°	Large events: 7–12	1260–1800	9.5	1470
	Small events: 3.4–4.5	300–480	3.8	370
2°	Large events: 7.4–11.6	1430–1700	9.8	1570
	Small events: 3.7–5.0	100–280	4.4	190
4°	Large events: 9.5–11.3	1450–1620	11.0	1500
	Small events: 0.6–1.6	260–300	1.0	290
6°	Large events: 7.4–8.3	1510–1540	7.8	1530
	Small events: 3.9–4.2	210–250	4.1	240
8°	Large events: 8.0–11.3	1720–1840	9.4	1800
	Small events: 1–4.8	10–94	3.4	50

Large and small slip events occur in the form of partial and through-going ruptures. Large events have longer recurrence intervals compared to small events

For all dip angles considered for Case 3 (velocity weakening, conditionally stable décollement), two types of characteristic earthquakes occur: large, through-going ruptures, and smaller ruptures confined to a single segment. Although the earthquake details vary, there is no clear dependence on the dip angle, and the slip and recurrence interval values for the largest earthquakes are within ∼ 20% of each other (Table 2). The average fault slip for the small events shows more variability, but as these events may not be detected by paleoseismological studies, we do not consider them during model selection.

3.2. Effect of Frictional Parameters on Earthquake Rupture Patterns

Compared to the dip of the décollement, the frictional parameters exert significantly more influence on the slip patterns. We investigate the effect of the three frictional cases on the range of slip and recurrence intervals (Fig. 4) and the patterns of earthquake rupture that occur across the ramp-décollement system (Fig. 3).

In Case 1 (velocity-strengthening décollement), most of the slip on the décollement is aseismic, with long-term creep and postseismic periods of accelerated slip. Earthquakes are limited to smaller events that rupture partially either below or above the décollement, with no through-going ruptures (Fig. 3, Case 1). In contrast, a velocity-weakening décollement (Cases 2 and 3) produces earthquake ruptures of much greater complexity, with both partial and through-going ruptures. These larger, through-going ruptures are more consistent with the behavior interpreted for the Ventura fault system.

In Case 2 (velocity-weakening, unstable décollement), earthquakes generally start at the fault bends and rupture in both directions. These earthquakes often propagate through the décollement and then either stop at the next fault bend (updip or downdip), or propagate across the second fault bend as well, resulting in a through-going rupture. Most of the fault system is locked between earthquakes, although there is a small amount of creep centered around the fault bends. Small earthquakes occasionally nucleate at the bottom of the velocity-weakening region. In Case 3 (velocity-weakening, conditionally stable décollement), the décollement creeps part of the time. This relieves some of the stresses on the décollement, which consequently participates in fewer earthquakes. Earthquakes nucleate on the ramps and either stop at the bends or rupture all the way through. Unlike Case 2, there are no partial earthquakes that rupture the décollement.

3.3. Effect of Fault Curvature

We investigate whether the degree of curvature at the fault bend impacts the seismic cycle significantly. To do this, we run all three frictional cases for our curved and kinked fault geometries, and observe the differences in the earthquake cycles and types of earthquakes produced (Fig. 6). In the kinked fault geometry, smaller earthquakes occur on the upper ramp, with some additional earthquakes occurring at shallow depths close to the first kink or at the lower kink between the décollement and the lower ramp (Fig. 6a). In contrast, the curved fault geometry shows earthquakes nucleating at the transition zone between the velocity-weakening and velocity-strengthening zones at 20 km depth; these earthquakes either propagate throughout the fault and break at the surface, or cause partial ruptures of the lower ramp and/or décollement (Fig. 6b). However, the overall earthquake patterns are generally similar. We conclude that the modeling is generally insensitive to the curvature of the fault bends with regards to earthquake style.

Despite this overall similarity, there is a measurable difference between the two geometries. Everything else being the same, the curved model produces larger earthquakes following longer interseismic periods. For example, for an input stress of 100 MPa and with Case 3 (velocity-weakening, conditionally stable décollement), the kinked fault produces events with 12 m of slip every 1720–1860 years, compared to 15–16 m of slip every 2140–2270 years for the curved fault geometry (Table 3). Thus, if geological faults are truly kinked, using a curved fault geometry for modeling might result in an over-estimate of slip and recurrence interval (Fig. 6). However, as shown in the previous figures, the effective confining pressure has a similar effect and therefore these parameters trade off with each other. Because we do not have strong constraints on the physical parameters of the fault geometry, we conclude that we cannot use this effect to distinguish between fault geometries.

We also note that for Case 3, the kinked geometry produces significantly more creep on the décollement

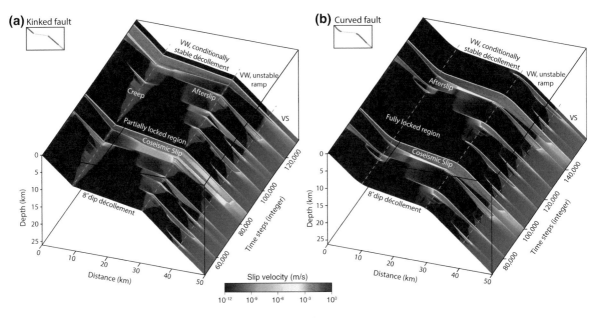

Figure 6
Seismic cycle simulations for **a** kinked versus **b** curved fault geometries for case 3 (velocity weakening (VW), conditionally stable décollement) and a décollement at 8° dip. Velocity-strengthening (VS) regions display continuous creep at depths below 20 km. The x-axis indicates the horizontal distance from the surface trace. The y-axis indicates the computational time steps in the simulation. The vertical axis is depth. Earthquake cycles are generally similar, with more creep on the upper ramp in the kinked fault geometry. The amount of slip and recurrence intervals vary by ~ 20%, with the kinked geometry exhibiting slightly smaller, more frequent events (Table 3) while the curved geometry exhibits more frequent and higher magnitude earthquakes

Table 3

Comparison of slip and recurrence intervals for the kinked, slightly curved and curved fault geometries for Case 3 (velocity weakening, conditionally stable décollement)

Fault geometry	Fault slip amount (m)	Recurrence interval range (year)	Stress drop (MPa)
Kinked	Large event: 12.0–12.2	1720–1860	6.8–7.9
	Small event: 4.2	50–110 years before large event	
Slightly curved	Large event: 15.5	2270	6–7
	Small event: 1–2	380–500	
More curved	Large event: 16	2140–2270	10
	Small event: 3	87–900	

The curved geometry produces events with larger amounts of slip and longer recurrence intervals compared to the kinked fault geometry. Stress drop is higher in the curved fault geometry

than the curved fault geometry (Fig. 6). The creep at the kinks helps to relieve the stress that builds up during the interseismic period. This may explain why there is less stress drop in earthquakes on the kinked fault geometry (6.8–7.9 MPa) compared to the curved geometry (10 MPa). However, the overall patterns are similar for the two geometries, and both can provide an acceptable fit to paleoseismic records.

4. Discussion and Conclusion

In this study, we adopt a two-dimensional model of the Ventura Thrust system to assess the sensitivity of dynamic earthquake models to changes in geometry and frictional properties of a ramp/décollement system. We compare the modeled slip at the top of the superficial ramp to slip inferred from observations of uplifted Holocene marine terraces associated with the Ventura fault. By testing three frictional parameterizations for the décollement (Case 1—velocity-strengthening; Case 2—unstable velocity-weakening, and Case 3—conditionally stable velocity-weakening), we evaluate which frictional parameters produce a better representation of the paleoseismic record along this fault. This paleoseismic record indicates that the Ventura fault likely ruptures periodically in large, through-going ruptures (Hubbard et al. 2014), and therefore models that produce only smaller single-segment ruptures are unlikely to properly represent the physics of the system.

Our results show that varying the fault geometry (décollement dip from 0° to 8°, and curvature of the fault bends) leads to small variations in earthquake cycles in terms of the rupture extent, the amount of slip and the recurrence intervals for small and large earthquakes. Changing the dip does not affect the overall earthquake rupture patterns, although slip amounts and recurrence intervals vary within 20% of each other. Changing the curvature of the fault bends results in relatively small differences in slip amounts and recurrence intervals, without significantly altering the earthquake rupture pattern as a whole. We conclude that a curved fault geometry can reasonably be used to simplify a model to allow for smoother rupture propagation and ground deformation. However, this may lead to an overestimation of slip amounts and recurrence intervals if the fault is truly kinked.

In contrast, the effective normal stress and frictional properties significantly affect the modeled slip patterns. Slip in earthquakes approximately doubles when the effective normal stress is increased from 100 to 200 MPa, with a corresponding doubling of recurrence intervals for all cases tested. Varying the friction changes the recurrence intervals and magnitude of slip, with earthquake slip of approximately 5 m for Case 1 (velocity-strengthening), compared to a combination of two earthquake types (large events with 7–20 m slip and small events with 4 m slip for Cases 2 and 3). Table S1 presents the differences in earthquake slip and recurrence intervals for the five fault geometries and three sets of frictional parameters that we tested.

Overall, because the model results are not overly sensitive to small changes in fault geometry, we conclude that fault geometries constrained by geological measurements, like the Ventura fault geometry, are likely sufficient to inform physics-based numerical models, without the concern that a small change in geometry will lead to significant differences in slip behavior. However, there are major uncertainties that do significantly alter the model outcome. Specifically, the effective normal stress is not well known, limiting the utility of rupture models. In addition, the friction, which depends directly on the fault geometry (thrust ramp versus décollement), is not known. However, with further studies it may be possible to constrain the friction parameters of different fault types based on known rupture histories. For instance, here we observe that one set of frictional parameters does not replicate the known behavior of the Ventura fault system: Case 1 produces only smaller earthquakes that rupture only the updip or downdip ramp; no rupture is capable of crossing the décollement. This is inconsistent with the geological record, which suggests larger, through-going ruptures. Within the limits of this two-dimensional study, we interpret that it is inappropriate to model the décollement as velocity-strengthening. Cases 2 and 3 remain possible, as they are capable of producing both partial and through-going ruptures.

We expect that the frictional behavior of fault ramps and décollements should have broad applicability, and as such it may be useful to study similar fault geometries in other settings. For instance, the 1999 $M_w = 7.6$ Chi–Chi earthquake in Taiwan was the result of slip on the Chelungpu thrust and part of the downdip décollement; such events can be reasonably replicated with all three of our friction parameterizations (Dominguez et al. 2003; Yue et al. 2005; Rousset et al. 2012). In contrast, the 2015 $M_w = 7.8$ Gorkha earthquake in Nepal initiated at a downdip fault bend, with slip almost entirely propagating updip onto the décollement; there was no significant slip on ramps (Diao et al. 2015; Bai et al. 2016; Hubbard et al. 2016, Sreejith et al. 2016). In our modeling, all earthquakes involve significant ramp slip; we therefore do not reproduce such

behavior. This slip pattern indicates that either we are not properly representing the frictional fault behavior, and/or that the geometric or geological differences between the Gorkha and Ventura systems (e.g., depth, dip angle, 3D effects) produce different results.

Our study, which combines fault geometry, frictional parameters, slip modeling, and geological observations, is an important step towards the development of future realistic rupture scenarios. However, reaching this goal will require considerable work in evaluating how friction depends on fault type, as well as the inclusion of three dimensional variations and interactions among fault segments (e.g., Li and Liu 2016; Yu et al. 2018). Some of the largest variations in dynamic rupture models still come from properties that are not well known, such as the frictional properties of décollements and appropriate values for effective normal stress. Ultimately, accurate modelling of earthquake dynamics for potential earthquake forecasting may help mitigate the consequences of earthquakes on the affected communities.

Acknowledgements

This work comprises Earth Observatory of Singapore contribution no. 200. This research is supported by the National Research Foundation of Singapore under the NRF Fellowship scheme (National Research Fellow Awards no. NRF-NRFF2013-04 and no. NRF-NRFF2013-06) and by the Earth Observatory of Singapore, the National Research Foundation, and the Singapore Ministry of Education under the Research Centres of Excellence initiative.

Publisher's Note Springer Nature remains neutral with regard to jurisdictional claims in published maps and institutional affiliations.

REFERENCES

Ampuero, J. P., & Rubin, A. M. (2008). Earthquake nucleation on rate and state faults—aging and slip laws. *Journal of Geophysical Research Solid Earth, 113,* B1.

Anderson, G., Aagaard, B., & Hudnut, K. (2003). Fault interactions and large complex earthquakes in the Los Angeles area. *Science, 302*(5652), 1946–1949.

Ariyoshi, K., Matsuzawa, T., Yabe, Y., Kato, N., Hino, R., Hasegawa, A., et al. (2009). Character of slip and stress due to interaction between fault segments along the dip direction of a subduction zone. *Journal of Geodynamics, 48*(2), 55–67.

Bai, L., Liu, H., Ritsema, J., Mori, J., Zhang, T., Ishikawa, Y., et al. (2016). Faulting structure above the Main Himalayan Thrust as shown by relocated aftershocks of the 2015 M_w 7.8 Gorkha, Nepal, earthquake. *Geophysical Research Letters, 43*(2), 637–642.

Barbot, S. D. (2018). Asthenosphere flow modulated by megathrust earthquake cycles. *Geophysical Research Letters, 2015,* 45.

Barbot, S., Lapusta, N., & Avouac, J. P. (2012). Under the hood of the earthquake machine: Toward predictive modeling of the seismic cycle. *Science, 336*(6082), 707–710.

Ben-Zion, Y., & Rice, J. R. (1997). Dynamic simulations of slip on a smooth fault in an elastic solid. *Journal of Geophysical Research Solid Earth, 102*(B8), 17771–17784.

Biasi, G. P., & Weldon, R. J. (2006). Estimating surface rupture length and magnitude of paleoearthquakes from point measurements of rupture displacement. *Bulletin of the Seismological Society of America, 96*(5), 1612–1623.

Biemiller, J., & Lavier, L. (2017). Earthquake supercycles as part of a spectrum of normal fault slip styles. *Journal of Geophysical Research Solid Earth, 122*(4), 3221–3240.

Bilham, R., Gaur, V. K., & Molnar, P. (2001). Himalayan seismic hazard. *Science, 293*(5534), 1442–1444.

Bollinger, L., Tapponnier, P., Sapkota, S. N., & Klinger, Y. (2016). Slip deficit in central Nepal: Omen for a repeat of the 1344 AD earthquake? *Earth Planets and Space, 68*(1), 12.

Brune, J. N. (1970). Tectonic stress and the spectra of seismic shear waves from earthquakes. *Journal of Geophysical Research, 75*(26), 4997–5009.

Cubas, N., Lapusta, N., Avouac, J. P., & Perfettini, H. (2015). Numerical modeling of long-term earthquake sequences on the NE Japan megathrust: Comparison with observations and implications for fault friction. *Earth and Planetary Science Letters, 419,* 187–198.

Dahlen, F. A. (1990). Critical taper model of fold-and-thrust belts and accretionary wedges. *Annual Review of Earth and Planetary Sciences, 18*(1), 55–99.

Dahlen, F. A., Suppe, J., & Davis, D. (1984). Mechanics of fold-and-thrust belts and accretionary wedges: Cohesive Coulomb theory. *Journal of Geophysical Research Solid Earth, 89*(B12), 10087–10101.

Daout, S., Barbot, S., Peltzer, G., Doin, M. P., Liu, Z., & Jolivet, R. (2016). Constraining the kinematics of metropolitan Los Angeles faults with a slip-partitioning model. *Geophysical Research Letters, 43,* 21.

Daub, E. G., & Carlson, J. M. (2008). A constitutive model for fault gouge deformation in dynamic rupture simulations. *Journal of Geophysical Research Solid Earth, 113,* B12.

Davis, T. L., & Namson, J. S. (1994). A balanced cross-section of the 1994 Northridge earthquake, southern California. *Nature, 372*(6502), 167.

Davis, D., Suppe, J., & Dahlen, F. A. (1983). Mechanics of fold-and-thrust belts and accretionary wedges. *Journal of Geophysical Research Solid Earth, 88*(B2), 1153–1172.

Diao, F., Walter, T. R., Motagh, M., Prats-Iraola, P., Wang, R., & Samsonov, S. V. (2015). The 2015 Gorkha earthquake investigated from radar satellites: Slip and stress modeling along the MHT. *Frontiers in Earth Science, 3,* 65.

Dieterich, J. H. (1978). Time-dependent friction and the mechanics of stick–slip. *Pure and Applied Geophysics, 116*(4–5), 790–806.

Dieterich, J. H. (2007). Applications of rate-and state-dependent friction to models of fault slip and earthquake occurrence-4.04.

Dolan, J. F., & Rockwell, T. K. (2001). Paleoseismologic evidence for a very large ($M_w > 7$), post-AD 1660 surface rupture on the Eastern San Cayetano Fault, Ventura County, California: Was this the elusive source of the damaging 21 December 1812 Earthquake? *Bulletin of the Seismological Society of America, 91*(6), 1417–1432.

Dominguez, S., Avouac, J. P., & Michel, R. (2003). Horizontal coseismic deformation of the 1999 Chi-Chi earthquake measured from SPOT satellite images: Implications for the seismic cycle along the western foothills of central Taiwan. *Journal of Geophysical Research Solid Earth, 108,* B2.

Duan, B., Liu, D., & Yin, A. (2017). Seismic shaking in the North China Basin expected from ruptures of a possible seismic gap. *Geophysical Research Letters, 44*(10), 4855–4862.

Elliott, J. R., Jolivet, R., González, P. J., Avouac, J. P., Hollingsworth, J., Searle, M. P., et al. (2016). Himalayan megathrust geometry and relation to topography revealed by the Gorkha earthquake. *Nature Geoscience, 9*(2), 174.

Feng, G., Li, Z., Shan, X., Zhang, L., Zhang, G., & Zhu, J. (2015). Geodetic model of the 2015 April 25 M_w 7.8 Gorkha Nepal Earthquake and M_w 7.3 aftershock estimated from InSAR and GPS data. *Geodetic journal international, 203*(2), 896–900.

Gabriel, A. A., Ampuero, J. P., Dalguer, L. A., & Mai, P. M. (2012). The transition of dynamic rupture styles in elastic media under velocity-weakening friction. *Journal of Geophysical Research Solid Earth, 117,* B9.

Goswami, A., & Barbot, S. (2018). Slow-slip events in semi-brittle serpentinite fault zones. *Scientific Reports, 8*(1), 6181.

Hauksson, E., Jones, L. M., & Hutton, K. (1995). The 1994 Northridge earthquake sequence in California: Seismological and tectonic aspects. *Journal of Geophysical Research Solid Earth, 100*(B7), 12335–12355.

Hori, T., & Miyazaki, S. I. (2011). A possible mechanism of M 9 earthquake generation cycles in the area of repeating M 7–8 earthquakes surrounded by aseismic sliding. *Earth Planets and Space, 63*(7), 48.

Hubbard, J., Almeida, R., Foster, A., Sapkota, S. N., Bürgi, P., & Tapponnier, P. (2016). Structural segmentation controlled the 2015 M_w 7.8 Gorkha earthquake rupture in Nepal. *Geology, 44*(8), 639–642.

Hubbard, J., Barbot, S., Hill, E. M., & Tapponnier, P. (2015). Coseismic slip on shallow décollement megathrusts: Implications for seismic and tsunami hazard. *Earth-Science Reviews, 141,* 45–55.

Hubbard, J., Shaw, J. H., Dolan, J., Pratt, T. L., McAuliffe, L., & Rockwell, T. K. (2014). Structure and seismic hazard of the Ventura Avenue Anticline and Ventura Fault, California:

Prospect for large, multisegment ruptures in the Western Transverse Ranges. *Bulletin of the Seismological Society of America, 104*(3), 1070. https://doi.org/10.1785/0120130125.

Ikari, M. J., Carpenter, B. M., & Marone, C. (2016). A microphysical interpretation of rate-and state-dependent friction for fault gouge. *Geochemistry Geophysics Geosystems, 17*(5), 1660–1677.

Ikari, M. J., Marone, C., & Saffer, D. M. (2011). On the relation between fault strength and frictional stability. *Geology, 39*(1), 83–86.

Johnson, K. M., Hsu, Y. J., Segall, P., & Yu, S. B. (2001). Fault geometry and slip distribution of the 1999 Chi–Chi, Taiwan earthquake imaged from inversion of GPS data. *Geophysical Research Letters, 28*(11), 2285–2288.

Kaneko, Y., Avouac, J. P., & Lapusta, N. (2010). Towards inferring earthquake patterns from geodetic observations of interseismic coupling. *Nature Geoscience, 3*(5), 363–369.

Kato, N. (2003). Repeating slip events at a circular asperity: Numerical simulation with a rate-and state-dependent friction law. *Bulletin of the Earthquake Research Institute University of Tokyo, 78,* 151–166.

Kato, N. (2014). Deterministic chaos in a simulated sequence of slip events on a single isolated asperity. *Geophysical Journal International, 198*(2), 727–736.

Lambert, V., & Barbot, S. (2016). Contribution of viscoelastic flow in earthquake cycles within the lithosphere–asthenosphere system. *Geophysical Research Letters, 43,* 19.

Lapusta, N., & Barbot, S. (2012). Models of earthquakes and aseismic slip based on laboratory-derived rate-and-state friction laws. *The Mechanics of Faulting From Laboratory to Real Earthquakes, 2012,* 153–207.

Lapusta, N., & Rice, J. R. (2003). Nucleation and early seismic propagation of small and large events in a crustal earthquake model. *Journal of Geophysical Research Solid Earth, 108,* B4.

Lapusta, N., Rice, J. R., Ben-Zion, Y., & Zheng, G. (2000). Elastodynamic analysis for slow tectonic loading with spontaneous rupture episodes on faults with rate-and state-dependent friction. *Journal of Geophysical Research Solid Earth, 105*(B10), 23765–23789.

Lavé, J., & Burbank, D. (2004). Denudation processes and rates in the Transverse Ranges, southern California: Erosional response of a transitional landscape to external and anthropogenic forcing. *Journal of Geophysical Research Earth Surface, 109,* F1.

Lavé, J., Yule, D., Sapkota, S., Basant, K., Madden, C., Attal, M., et al. (2005). Evidence for a great Medieval earthquake ($\sim$ 1100 AD) in the central Himalayas, Nepal. *Science, 307*(5713), 1302–1305.

Lee, S. J., & Ma, K. F. (2000). Rupture process of the 1999 Chi–Chi, Taiwan, earthquake from the inversion of teleseismic data. *Terrestrial Atmospheric and Oceanic Sciences, 11*(3), 591–608.

Li, D., & Liu, Y. (2016). Spatiotemporal evolution of slow slip events in a nonplanar fault model for northern Cascadia subduction zone. *Journal of Geophysical Research Solid Earth, 121*(9), 6828–6845.

Liu, Y., & Rice, J. R. (2005). Aseismic slip transients emerge spontaneously in three-dimensional rate and state modeling of subduction earthquake sequences. *Journal of Geophysical Research Solid Earth, 110,* B8.

Liu, Y., & Rice, J. R. (2007). Spontaneous and triggered aseismic deformation transients in a subduction fault model. *Journal of Geophysical Research Solid Earth, 112,* B9.

Liu, Y., & Rice, J. R. (2009). Slow slip predictions based on granite and gabbro friction data compared to GPS measurements in northern Cascadia. *Journal of Geophysical Research Solid Earth, 114,* B9.

Marone, C. J., Scholtz, C. H., & Bilham, R. (1991). On the mechanics of earthquake afterslip. *Journal of Geophysical Research Solid Earth, 96*(B5), 8441–8452.

Marshall, S. T., Funning, G. J., & Owen, S. E. (2013). Fault slip rates and interseismic deformation in the western Transverse Ranges, California. *Journal of Geophysical Research Solid Earth, 118*(8), 4511–4534.

McAuliffe, L. J., Dolan, J. F., Rhodes, E. J., Hubbard, J., Shaw, J. H., & Pratt, T. L. (2015). Paleoseismologic evidence for large-magnitude (M_w 7.5–8.0) earthquakes on the Ventura blind thrust fault: Implications for multifault ruptures in the Transverse Ranges of southern California. *Geosphere, 11*(5), 1629–1650. https://doi.org/10.1130/ges01123.1.

McPhillips, D., & Scharer, K. M. (2018). Quantifying uncertainty in cumulative surface slip along the Cucamonga Fault, a crustal thrust fault in southern California. *Journal of Geophysical Research Solid Earth, 123*(10), 9063–9083.

Mele Veedu, D., & Barbot, S. (2016, April). *The Parkfield Tremors: Slow and fast ruptures on the same asperity.* In EGU General Assembly Conference Abstracts (vol. 18, p. 16555).

Michel, S., Avouac, J. P., Lapusta, N., & Jiang, J. (2017). Pulse-like partial ruptures and high-frequency radiation at creeping-locked transition during megathrust earthquakes. *Geophysical Research Letters, 44*(16), 8345–8351.

Noda, H., & Hori, T. (2014). Under what circumstances does a seismogenic patch produce aseismic transients in the later interseismic period? *Geophysical Research Letters, 41*(21), 7477–7484.

Noda, H., & Lapusta, N. (2013). Stable creeping fault segments can become destructive as a result of dynamic weakening. *Nature, 493*(7433), 518–521.

Qiu, Q., Hill, E. M., Barbot, S., Hubbard, J., Feng, W., Lindsey, E. O., et al. (2016). The mechanism of partial rupture of a locked megathrust: The role of fault morphology. *Geology, 44*(10), 875–878.

Rockwell, T. K., Clark, K., Gamble, L., Oskin, M. E., Haaker, E. C., & Kennedy, G. L. (2016). Large Transverse Range earthquakes cause coastal upheaval near Ventura, southern California. *Bulletin of the Seismological Society of America, 106*(6), 2706–2720.

Rockwell, T. K., Keller, E. A., Clark, M. N., & Johnson, D. L. (1984). Chronology and rates of faulting of Ventura River terraces, California. *Geological Society of America Bulletin, 95*(12), 1466–1474.

Rollins, C., Avouac, J. P., Landry, W., Argus, D. F., & Barbot, S. (2018). Interseismic strain accumulation on faults beneath Los Angeles, California. *Journal of Geophysical Research Solid Earth, 123*(8), 7126–7150.

Romanet, P., Bhat, H. S., Jolivet, R., & Madariaga, R. (2018). Fast and slow slip events emerge due to fault geometrical complexity. *Geophysical Research Letters, 45*(10), 4809–4819.

Rousset, B., Barbot, S., Avouac, J. P., & Hsu, Y. J. (2012). Postseismic deformation following the 1999 Chi-Chi earthquake, Taiwan: Implication for lower-crust rheology. *Journal of Geophysical Research Solid Earth, 117,* B12.

Rubin, C. M., Lindvall, S. C., & Rockwell, T. K. (1998). Evidence for large earthquakes in metropolitan Los Angeles. *Science, 281*(5375), 398–402.

Ruina, A. (1983). Slip instability and state variable friction laws. *Journal of Geophysical Research Solid Earth, 88*(B12), 10359–10370.

Ryan, K. J., Geist, E. L., Barall, M., & Oglesby, D. D. (2015). Dynamic models of an earthquake and tsunami offshore Ventura, California. *Geophysical Research Letters, 42*(16), 6599–6606.

Sapkota, S. N., Bollinger, L., Klinger, Y., Tapponnier, P., Gaudemer, Y., & Tiwari, D. (2013). Primary surface ruptures of the great Himalayan earthquakes in 1934 and 1255. *Nature Geoscience, 6*(1), 71–76.

Sarna-Wojcicki, A. M., Williams, K. M., & Yerkes, R. F. (1976). *Geology of the Ventura Fault, Ventura County, California (No. 781)*.

Sarna-Wojcicki, A. M., & Yerkes, R. F. (1982). *Comment on Article by RS Yeats on "Low-Shake Faults of the Ventura Basin, California"*.

Scholz, C. H. (1998). Earthquakes and friction laws. *Nature, 391*, 37–42.

Segall, P. (2012). Understanding earthquakes. *Science, 336*(6082), 676–677.

Shaw, J. H., & Shearer, P. M. (1999). An elusive blind-thrust fault beneath metropolitan Los Angeles. *Science, 283*(5407), 1516–1518.

Shibazaki, B., Matsuzawa, T., Tsutsumi, A., Ujiie, K., Hasegawa, A., & Ito, Y. (2011). 3D modeling of the cycle of a great Tohoku-Oki earthquake, considering frictional behavior at low to high slip velocities. *Geophysical Research Letters, 38*, 21.

Sreejith, K. M., Sunil, P. S., Agrawal, R., Saji, A. P., Ramesh, D. S., & Rajawat, A. S. (2016). Coseismic and early postseismic deformation due to the 25 April 2015, M_w 7.8 Gorkha, Nepal, earthquake from InSAR and GPS measurements. *Geophysical Research Letters, 43*(7), 3160–3168.

Suppe, J. (2007). Absolute fault and crustal strength from wedge tapers. *Geology, 35*(12), 1127–1130.

Thomas, M. Y., Avouac, J. P., & Lapusta, N. (2017). Rate-and-state friction properties of the Longitudinal Valley Fault from kinematic and dynamic modeling of seismic and aseismic slip. *Journal of Geophysical Research Solid Earth, 122*(4), 3115–3137.

Tse, S. T., & Rice, J. R. (1986). Crustal earthquake instability in relation to the depth variation of frictional slip properties. *Journal of Geophysical Research Solid Earth, 91*(B9), 9452–9472.

Wang, K., & Fialko, Y. (2015). Slip model of the 2015 M_w 7.8 Gorkha (Nepal) earthquake from inversions of ALOS-2 and GPS data. *Geophysical Research Letters, 42*(18), 7452–7458.

Wesnousky, S. G. (2006). Predicting the endpoints of earthquake ruptures. *Nature, 444*(7117), 358.

Wu, Y., & Chen, X. (2014). The scale-dependent slip pattern for a uniform fault model obeying the rate-and state-dependent friction law. *Journal of Geophysical Research Solid Earth, 119*(6), 4890–4906.

Yu, S. B., Kuo, L. C., Hsu, Y. J., Su, H. H., Liu, C. C., Hou, C. S., et al. (2001). Preseismic deformation and coseismic displacements associated with the 1999 Chi–Chi, Taiwan, earthquake. *Bulletin of the Seismological Society of America, 91*(5), 995–1012.

Yu, H., Liu, Y., Yang, H., & Ning, J. (2018). Modeling earthquake sequences along the Manila subduction zone: Effects of three-dimensional fault geometry. *Tectonophysics, 733*, 73–84.

Yue, L. F., Suppe, J., & Hung, J. H. (2005). Structural geology of a classic thrust belt earthquake: The 1999 Chi-Chi earthquake Taiwan (M_w = 7.6). *Journal of Structural Geology, 27*(11), 2058–2083.

Zhao, W. L., Davis, D. M., Dahlen, F. A., & Suppe, J. (1986). Origin of convex accretionary wedges: Evidence from Barbados. *Journal of Geophysical Research Solid Earth, 91*(B10), 10246–10258.

(Received September 5, 2018, revised January 10, 2019, accepted January 17, 2019, Published online February 5, 2019)

Pure Appl. Geophys. 176 (2019), 4009–4041
© 2018 Springer Nature Switzerland AG
https://doi.org/10.1007/s00024-018-1990-y

▌Pure and Applied Geophysics

Fully Coupled Simulations of Megathrust Earthquakes and Tsunamis in the Japan Trench, Nankai Trough, and Cascadia Subduction Zone

GABRIEL C. LOTTO,[1] TAMARA N. JEPPSON,[2] and ERIC M. DUNHAM[1,3]

Abstract—Subduction zone earthquakes can produce significant seafloor deformation and devastating tsunamis. Real subduction zones display remarkable diversity in fault geometry and structure, and accordingly exhibit a variety of styles of earthquake rupture and tsunamigenic behavior. We perform fully coupled earthquake and tsunami simulations for three subduction zones: the Japan Trench, the Nankai Trough, and the Cascadia Subduction Zone. We use data from seismic surveys, drilling expeditions, and laboratory experiments to construct detailed 2D models of the subduction zones with realistic geometry, structure, friction, and prestress. Greater prestress and rate-and-state friction parameters that are more velocity-weakening generally lead to enhanced slip, seafloor deformation, and tsunami amplitude. The Japan Trench's small sedimentary prism enhances shallow slip, but has only a small effect on tsunami height. In Nankai where there is a prominent splay fault, frictional parameters and off-fault material properties both influence the choice of rupture pathway in complex ways. The splay generates tsunami waves more efficiently than the décollement. Rupture in Cascadia is buried beneath the seafloor, but causes a tsunami that is highly complex due to the rough seafloor bathymetry. Neglecting compliant sediment layers leads to substantially different rupture behavior and tsunami height. We demonstrate that horizontal seafloor displacement is a major contributor to tsunami generation in all subduction zones studied. We document how the nonhydrostatic response of the ocean at short wavelengths smooths the initial tsunami source relative to commonly used approach for setting tsunami initial conditions. Finally, we determine self-consistent tsunami initial conditions by isolating tsunami waves from seismic and acoustic waves at a final simulation time and backpropagating them to their initial state using an adjoint method. We find no evidence to support claims that horizontal momentum transfer from the solid Earth to the ocean is important in tsunami generation.

Key words: Tsunami, megathrust earthquake, subduction zone, Japan Trench, Nankai Trough, Cascadia subduction zone, tsunami modeling, initial conditions, dynamic rupture.

1. Introduction

Subduction zones span the surface of the Earth and host the world's largest earthquakes and tsunamis. Despite their ubiquity in convergent margins, subduction zones are qualitatively and quantitatively distinct from one another in several ways that greatly influence earthquake rupture and tsunamigenesis. To this point, most models of subduction zone earthquakes have been highly idealized, and none have attempted to couple dynamic rupture to tsunami generation in a realistic setting. In this study, we perform fully coupled simulations of subduction zone earthquakes and tsunamis in order to gain insight on the following: the influence of geometry, geologic structure, friction, and stress on the megathrust rupture process; the role of horizontal and vertical seafloor motion in contributing to tsunami height; and the extent to which the standard tsunami modeling procedure makes justifiable assumptions, especially with respect to initial conditions.

We begin by highlighting some of the areas of greatest difference between subduction zones, motivating our interest in modeling specific detail in addition to general subduction features. Decades of seismic imaging has revealed major differences in geometry from one subduction zone to the next (e.g., Fleming and Tréhu 1999; Kopp and Kukowski 2003; Miura et al. 2005; Nakamura et al. 2014; Nakanishi et al. 2002). Geometrical differences include the presence or absence of splay faults, which have been

Electronic supplementary material The online version of this article (https://doi.org/10.1007/s00024-018-1990-y) contains supplementary material, which is available to authorized users.

[1] Department of Geophysics, Stanford University, Stanford, CA, USA. E-mail: glotto@stanford.edu

[2] Department of Geology and Geophysics, Texas A and M University, College Station, TX, USA.

[3] Institute for Computational and Mathematical Engineering, Stanford University, Stanford, CA, USA.

observed since the 1970s, most notably in Alaska (Plafker 1972), Costa Rica (Shipley et al. 1992), and Nankai (Park et al. 2002a). Virtually all subduction zones feature compliant prisms of weakly consolidated sediments, though these vary widely in their landward extent between different subduction zones (Huene et al. 2009) and along-strike within the same convergent margin (e.g., Nakanishi et al. 2002). Various ocean drilling projects have demonstrated the extreme elastic compliance of sedimentary prisms in Cascadia (Tobin et al. 1995), Nankai (Raimbourg et al. 2011), Costa Rica (Gettemy and Tobin 2003), Barbados (Tobin and Moore 1997), and Tohoku (Nakamura et al. 2014; Jeppson et al. 2018), and while exact values of elastic moduli of sediments vary from margin to margin, they are typically one or more orders of magnitude smaller than elastic moduli from deeper parts of the subduction zone. Lab and drilling results are consistent with estimates of rigidity from subduction zone earthquake data, which pin on-fault shear moduli between 1 and 10 GPa at shallow depths (Bilek and Lay 1999).

In addition to prism size and material compliance, the thickness of sediments on the incoming crust varies between subduction zones. Sediment thickness, along with crustal age and convergence rate, helps to control the temperature on the fault and thus the updip and downdip limits of velocity-weakening friction and possibly seismicity (Oleskevich et al. 1999). The updip limit of seismicity has generally been associated with the dehydration of stable-sliding smectite clays to velocity-weakening illite and chlorite as temperature increases with depth (Hyndman and Wang 1993; Hyndman et al. 1997; Wang 1980). Field and laboratory data indicate that this transformation occurs at temperatures between 100 and 150 °C (e.g., Hower et al. 1976; Jennings and Thompson 1986; Moore and Vrolijk 1992). These transitional temperatures occur at depths between 2 and 10 km, depending on the subduction zone (Oleskevich et al. 1999). The downdip limit of seismicity may also be determined in part by a velocity-weakening to velocity-strengthening transition, relating to increasing temperatures or mineral transformations (Blanpied et al. 1995; Tse and Rice 1986; Mitchell et al. 2015).

Plate geometry and tectonic forces, which determine the absolute state of stress in a subduction zone, vary around the world. Excess pore pressure is another major determinant of effective stress (Saffer and Tobin 2011). As initially saturated seafloor sediments subduct and pressurize, they expel water through porosity reduction and mineral dehydration. Here, too, we observe differences between subduction zones, where fluid pressures in sedimentary prisms are often elevated but can fall anywhere in the range between hydrostatic and lithostatic (Moore and Vrolijk 1992). Fluid expulsion rate can vary depending on whether a margin is accretionary or nonaccretionary, and on the overall permeability of a fault zone (Saffer and Tobin 2011).

Given such a diversity in qualitative and quantitative subduction zone characteristics, it should not surprise us to find substantial differences in rupture style and tsunamigenic efficiency. Tsunamis from great megathrust events like the M_w 9.1–9.3 2004 Sumatra earthquake (Stein and Okal 2005) result in significant loss of life and cause massive damage to property and infrastructure, but so do those from more-efficient tsunami earthquakes that release hundreds of times less energy, such as the M_w 7.7 2010 Mentawai event (Lay et al. 2011). Meanwhile, other major significant subduction zone earthquakes like the M_w 8.6 2005 Nias–Simeulue event (Briggs et al. 2006) generate only small tsunamis that do limited damage. The non-monotonicity of the relationship between earthquake magnitudes and tsunami heights should cause us to ask which specific properties of subduction zones influence tsunami generation, and to what extent they do so.

The first major focus of this study is how geometry, friction, stress, and material structure influence rupture and the ensuing tsunami. We can link several subduction zone characteristics to differences in earthquake rupture and tsunamigenic behavior. For instance, fault and seafloor geometry is a prominent factor in determining tsunami amplitude. A steeply dipping décollement or splay fault will produce a larger portion of vertical uplift, which directly corresponds to tsunami height (e.g., DeDontney and Hubbard 2012; Kame et al. 2003). The angle of a sloping seafloor, including the presence of seamounts or any rough bathymetry, also directly contributes to tsunami height even if seafloor motion is entirely horizontal (Lotto et al. 2017; Tanioka and Satake 1996). Several studies (DeDontney and Hubbard 2012; Kame et al. 2003; Wendt et al. 2009) have used dynamic rupture simulations of branching faults to demonstrate that prestress, frictional properties, and rupture velocity are important in determining rupture pathway, e.g., whether an earthquake ruptures along a steep splay, a flatter décollement, or both. The results of our Nankai simulations, presented below, bear out the importance of prestress and friction on the choice of rupture pathway.

In addition to determining rupture pathway, fault friction and prestress (particularly the contributions of excess pore pressure) help to control both rupture speeds and long-term deformation (Ma 2012; Saffer and Tobin 2011; Wang and Hu 2006). Our previous work shows that the choice of rate-and-state frictional parameter $b - a$ has a major influence on rupture velocity, total slip, and tsunami height for a simple subduction-zone-like geometry (Lotto et al. 2017). That same work demonstrates the effect of compliant prism materials on earthquake rupture and tsunami amplitude. In Lotto et al. (2017) we find that larger, more compliant prisms in a purely elastic medium lead to enhanced shallow slip and greater tsunami amplitudes. One can invoke Hooke's Law to simply account for this effect: a more compliant material will be more susceptible to elastic deformation for a given earthquake-induced stress change. Alternatively, Ma (2012) argues that sedimentary prisms are subject to significant inelastic deformation during shallow subduction zone earthquakes. Additionally, normal stress perturbations caused by slip on a bimaterial interface—commonly found in shallow subduction zone faults—can alter earthquake rupture and lead to unstable slip (Andrews and Ben-Zion 1997; Weertman 1980; Ma and Beroza 2008). Michael Aldam et al. (2017) find that several physical quantities, including slip velocity and normal stress drop, exhibit a nonmonotonic dependence on the bimaterial contrast.

Simulations and empirical results from real earthquakes support the idea that compliant prisms have a large effect on earthquake rupture. Numerical experiments by Tamura and Ide (2011) of a branching fault system in a heterogeneous medium with a free surface show that when the upper medium is more compliant, rupture is encouraged on the branching fault. Several slow-rupturing tsunami earthquakes have been associated with shallow slip through subducted sediments (Kanamori and Kikuchi 1993; Polet and Kanamori 2000; Satake 1994; Tanioka and Satake 1996). Gulick et al. (2011) argue that dewatering and lithification of shallow sediments in the Sumatra–Andaman margin made them strong enough to enable shallow rupture during the 2004 Sumatra earthquake.

No single conceptual model fully captures the variations between subduction zones (Huene et al. 2009). And while many numerical studies of dynamic rupture focus on the effects of varying one parameter, recent publications have shown that dynamic rupture experiments produce nonmonotonic or counterintuitive results, especially when multiple variables are introduced (Michael Aldam et al. 2017; Lotto et al. 2017). Given the extensive qualitative and quantitative differences between subduction zones and the extent to which those differences affect rupture and tsunamigenesis, it is necessary to model individual subduction zones in some sufficient level of detail in order to make reasonable conclusions about tsunami hazard. In this study, we consider earthquake rupture and tsunami generation for three distinct subduction zones: the Japan Trench, the Nankai Trough, and the Cascadia subduction zone. We model each subduction zone in 2D with realistic geometry and material properties, and make reasonable choices for friction, prestress, and pore pressure.

In addition to exploring the effects of using realistic frictional and structural parameters, we are also motivated by some more fundamental questions about the physics of tsunami generation. To what extent does horizontal deformation of the seafloor contribute to sea surface uplift? Do common tsunami modeling techniques make valid assumptions about tsunami generation and propagation?

We are presently aware of only three modeling approaches—ours (Lotto and Dunham 2015) and two others (Maeda and Furumura 2013; Saito and Tsushima 2016)—that attempt the fully coupled problem of dynamic rupture and tsunami generation. Typical tsunami generation approaches separate the problem into several distinct steps: determination of fault slip, calculation of seafloor deformation, and translation of that deformation into the ocean to provide tsunami initial conditions. Breaking the problem up this way requires making several approximations, not all of which may be justified.

Tanioka and Satake (1996) first theorized the role that horizontal motion of a sloping seafloor plays in determining the initial height of a tsunami. The Tanioka and Satake initial tsunami height, η_{ts}, superimposes the obvious effect of vertical seafloor displacement, u_y, with the kinematic effect of horizontal displacement, u_x, of a seafloor with slope m, and translates that motion to the sea surface under the assumption of hydrostatic ocean response:

$$\eta_{ts}(x) = u_y(x) - m(x)u_x(x). \tag{1}$$

Though the contribution of horizontal displacement to ocean uplift (Fig. 1) was recognized over two decades ago, most tsunami models continue to neglect the second term in Eq. (1). Likewise, most tsunami models also do not account for Kajiura (1963) nonhydrostatic correction to the initial tsunami height, which acts as a low-pass filter, reducing the contribution of short-wavelength seafloor deformation to ocean uplift. Additionally, tsunami propagation models based on the shallow water wave equation neglect dispersion, which is another consequence of nonhydrostatic ocean response at wavelengths comparable to or less than the ocean depth. Numerical methods that decouple the

earthquake rupture and tsunami generation process also inherently ignore the compressibility of the ocean. Many also neglect the time-dependent nature of the rupture process.

In addition to the kinematic effects of horizontal seafloor motion described above, Song and others (e.g., Song et al. 2008, 2017) have argued that horizontal momentum transfer from the solid Earth to the ocean is a major contributor to tsunami height. We have previously found that such an effect is negligible for subduction zone earthquakes (Lotto et al. 2017), at least for the specific geometries studied in that work. Fully coupled methods like ours naturally account for momentum transfer, whereas typical tsunami models neglect it by setting initial horizontal velocity in the ocean to zero.

To what extent are these sundry assumptions valid for real subduction zone earthquakes? Through our simulations we aim to determine the relevance of the following factors on tsunami generation and propagation: horizontal seafloor displacement, nonhydrostatic corrections to η_{ts}, tsunami dispersion, ocean compressibility, and a time-dependent rupture process.

2. Modeling Framework

We perform fully coupled simulations of earthquakes and tsunamis at three subduction zones using

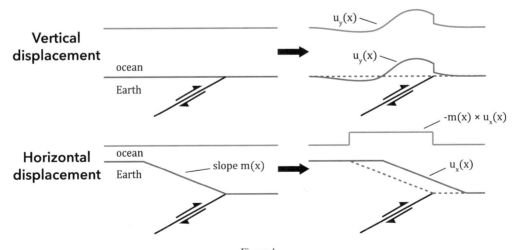

Figure 1
The role of vertical and horizontal seafloor deformation in setting initial sea surface height, according to the theory of Tanioka and Satake (1996) and Eq. (1). Figure redrawn from that publication

2D profiles centered at the oceanic trench and extending hundreds of kilometers both landward and seaward. Seafloor and fault geometries as well as material properties are based upon seismic surveys and ocean drilling experiments. Our models reflect the complexities of seafloor bathymetry and include major fault bends, although we do not explicitly consider short wavelength fault roughness or a finite-width fault damage zone. The subduction zone models also capture complexities in geology where they are known to exist, including smoothly varying material properties, compliant prisms, subducted sediments, and mantle wedges.

We simulate coupled earthquakes and tsunamis using a provably stable and high-order accurate finite difference code that couples an acoustic ocean in the presence of gravity to an elastodynamic Earth (Kozdon et al. 2013). Surface gravity waves are incorporated via a linearized boundary condition that imposes gravity on perturbations about an ocean initially in hydrostatic balance (Lotto and Dunham 2015). This approach allows us to model the full seismic, ocean acoustic and tsunami wavefield in one self-consistent framework. Tsunami waves are generated by time-dependent seafloor deformation in response to dynamic earthquake rupture, and propagate dispersively in a compressible ocean.

Dynamic rupture is modeled on the plate boundary fault—and on a splay fault, for Nankai—using a differential form of rate-and-state friction described by Kozdon and Dunham (2013) that alleviates issues of instability and ill-posedness associated with bimaterial interfaces:

$$\frac{d\tau}{dt} = \frac{a\bar{\sigma}}{V} \tanh\left(\frac{\tau}{a\bar{\sigma}}\frac{dV}{dt}\right) - \frac{|V|}{L}\left[\tau - \bar{\sigma}f_{ss}(V)\right], \quad (2)$$

for time t, shear strength τ, slip velocity V, and state evolution distance L. Effective normal stress $\bar{\sigma}$ is defined as the difference between total normal stress and pore pressure,

$$\bar{\sigma} = \sigma - p. \quad (3)$$

The steady-state friction coefficient, f_{ss}, is dependent on V as

$$f_{ss}(V) = f_0 - (b-a)\ln(V/V_0), \quad (4)$$

where V_0 is a reference velocity and $b - a$ is a dimensionless variable that determines the extent to which friction increases or decreases with increasing slip velocity ($b - a > 0$ is velocity-weakening and $b - a < 0$ is velocity-strengthening). The parameter f_0 is the friction coefficient for steady sliding at V_0. For all simulations of all subduction zones, we set a constant $V_0 = 10^{-6}$ m/s and $L = 0.8$ m.

We account for undrained poroelastic changes in pore pressure, Δp, in response to changes in total normal stress, $\Delta\sigma$, with a linear relation of the form $\Delta p = B\Delta\sigma$. Considering the poroelastic effect lets us rewrite effective normal stress as Kozdon et al. (2013)

$$\bar{\sigma} = \bar{\sigma}_0 + (1 - B)\Delta\sigma, \quad (5)$$

which acts to partially buffer changes to effective normal stress. Equation (5) is a limiting case of a model (Cocco and Rice 2002) for a fault that is bounded by damaged material, where B is Skempton's coefficient. Across all simulations, we choose a moderate value of $B = 0.6$ (where in the $B \to 1$ limit, effective stress would remain constant).

At the coast of each of our subduction zone models is a vertical "cliff", a nonphysical feature that allows us to include a clear ocean-Earth boundary and has no effect on earthquake rupture or the first few hundred seconds of tsunami propagation in the open ocean. The cliff tends to produce its own tsunami signal as it moves horizontally in response to seismic waves, but this tsunami is not a realistic feature.

For each subduction zone simulation, grid points along the fault are separated by about 200 m, though this varies because of the curved geometry. The time step size is 5×10^{-4} s. This allows us to accurately capture wave frequencies lower than ~ 0.5 Hz, though the frequency resolution is reduced in low-velocity layers. At distances hundreds of kilometers away from the fault, we progressively stretch the grid to place the outer boundaries far from the region of interest so that numerical reflections are unnoticeable.

In the following sections, we introduce our three subduction zones of interest, justify our choices of physical parameters, and present results for each subduction zone. This is followed by a comprehensive discussion of the more general issues raised in this introduction.

3. Japan Trench: Background

The 2011 M_w 9.0 Tohoku earthquake, Japan's most recent great megathrust event, ruptured a large portion of the Japan Trench and generated a devastating tsunami, several meters high in the open ocean (Fujii et al. 2011; Ozawa et al. 2011). Observational and modeling evidence supports the idea that the earthquake ruptured to the trench, rapidly deforming the seafloor by $\sim$ 20 to 50 m and contributing to the extreme tsunami amplitude (Fujiwara et al. 2011; Sato et al. 2011; Kodaira et al. 2012; Kozdon and Dunham 2013).

Our model of the Japan Trench uses seafloor and fault geometry that is essentially identical to that of our previous efforts to model the Tohoku event (Kozdon and Dunham 2013, 2014; Lotto et al. 2017), based on seismic lines from Miura et al. (2005), but with somewhat smoother interfaces and bathymetry. The material properties are derived from the structural models of Miura et al. (2005) and Nakamura et al. (2014). The structure (Fig. 2) features a fairly small accretionary prism that extends 22 km downdip along the décollement, and a 2 km layer of sediments along the seafloor landward of the oceanic trench. The depth of the ocean at the trench is 7.5 km. At the coastline ($x = -220$ km, where $x = 0$ is the trench), the water meets the land in a nonphysical "cliff", as

mentioned in Sect. 2. We nucleate the earthquake at a depth of 20.8 km, by quickly increasing shear stress over a small region along the fault.

Dynamic rupture simulations on faults in elastic media require setting several parameters relating to friction and prestress. In order to select a, b, and f_0, we appeal to several experimental results obtained from the Japan Trench Fast Drilling Project (JFAST), conducted during Integrated Ocean Drilling Program Expeditions 343/343T (Mori et al. 2012). The JFAST project penetrated a part of the plate boundary fault that had experienced large shallow slip during the 2011 Tohoku earthquake and recovered a core of clay-rich material from 822 m below the seafloor (Chester et al. 2011). Additionally, a temperature observatory was installed in the drilled hole through the plate boundary, which recorded temperature observations over a 9-month period (Fulton et al. 2013). The observed temperature anomaly in the vicinity of the fault corresponded to an apparent friction coefficient of 0.08. High-velocity (1.3 m/s) friction experiments on JFAST fault samples showed very low shear stress and a correspondingly small stress drop (Ujiie et al. 2013). More recent laboratory experiments on Japan Trench material at sub-seismic velocities have found a range of somewhat higher values for the friction coefficient, from 0.17 (Hirono et al. 2016) to 0.20 (Ikari et al. 2015) to $\sim$0.35 (Sawai et al. 2017). Inspired by these data, we run two sets of simulations of the Japan Trench, one using $f_0 = 0.15$ and the other using $f_0 = 0.35$ at shallow depths. At $x < -22$ km the décollement is

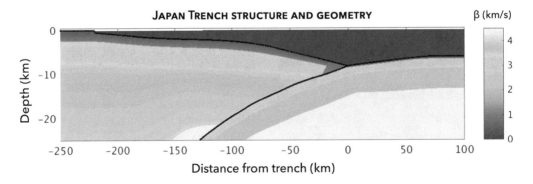

Figure 2

Material structure and geometry of the Japan Trench used for our simulations, colored by shear wave speed β. Bold black lines trace the plate boundary fault and the seafloor

no longer bounded by a highly compliant sedimentary prism and we transition f_0 from the lower values above to a more moderate value of 0.5 over several kilometers.

There are also several experimental results that measured the rate-and-state parameters a and b for various Japan Trench samples. It is often assumed that clay minerals, which comprise the shallow parts of subduction zone faults, exhibit velocity-strengthening behavior. However, laboratory experiments have shown that the frictional behavior of smectite-rich sediments is not always stable; for example, Saffer and Marone (2003) observed velocity-weakening behavior in smectite at room temperature under low slip velocities. Tests of smectite-rich materials recovered from the Japan Trench showed frictional behavior that is mostly velocity-strengthening but sometimes velocity-weakening or velocity-neutral (Ikari et al. 2015). More recently, Sawai et al. (2017) showed that JFAST core samples exhibited both velocity-weakening and velocity-strengthening behavior under a range of combinations of temperature and slip velocity, and that $b - a$ often responded non-monotonically to those variables. Given the poorly constrained nature of shallow friction in subduction zones including the Japan Trench, we run simulations with four different values of $b - a$ in the region adjacent to the prism: $b - a = -0.004, -0.002, 0.000,$ or 0.002.

Deeper in the subduction zone, we set $b - a = 0.004$, a solidly velocity-weakening value that allows for earthquake nucleation and propagation at depth. Velocity-weakening friction is required to nucleate an earthquake but this choice of $b - a$ is also supported by results from Sawai et al. (2016), who report that blueschist fault rocks, likely present at seismogenic depths at the Japan Trench, exhibit velocity-weakening behavior at temperatures relevant to hypocentral depths. Similarly to f_0, $b - a$ transitions from one of the values above to $b - a = 0.004$ as the fault descends past the prism. Downward along the décollement ($x = -180$ km), in order to smoothly cease rupture propagation, we transition the fault from velocity-weakening to strongly velocity-strengthening by gradually increasing the value of a by an order of magnitude.

Since our simulations consider only linear elastic deformation of the solid, stress and pore pressure need only be initialized on the fault. Total normal stress is assumed to increase lithostatically as a function of fault depth below seafloor d as $\sigma = \rho g d$, with $g = 9.8$ m/s^2 and density at a constant, nominal value of $\rho = 2000$ kg/m^3. This formulation approximates a state of stress in which the minimum principal stress is vertical and the décollement is nearly horizontal.

As Eq. (3) indicates, pore fluid pressure plays a major role in determining stress on the fault. We introduce the Hubbert–Rubey fluid pressure ratio (Hubbert and Rubey 1959), λ, to quantify pore pressure relative to lithostatic pressure for a submarine fault:

$$\lambda = \frac{p}{\rho g d}. \qquad (6)$$

A pore pressure ratio of $\lambda = 1$ corresponds to pore pressure that increases lithostatically. For our nominal density of $\rho = 2000$ kg/m^3 in the earth, $\lambda = 0.5$ means pore pressure increases hydrostatically. Anywhere in between, pore fluid are overpressured. Both Seno (2009) and Kimura et al. (2012) argue that pore fluid pressure is highly elevated in the Japan Trench, with $\lambda > 0.9$ everywhere along the décollement. We ran sets of simulations with $\lambda = 0.7$ and $\lambda = 0.9$, though we prefer the latter value.

To inhibit excess slip at depth, we follow the example of Rice (Rice 1992) and assume that below some depth, pore pressure begins to increase lithostatically, i.e., $\lambda = 1$, and effective normal stress no longer grows. This occurs at a depth such that the maximum effective normal stress is $\bar{\sigma}_{max} = 40$MPa. We also assume that the fault has some minimum strength near the trench, such that $\bar{\sigma}_{min} = 1$MPa. Initial shear stress along the entire fault is calculated as $\tau = f_0 \bar{\sigma}$.

4. Japan Trench: Results

The Japan Trench serves as something of a base case scenario within this study, given its relatively simple geometry and structure and our prior modeling of the 2011 Tohoku earthquake and tsunami (Kozdon

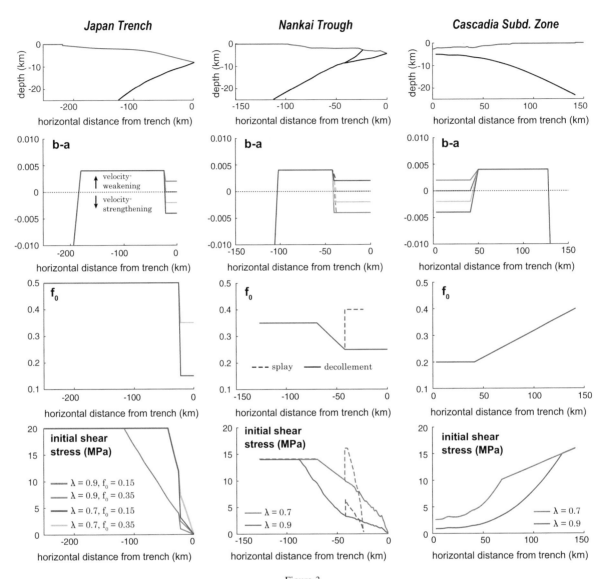

Figure 3

Initial conditions on frictional parameters $b - a$ and f_0, and shear stress τ, for the three subduction zones considered in this study. For our Japan Trench simulations, we run two sets of simulations, one with $f_0 = 0.15$ and one with $f_0 = 0.35$ near the trench. For each subduction zone, we vary $b - a$ among four options: $b - a = -0.004, -0.002, 0.000, \text{and} 0.002$. We also vary the pore pressure ratio, λ, between 0.7 and 0.9. This leads to two different versions of initial shear stress for Nankai and Cascadia (and four for the Japan Trench, given the two parameterizations of f_0 for that subduction zone). Dashed lines for the Nankai Trough refer to conditions on the splay rather than the plate boundary fault

and Dunham 2013, 2014; Lotto et al. 2017). In this section, we present a range of numerical experiments that explore the variability produced by adjusting friction and initial stress.

We first examine the effects of varying the rate-and-state friction parameter $b - a$ in the shallow part of the subduction zone, as in the upper left plot of Fig. 3. For the Japan Trench, shallow $b - a$ has a

clear but limited effect on slip and on final tsunami height (at $t = 500$ s). Figure 4 shows results for two extreme conditions of stress, $\lambda = 0.9$, $f_0 = 0.15$ and $\lambda = 0.7$, $f_0 = 0.35$. In both cases, smaller values of $b - a$ (more velocity-strengthening friction) lead to less slip across the whole fault but especially at the prism, as unstable slip is inhibited near the trench. The effect of changing $b - a$ is greater, as is total

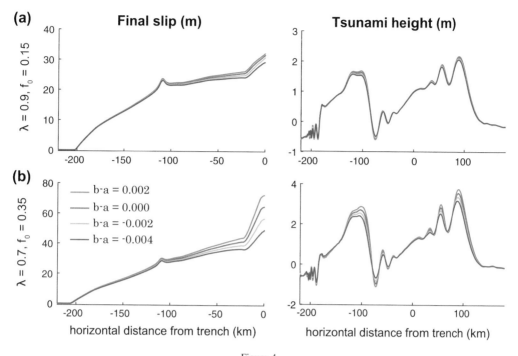

Figure 4
The effect of modifying rate-and-state parameter $b - a$ on total slip and final tsunami height (time $t = 500$ s) for the Japan Trench, with **a** $\lambda = 0.9, f_0 = 0.15$ and **b** $\lambda = 0.7, f_0 = 0.35$. Increasing $b - a$ makes the shallow part of the fault more susceptible to unstable slip, increasing slip and tsunami height everywhere. This effect is more pronounced for **b**, where prestress is higher

slip, for the case with greater initial stress (i.e., $\lambda = 0.7$, $f_0 = 0.35$). Tsunami height varies in accordance with slip, although in a less noticeable way.

These results fit reasonably well in the context of our previous work on compliant prisms (Lotto et al. 2017), for which we varied compliance, prism size, and shallow $b - a$ in an idealized subduction zone geometry. In that work, we used a structural model where material properties were piecewise constant, having one set of values outside the prism and another set inside the compliant prism. We characterized the compliance of the prism by a nondimensional parameter r, defined as the ratio of shear wave speeds β in the prism and elsewhere in the Earth, $r = \beta_{\text{prism}}/\beta_{\text{earth}}$. Prism size W was characterized as the downdip extent of the prism. Though the Japan Trench geometry in this study has a more complex structure and initial stress distribution, we can make rough estimates of $W = 22$ km and $r \approx 4$. For those sets of parameters, our previous work shows that $b - a$ has a similarly modest effect on slip and tsunami height (see Fig. 5c, d of Lotto et al. 2017).

We now focus on the effects of the friction coefficient and pore pressure ratio, selecting a moderate value of $b - a = -0.002$ (slightly velocity-strengthening behavior) near the trench and modifying shallow f_0 and λ. In this set of simulations, presented in Fig. 5, initial stress (as modified via pore pressure ratio) is the single greatest factor affecting slip, horizontal and vertical seafloor displacement, and tsunami height. Lower initial shear stresses (see lower left plot of Fig. 3) lead to lower stress drops (see Supplementary Figure 21) and thus less slip throughout the fault. The effect of f_0 on slip is apparent only beneath the prism where higher f_0 means higher initial shear stress. Changes to f_0 and λ have similar effects on horizontal and vertical seafloor displacement. Differences in tsunami heights are also largely determined by pore pressure ratio, and despite the enhanced slip near the trench with $f_0 = 0.35$, the tsunami—especially its landward wave—barely registers that difference. (Note that the tsunami signal at the far left of Figs. 4 and 5e at $x < -175$ km is a wave propagating from the coast toward the ocean, caused by horizontal displacement

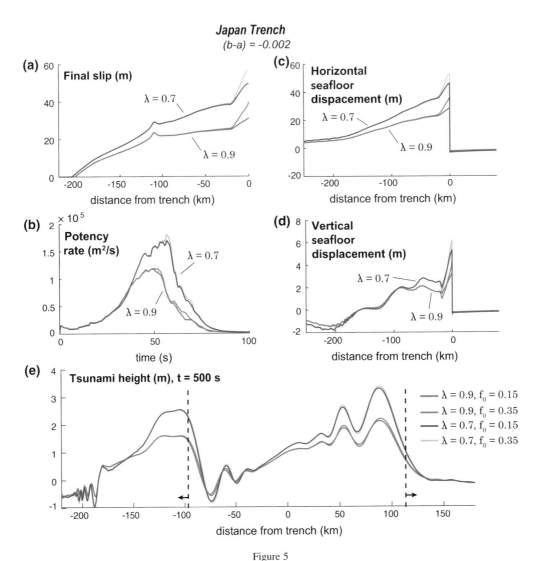

Figure 5

Modeling results from the Japan Trench, varying f_0 and λ but keeping a constant shallow $b - a = -0.002$ across simulations. Pore pressure ratio, which controls stress drop and hence overall slip, has the greatest influence on **a** slip, **b** potency rate, **c** horizontal seafloor displacement, **d** vertical seafloor displacement, and **e** tsunami height at $t = 500$ s, while shallow f_0 is of secondary importance. Vertical dashed lines mark the distance the fastest tsunami waves produced at $x = 0$ would travel by the final time

of the artificial ocean-earth "cliff" during seismic deformation. This byproduct of our structural model is not a realistic feature and would not be observed.)

We now introduce another way to quantify earthquake rupture, seismic potency per distance along strike:

$$P(t) = \int_{l_1}^{l_2} s(l,t)\, \mathrm{d}l, \qquad (7)$$

where s is slip and the variable l follows the downdip fault axis from l_1 to l_2. Potency is defined as

earthquake moment per unit rigidity; we use potency rather than moment to avoid ambiguities in the choice of rigidity on a bimaterial fault plane. We further define potency rate per distance along strike, $\dot{P}$, as the rate of change in potency over time,

$$\dot{P}(t) = \frac{\mathrm{d}P}{\mathrm{d}t}. \qquad (8)$$

We use potency rate to compare and contrast the time-dependent earthquake source process between different simulations. For example, in Fig. 5b we see that earthquakes with $\lambda = 0.7$ rupture for longer and have more total potency than earthquakes with $\lambda = 0.9$, but that going from $f_0 = 0.15$ to $f_0 = 0.35$ only affects the rupture process marginally.

5. Nankai Trough: Background

The Nankai Trough in southern Japan has hosted five M_w 8+ earthquakes since the beginning of the eighteenth century (Masataka Ando 1975). Each event launched a significant tsunami and caused over 1000 fatalities. The last pair of tsunamigenic earthquakes were the 1944 Tonankai (M_w 8.1) and 1946 Nankaido (M_w 8.3) events. Consequently, the Nankai Trough has been closely studied by earth scientists for decades.

Southeast of Shikoku Island and the Kii Peninsula, the Philippine Sea Plate subducts beneath the Eurasian Plate and produces a very large accretionary prism (Park et al. 2002a). The prism, which has been developing since the Miocene, consists of compliant materials, mainly offscraped and underplated from turbidites and Shikoku Basin sediments (Park et al. 2002b). The Nankai Trough is distinguished from other subduction zones by its prominent splay fault, a thrust fault that bends up from the megathrust to the seafloor at a higher angle than the décollement (Park et al. 2002a).

A geodetic slip inversion of the 1944 and 1946 earthquakes by Sagiya and Thatcher (1999) suggested a highly complex source process for the two events, which they associated in part with coseismic slip on splay faults. The presence of splay faults has major implications for tsunami generation; an earthquake rupture that branches off of the main megathrust and continues along a steeper splay could possibly generate a tsunami with a higher amplitude and certainly with a different wave profile than one that continues along the décollement. Though Tanioka and Satake (2001), drawing from an inversion of tsunami waveforms from the 1946 event, found that large slip on splay faults was not required to fit available data, other publications have reached the opposite conclusion. Cummins and Kaneda (2000) argued that the tsunami data could be matched equally well if all slip was confined to a splay rather than the plate boundary fault.

Scientific opinion is more unified with respect to the 1944 Tonankai event. A topographical argument was made by Kikuchi et al. (2003), who reasoned that the repetition of fault motions must be responsible for the rough profile of the Kumano Basin above the inferred slip distribution of the 1944 earthquake. Moore et al. (2007), analyzing the results of 2D and 3D seismic reflection surveys, showed that there has been more activity on the splay than on the décollement and concluded that the splay fault likely contributed to the 1944 tsunami. Geothermometric measurements on core samples from the Nankai Trough revealed evidence of frictional heating on shallow portions of both the splay fault and the plate boundary fault, implying that coseismic rupture has occurred on both segments (Sakaguchi et al. 2011).

Several studies, inspired by the splay fault at Nankai, have used numerical models to explore various parameters that influence rupture in the vicinity of a branching fault system. Kame et al. (2003) focused on the effects of prestress, rupture velocity, and branch angle (in an unbounded, homogeneous, elastic medium) and found that the prestress has a significant effect on the favored rupture direction, though enhanced dynamic stressing sometimes resulted in slip on a less-favorably oriented segment. Collapsing the geometry of the Nankai Trough into the context of their parameter space study, they predicted that coseismic rupture would follow the splay fault and cause negligible slip along the plate boundary fault. DeDontney and Hubbard (2012) expanded on that work, using realistic initial stress states derived from elastic wedge theory and a range of values for friction and dip angle to produce a variety of simulations where rupture propagated along the main fault, a branching fault, or both. Simulating dynamic rupture in 3D, Wendt et al. (2009) observed rupture remaining on the plate boundary thrust for a homogeneous prestress, but found that the introduction of a stress barrier leads to activation of splay faults. Tamura and Ide (2011), modeling a branching fault system in a 2D bimaterial medium with a free surface, reported that for homogeneous prestress, rupture is enhanced on the branching fault when the upper material is more compliant, as would be likely the case in a subduction zone. These studies advance the hypothesis that

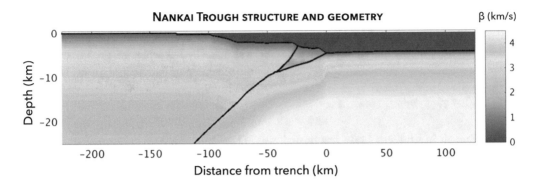

Figure 6

Material structure and geometry of the Nankai Trough used for our simulations, colored by shear wave speed β. Bold black lines trace the plate boundary fault, the splay fault, and the seafloor

rupture on the splay is plausible or even likely, but they greatly simplify the Nankai Trough; our simulations capture realistic geometric and geologic features, with spatially variable prestress and frictional properties.

Our geometric and structural model for Nankai is based upon two seismic profiles extending from the Kii Peninsula out past the trench axis: a wide-angle seismic survey (profile KR9806) interpreted by Nakanishi et al. (2002) and waveform tomography images from Kamei et al. (2012). Thus, our model (Fig. 6) is representative of the segment of the Nankai Trough through which the 1944 Tonankai earthquake ruptured. A compliant layer of sediments, 1–3 km in thickness, underlies the seafloor and fills the prism bounded by the décollement and splay. At the trench ($x = 0$), the ocean reaches its maximum depth of 5.1 km. The splay branches off the décollement at $x = -42$ km (a depth of 8.5 km) and breaches the seafloor at $x = -24$ km.

Our approach to assigning initial stresses on the fault is somewhat different for Nankai and Cascadia than for Tohoku. Rather than assuming a simple depth-dependent normal stress on the faults, we instead opt for the more self-consistent approach of prescribing a stress field everywhere in the solid and resolving it onto the faults. We draw on the work of DeDontney and Hubbard (2012), who use critical and elastic wedge theory to determine the stress state of a Nankai-like subduction zone geometry. They characterize the stress state in the prism by a ratio of principal stresses, σ_1/σ_3, and by the angle of σ_1 (the

more horizontal principal stress component) relative to the dip of the plate boundary fault, Ψ (Fig. 7). For the Nankai-like subduction zone geometry, they determine that $\Psi = 13.5°$ and $\sigma_1/\sigma_3 = 2.28$ are appropriate values. Our approach is to have σ_1 increase lithostatically with depth below seafloor and to scale σ_3 according to the ratio of principal stresses above. The resulting stress state is resolved onto the faults, and the ratio τ/σ on the fault is used to determine an approximate f_0. This approach yields values of $f_0 = 0.4$ on the splay, $f_0 = 0.25$ on the shallow plate boundary thrust, and $f_0 = 0.35$ on the deep plate boundary fault. These values are consistent with experimental results that pin friction on the splay fault between $f_0 = 0.36$ and $f_0 = 0.46$ (Hirono et al. 2016; Ikari and Saffer 2011). Theory and experimental data for the plate boundary thrust are also supportive but less precise—one estimate has

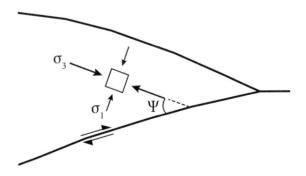

Figure 7

Principal stresses σ_1 and σ_3 oriented relative to the plate boundary fault by angle Ψ. This stress convention is used to determine the prestress for Nankai and Cascadia

$0.16 \leq f_0 \leq 0.26$ (Kopf and Brown 2003) and another finds $0.32 \leq f_0 \leq 0.40$ (Ikari and Saffer 2011). For a chosen value of λ, we use Eqs. (3) and (6) to calculate effective normal stress and $\tau = f_0 \bar{\sigma}$ to calculate initial shear stress.

With regard to the choice of λ for Nankai simulations, published estimates for pore pressure vary widely between and within studies. While Seno (2009) gives a pore pressure ratio as high as $\lambda \approx 0.98$ on the megathrust, other results yield more moderate values of $\lambda \approx 0.8$ and $0.68 \leq \lambda \leq 0.77$ (Skarbek and Saffer 2009; Tobin and Saffer 2009). Pore pressure on the splay is probably similarly elevated, with Tsuji et al. (2014) estimating that the pore pressure ratio lies roughly between $\lambda \approx 0.6$ and $\lambda \approx 0.9$. Given the wide spread of estimates of λ and the influence of prestress on the results of dynamic rupture experiments, we run two sets of simulations, one with $\lambda = 0.7$ on both splay and megathrust and another with $\lambda = 0.9$.

To choose rate-and-state parameters a and b, we appeal in part to Ikari and Saffer (2011), whose experiments on Nankai Trough sediments showed mostly velocity-strengthening friction, but with significant variability on both the splay and the megathrust. As with the Japan Trench, we run simulations with four possible values of $b - a$ $(-0.004, -0.002, 0.000,$ and $0.002)$, allowing that parameter to vary independently on the splay and the megathrust for a total of 16 permutations.

6. Nankai Trough: Results

The Nankai Trough is unique in this study, in that rupture nucleated at depth comes to a fault juncture, and can choose between propagating along the steeper splay, the flatter décollement, or both. Because the splay dips at a higher angle and breaks the seafloor surface closer to the coast, the rupture pathway is expected to have serious implications for tsunami hazard. Indeed, we find that the choice of rupture pathway has a qualitative effect on the the tsunami waveform and a quantitative effect on its amplitude.

We start by looking at simulation results where pore pressure ratio is constant at $\lambda = 0.9$ and shallow $b - a$ varies between $b - a = -0.004$ and

$b - a = 0.002$. We focus on simulations where $b - a$ is the same on both the splay and décollement; we ran several versions where $b - a$ was different on the two fault segments above the juncture, but there is no particular reason to believe that one segment is more or less velocity-weakening than the other, and we observe plenty of variability in outcomes even when $b - a$ is identical.

Figure 8a shows cumulative slip on the fault for both the splay (dashed lines) and the décollement (solid lines). We see that small changes in $b - a$ result in major changes to the choice of rupture pathway. The most velocity-strengthening model, with $b - a = -0.004$ (blue lines), ruptures very little along the décollement, producing almost zero slip at the trench. However, that same model produces greater slip on the splay than any other model, despite experiencing a relatively low stress drop on the splay (see Supplementary Figure 22). The other three choices of $b - a$ all produce about the same slip on the splay, but, unsurprisingly, the more velocity-weakening models produce greater slip on the décollement. Interestingly, all four simulations result in slip along the splay, and that slip is greater at the seafloor than at the fault junction. Potency rate (Fig. 8b)—defined in Eqs. (7) and (8) and summed over both the splay and décollement—shows that ruptures that continue along the décollement produce more slip for longer periods of time than ruptures with less slip at the trench.

Slip is directly responsible for horizontal and vertical seafloor deformation (Fig. 8c, d). Displacements peak at $x = 0$ and $x = -23.8$ km, where the décollement and splay reach the seafloor. Both peaks in vertical displacement correlate strongly with the shallow slip on the splay and décollement faults, with the $b - a = -0.004$ model causing maximum displacement on the splay peak and the $b - a = 0.002$ model topping the décollement peak. The splay peak sees greater displacement due to the higher angle of the splay fault compared to the décollement. The leftmost peak in horizontal seafloor displacement, however, is strictly greater for models with more velocity-weakening friction. In other words, models with greater $b - a$ more efficiently produce horizontal deformation even with less slip along the splay.

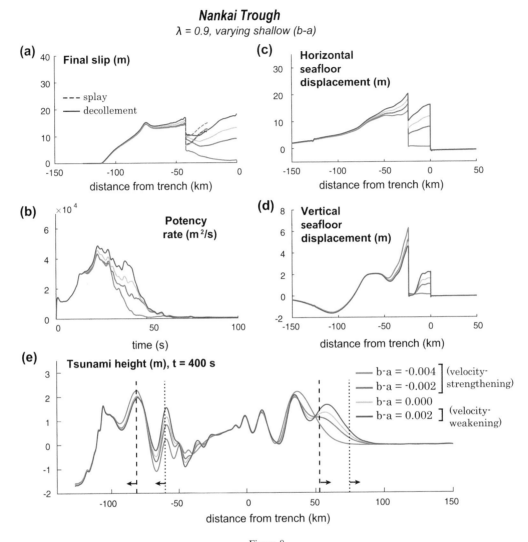

Nankai Trough
λ = 0.9, varying shallow (b-a)

Figure 8
Modeling results from the Nankai Trough, keeping pore pressure ratio constant at $\lambda = 0.9$ and varying $b - a$ identically on both the splay and décollement. Plots show **a** final slip on the splay and décollement, **b** potency rate of the earthquake, **c** horizontal and **d** vertical seafloor displacement at the final time step, with one peak each from motion on the splay and décollement, and **e** tsunami height at a time of $t = 400$ s. Vertical lines mark the distance the fastest tsunami waves produced at $x = 0$ (dotted) and $x = -23.8$ km (dashed) would travel by the final time step

When the seafloor is sloping, as it does in the subduction zones presented in this study, vertical and horizontal seafloor motion both directly contribute to tsunami generation. Tsunami height, which is plotted at a time of $t = 400$ s in Fig. 8e, shows the influence of both factors. The seaward wave (on the right side) and the landward wave each have two peaks, one each associated with the splay and the décollement. The leading seaward wave is caused by slip at the

trench, and thus it does not exist for the $b - a = -0.004$ model. The second seaward peak—the higher amplitude of the two—is caused by splay rupture, and is greatest for the model with the most velocity-strengthening friction. The landward tsunami (excluding the section where $x < -100$ km, which is a seaward-traveling byproduct of the unrealistic ocean "cliff" in our structural model) also has two major peaks. Unlike the seaward tsunami, the leading landward wave is larger due its association with the splay. The height of the secondary landward signal scales strongly with $b - a$.

We see several qualitative similarities for Nankai Trough simulations where pore pressure ratio is set to $\lambda = 0.7$ (Fig. 9), instead of $\lambda = 0.9$ (Fig. 8), though for $\lambda = 0.7$ the higher prestress leads to overall greater slip, seafloor displacements, and tsunami heights than for $\lambda = 0.9$. Fault displacement on the shallow décollement—and to a lesser extent on the deep plate boundary fault—scales strongly with shallow $b - a$, ranging from zero slip on the trench ($b - a = -0.004$) to nearly 40 m ($b - a = 0.002$). As with $\lambda = 0.9$, we see that the model with the most velocity-strengthening friction actually produces the most slip on the splay, though the other models all produce some slip there as well. Models that slip to the trench produce secondary peaks in potency rate due to the greater length of the décollement and the extended slip amidst low-velocity sediments.

Once again, vertical and horizontal seafloor displacement features two main peaks, one each associated with the splay and the décollement. Horizontal displacement is everywhere greater with higher $b - a$. Vertical displacement around $x = 0$ similarly scales with $b - a$, but the $x = -23.8$ km peak is largest for $b - a = -0.004$ and virtually identical for the other three simulations. The resulting tsunami also shows two peaks traveling toward land and toward the sea. For all simulations except that with $b - a = 0.002$, the tsunami wave peak produced by the splay is the larger of the two, and therefore more important for hazard. The $b - a = -0.004$ simulation produces the highest and broadest splay peak, despite having the most velocity-strengthening friction.

Throughout this study, we have used structural models that are as realistic as possible, using constraints from seismic and drilling data. Such detailed structural models are rarely used in earthquake or tsunami studies, however. Here, we explore the full significance of the subduction zone structure by comparing the realistic structural model in Fig. 6 to the two far simpler structural models in Fig. 10d. Such compromises or simplifications to structure are commonly used, for example when converting fault slip to seafloor displacement using analytical solutions for dislocations in homogeneous half-spaces.

The first simplified structural model employs uniform material properties in the solid, using properties typical of rock fairly deep along the subducting plate (density $\rho = 2800$ kg/m^3, s-wave speed $\beta = 4$ km/s, and p-wave speed $\alpha = 7$ km/s). The second model simplified structural uses those values at all points deeper than approximately $y = -8.51$ km, the depth of the fault branch junction, and has values representative of shallow sediments ($\rho = 2.2$ kg/m^3, $\beta = 1.5$ km/s, and $\alpha = 3.1$ km/s) above that junction.

Figure 10 compares the results of using the three structural models on simulations where $\lambda = 0.9$ and $b - a = -0.002$ on both the splay and shallow décollement. For this set of plausible parameters, the two-layer structural model decreases the magnitude of slip and tsunami height, and reduces the maximum potency rate relative to the realistic, heterogeneous structural model. However, the uniform structural model fundamentally changes the character of the earthquake rupture. With the uniform model, all slip is confined to the splay, potency rate is consistently smoother and reveals a shorter rupture process, and the tsunami (excluding the nonphysical effects at $x < -100$ km) has a lower amplitude than either of the other two structural models.

7. Cascadia Subduction Zone: Background

The Cascadia Subduction zone is distinct from the Japan Trench and the Nankai Trough in that it has a relative absence of interplate seismicity over the historical record (e.g., Wang et al. 2003; Wang and Tréhu 2016). Still, great megathrust earthquakes are a known feature of the Cascadia margin over the past several millennia (Witter et al. 2003). The most recent such event, inferred from evidence of coseismic subsidence in North America and an "orphan tsunami" in Japan, occurred on the evening of 26 January 1700 (Nelson et al. 1995; Satake et al. 1996). The approximately M_w 9.0 event was the latest in a series of full-margin ruptures that recur every 500–530 years (which, when combined with smaller events in southern Cascadia, leads to an average

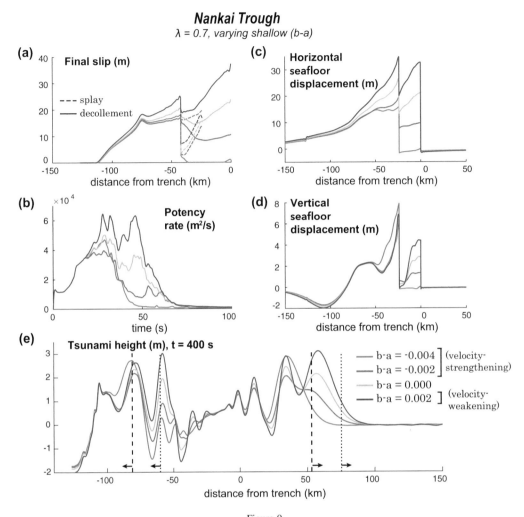

Figure 9
Modeling results from the Nankai Trough, keeping pore pressure ratio constant at $\lambda = 0.7$ and varying $b - a$ identically on both the splay and décollement. Plots show **a** final slip on the splay and décollement, **b** potency rate of the earthquake, **c** horizontal and **d** vertical seafloor displacement at the final time step, with one peak each from motion on the splay and décollement, and **e** tsunami height at a time of $t = 400$ s. Vertical lines mark the distance the fastest tsunami waves produced at $x = 0$ (dotted) and $x = -23.8$ km (dashed) would travel by the final time

recurrence period of $\sim$ 240 years in that section) (Goldfinger et al. 2012).

There have been few attempts to simulate dynamic earthquake rupture in the Cascadia Subduction Zone. Previous studies (Flück et al. 1997; Leonard et al. 2010; Satake et al. 2003; Wang et al. 2003, 2013) have primarily used elastic dislocation models to estimate coseismic seafloor deformation and coastal subsidence given some prescribed fault slip. Those seafloor deformation calculations have sometimes been used to simulate tsunamis, although

the process of converting earthquake slip to seafloor deformation to tsunami amplitude is not as straightforward as is often assumed for realistic earthquake

Figure 10 ▶
Modeling results from the Nankai Trough with $\lambda = 0.9$ and $b - a = -0.002$ on both the splay and shallow décollement, comparing our standard realistic structural models to two simpler ones with **a** uniform material properties in the Earth and a two-layer set of material properties. Plots show **b** final slip on the splay and décollement, **c** horizontal and **d** vertical seafloor displacements, and **e** tsunami height at a time of $t = 350$ s

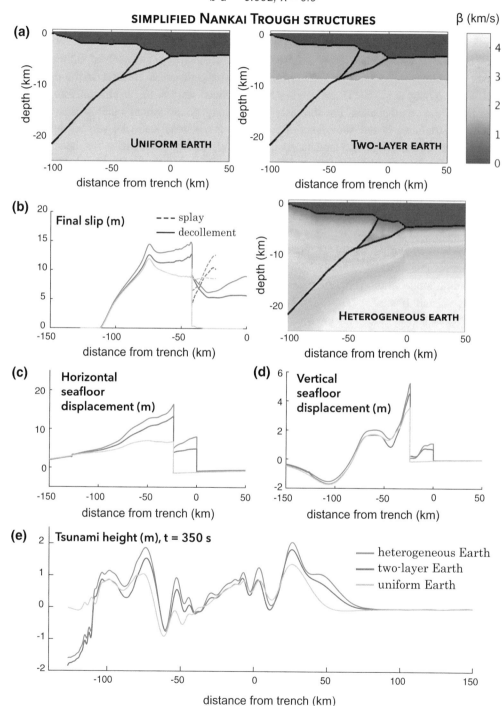

Nankai Trough - comparing structural models
b-a = -0.002, λ = 0.9

(a) SIMPLIFIED NANKAI TROUGH STRUCTURES   β (km/s)

UNIFORM EARTH   TWO-LAYER EARTH

(b) Final slip (m)   --- splay   — decollement

HETEROGENEOUS EARTH

(c) Horizontal seafloor displacement (m)

(d) Vertical seafloor displacement (m)

(e) Tsunami height (m), t = 350 s   heterogeneous Earth   two-layer Earth   uniform Earth

scenarios (Lotto et al. 2017). The elastic dislocation models for Cascadia have been constrained mostly by evidence of coseismic subsidence from paleoseismic

data, but also by thermal constraints, strain rate observations, and recurrence rates (Wang et al. 2003). The growing paleoseismic evidence has for many years shown that at several locations along the Pacific Northwest coast, subsidence between 0.5 and 1.5 m occurred (e.g., Benson et al. 2001; Guilbault et al. 1996; Shennan et al. 1996). More recent interpretations of subsidence data have confirmed the general magnitude of subsidence but inferred heterogeneous slip along-strike for the 1700 event (Kemp et al. 2018; Wang et al. 2013), as is typical for recorded megathrust earthquakes. For the dislocation models, uncertainties in the subsidence data lead to a wide variety of acceptable slip models. And, crucially for tsunami hazard applications, coastal subsidence estimates carry no information to constrain shallow slip or tsunami height (Wang et al. 2013). Earthquake rupture models that neither prescribe slip nor make too many assumptions about tsunami generation may prove useful for understanding local hazard in Cascadia.

Our 2D structural model draws (Fig. 11) on interpretations of two seismic transects: Line SO10 from the ORWELL Project (Flueh et al. 1998) and Line 4 from the COAST dataset (Holbrook et al. 2012; Webb 2017). The two seismic lines both cross the Cascadia margin just south of latitude 47°N, by Grays Harbor and Willapa Bay in Southwest Washington. Both profiles focus on the shallow portion of the subduction zone (depths less than 20 km), so we use the Juan de Fuca slab model of McCrory et al. (2012) to inform the deeper geometry of our plate boundary fault.

Compared with the other subduction zones in this study and in the world, Cascadia is unique in terms of its structure and seismic quiescence (Goldfinger et al. 2012; Webb 2017). Cascadia also has one of the hottest plate boundary thrusts, because of its young ($\sim$ 8 Ma) subducting plate and thick incoming sediments; its high temperature may contribute to its dearth of interplate seismicity (McCaffrey 1997). On the left side of our profile, there is a smooth, 2.5–3 km thick layer of undeformed sediments sitting above the oceanic crust. Starting at the deformation front (defined as $x = 0$), sediments form a large accretionary prism, whose outer wedge is interpreted to contain several landward vergent thrust faults (Webb 2017) (though these have not been included in our simulations). The décollement in this section dips at an extremely shallow angle, less than 4°. A bathymetric feature known as the Quinault Ridge (at $x \approx 38$ km) separates the outer wedge from the lower slope terrace, or midslope terrace, where the prism continues to thicken. To the right of this section, starting at $x \approx 50$ km, is an older section of the accretionary prism. The continental shelf begins at $x \approx 76$ km, where a shallow layer of ocean ($\sim$ 200 m depth) continues up to the coast, at $x = 142$ km. At its greatest depth, the ocean above the Cascadia margin is only 2.7 km deep, far shallower than the water above Nankai and the Japan Trench.

Neither seismic transect we use to build our Cascadia geometry shows evidence of a plate boundary thrust fault that breaches the seafloor at the deformation front. The lack of historic interplate earthquakes means that it is difficult to infer the rupture pathway at shallow depths. It is possible that the 1700 Cascadia earthquake ruptured to the trench, and elastic dislocation models that allow for trench-breaking rupture can sufficiently fit the subsidence

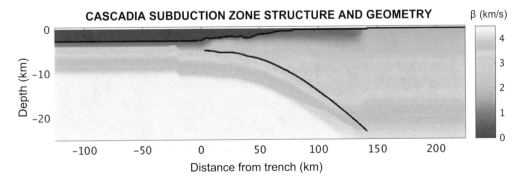

Figure 11
Material structure and geometry of the Cascadia Subduction Zone used for our simulations, colored by shear wave speed β. Bold black lines trace the plate boundary fault and the seafloor

data (Wang et al. 2013). However, seeing no direct evidence of a décollement that meets the seafloor, we use a buried thrust fault in our simulations.

Prestress on the fault is set using the same procedure as in the Nankai Trough simulations. We calculate a depth-dependent pair of principal stresses σ_1 and σ_3 and project them onto the plate boundary thrust. The ratio of shear to normal stress projected on the fault allows us to estimate the friction coefficient f_0, which in this case is set to $f_0 = 0.2$ below the accretionary prism and linearly ramped up to 0.4 at the deepest part of the fault (Fig. 3).

There is a good amount of evidence that the pore fluid along the fault is, as with Nankai and the Japan Trench, highly overpressured. Shear stress on the Cascadia subduction fault has been shown to be very low from heat flow measurements and focal mechanism solutions, and near-lithostatic pore pressure in the fault zone has been deemed to be the cause (Wang et al. 1995). Others have invoked near-lithostatic pore pressure, especially at depth, to explain high Vp/Vs ratios beneath southern Vancouver Island (Pascal Audet et al. 2009; Peacock et al. 2011). Seno (2009) gives a quantitative estimate for the pore pressure ratio as $\lambda = 0.93$ for Washington and $\lambda = 0.895$ for South Vancouver Island. Once again, we run two sets of simulations, one with $\lambda = 0.7$ and the other with $\lambda = 0.9$.

We vary the shallow rate-and-state friction parameter $b - a$ in the same way we do for the Japan Trench, running various simulations with $b - a = -0.004, -0.002, 0.000,$ or 0.002 under the accretionary prism, and $b - a = 0.004$ at depth (Fig. 3).

8. Cascadia Subduction Zone: Results

The Cascadia Subduction Zone is the only one in our study with a completely buried plate boundary thrust. Its rupture behavior is, in many ways, much simpler than that of the Japan Trench or Nankai Trough, because it neither encounters a junction nor breaches the seafloor. However, we find that tsunamis generated in Cascadia have a surprisingly complex signal and source.

In Fig. 12, we feature a set of Cascadia simulations with pore pressure ratio $\lambda = 0.9$, with shallow

rate-and-state parameter $b - a$ varied between $b - a = -0.004$ and $b - a = 0.002$. Though we only plot the results with $\lambda = 0.9$, simulations with $\lambda = 0.7$ produce qualitatively similar results but with greater slip (30–80% more for models with $\lambda = 0.7$), tsunami heights, etc. As expected, the more velocity-weakening models experience greater stress drop (Supplementary Figure 23) and yield greater overall slip, especially in the region where $b - a$ varies ($x < 50$ km, as per the upper right plot of Fig. 3). Cumulative slip is independent of shallow friction at the deepest and shallowest sections of the fault. Potency rate has a very similar history between simulations; the earthquake behaves in a qualitatively similar manner regardless of shallow $b - a$.

Horizontal seafloor displacement (Fig. 12c) scales very clearly with slip, and shows a dependence on shallow $b - a$ for the region where $0 < x < 100$ km. Vertical seafloor displacement, however, shows the effects of varying slip to a lesser degree and in a more limited region, $x < 50$ km, with a maximum near $x = 0$ where the fault is closest to the surface. The final tsunami profile is more complex than those from the other two subduction zones due to the shallow depth of much of the fault slip and the roughness of the seafloor bathymetry. The details of the tsunami process will be discussed further in the next section, but here we note that the differences in the tsunami profiles between the various models in Fig. 12e stem from differences in shallow slip and vertical seafloor displacement.

9. Comparison of Subduction Zones

The three subduction zones we focus on in this study show great variation in their geologic structure, seafloor bathymetry, and fault geometry. Thus far we have compared differences in results within each individual subduction zone, but in this section we make some comparisons across subduction zones. For each one, we select one reasonable parameter set and explore how unique features of each subduction zone influence earthquake rupture and tsunami generation and propagation. For the Japan Trench, we use the case with $\lambda = 0.9, f_0 = 0.35,$ and $b - a = -0.002$; for

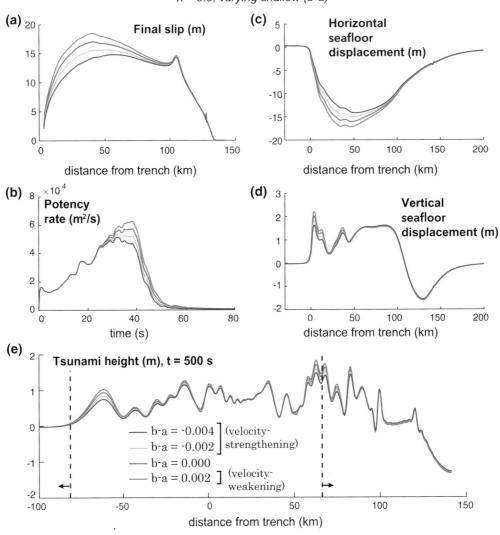

Figure 12
Modeling results from the Cascadia Subduction Zone, keeping pore pressure ratio constant at $\lambda = 0.9$ and varying $b - a$ identically on both the splay and décollement. Plots show **a** final slip on the splay and décollement, **b** potency rate of the earthquake, **c** horizontal and **d** vertical seafloor displacement at the final time step, and **e** tsunami height at a time of $t = 500$ s. Vertical dashed lines mark the distance the fastest tsunami waves produced at $x = 0$ would travel by the final time step

Nankai, we use $\lambda = 0.9$ and $b - a = -0.002$; and for Cascadia we use $\lambda = 0.9$ and $b - a = 0.000$.

9.1. Rupture Process

At each subduction zone, earthquake rupture begins with nucleation at depth and proceeds in both updip and downdip directions. Supplementary

Figure 24 shows potency rate functions for the three subduction zones on the same plot, giving a relative sense of the time history of each rupture.

The Japan Trench rupture proceeds in a very similar way to that of the ruptures in our prior study of idealized subduction zones with compliant prisms (Lotto et al. 2017). We see in Fig. 13a that rupture velocity gradually increases as the rupture propagates

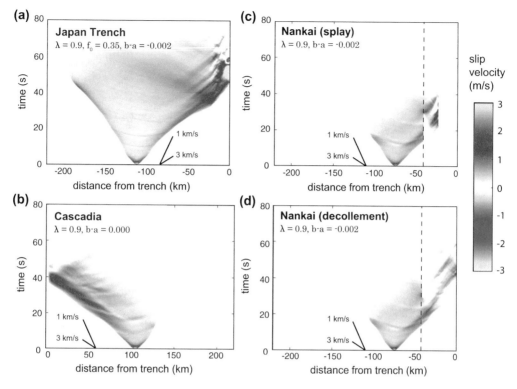

Figure 13
Slip velocity over time and space for selected simulations on the following subduction zone faults: **a** Japan Trench, **b** Cascadia Subduction Zone, **c** Nankai Trough including the splay, and **d** Nankai including the décollement. Vertical dashed lines indicate location of splay–décollement junction

toward the trench then slows as it reaches the accretionary prism. There, slip velocity accelerates despite friction becoming more velocity-strengthening at the trench, leaving enhanced slip profiles like those of Fig. 5a. This is caused by interactions with the seafloor and free surface and wave energy being trapped within the compliant prism.

In the Cascadia Subduction Zone (Fig. 13b), rupture propagates more smoothly. As the earthquake travels updip, through moderately compliant materials but not through extremely weak sediments, slip velocity gradually increases around the rupture front. Since we do not allow rupture to continue up to the seafloor, the slip velocity does not increase explosively and rupture ceases near the deformation front, causing only limited reflection of slip downdip.

For the Nankai Trough, we visualize slip separately for the splay fault and the décollement. Downdip of $x = -42$ km, the junction between the

splay and décollement, Fig. 13c and d are identical because they show the same fault segment. Though not all Nankai simulations result in identical patterns of slip on the splay and décollement, the case shown here is typical in that rupture first continues to propagate along the décollement but later jumps to the middle of the splay fault. Slip on the décollement is patchier than it is for the other subduction zones. Slip on the splay tends to proceed from the middle outward, increasing at the trench but slipping less near the junction, leading to the final slip pattern shown in dashed lines in Fig. 8.

9.2. Tsunami Generation and Propagation

One of the primary benefits of modeling subduction zone earthquakes and tsunamis in a fully coupled framework is the ability to observe the entire tsunami generation process without relying on various

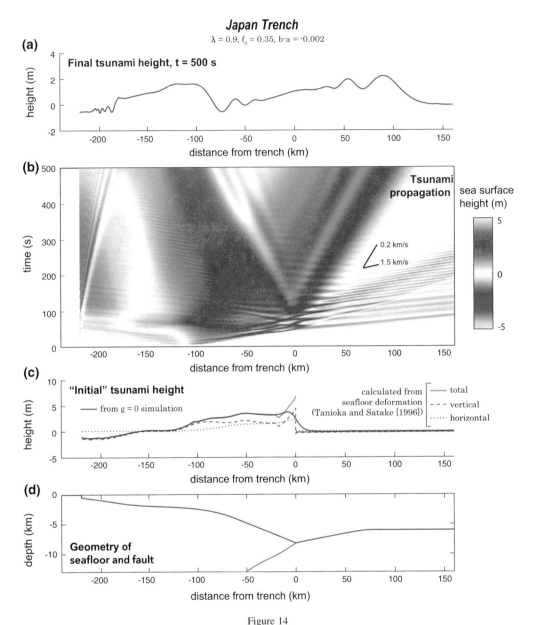

Figure 14

The life cycle of a tsunami in the Japan Trench: **a** tsunami height at the final time step; **b** sea surface height, reflecting earthquake rupture from nucleation to the trench as well as tsunami propagation; **c** two estimates of initial tsunami height, one from a zero-gravity simulation and another calculated from horizontal and vertical seafloor displacements; and **d** geometry of the seafloor and fault, which help determine seafloor motion.

approximations typically used to convert seafloor displacement to tsunami initial conditions. Whereas typical tsunami models must convert seafloor displacements to sea surface height before running a tsunami simulation, our approach takes into account the full physics of the problem, including any possible contributions from momentum transfer between the Earth

and ocean (Song et al. 2008; Lotto et al. 2017). In this section, we track the entire tsunami process from generation to propagation, and compare between the three subduction zones of interest.

In Fig. 14 we plot the whole life cycle of a Japan Trench tsunami, for a simulation with parameters $\lambda = 0.9$, $f_0 = 0.35$ and shallow $b - a = -0.002$.

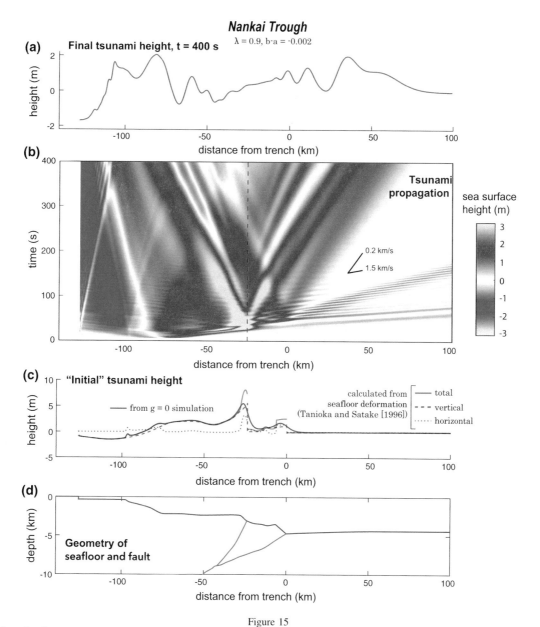

Figure 15
The life cycle of a tsunami in the Nankai Trough: **a** tsunami height at the final time step; **b** sea surface height, reflecting earthquake rupture from nucleation to the trench as well as tsunami propagation; **c** two estimates of initial tsunami height, one from a zero-gravity simulation and another calculated from horizontal and vertical seafloor displacements; and **d** geometry of the seafloor and faults, which help determine seafloor motion. Vertical dashed line marks where the splay intersects the seafloor

Figure 14a shows the sea surface height at time $t = 500$ s, as in Fig. 5e. This tsunami profile is the final slice of the space–time plot of sea surface height (Fig. 14b). Earthquake nucleation occurs at about $x = -110$ km at time $t = 0$, with subsequent fault slip generating various guided waves including oceanic Rayleigh waves and leaking P-wave modes (or oceanic PL waves) (Kozdon and Dunham 2014) that are apparent on the ocean surface. The rupture continues to the trench at $x = 0$, where shallow slip leads directly to seafloor deformation, causing the beginnings of a tsunami at around $t = 50$ s. Rupture to the trench generates additional ocean-guided waves, especially in the seaward direction due to

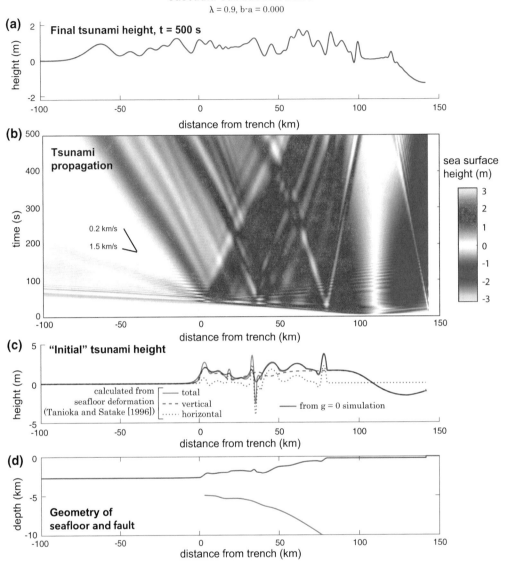

Figure 16
The life cycle of a tsunami in the Cascadia Subduction Zone: **a** tsunami height at the final time step; **b** sea surface height, reflecting earthquake rupture from nucleation to the trench as well as tsunami propagation; **c** two estimates of initial tsunami height, one from a zero-gravity simulation and another calculated from horizontal and vertical seafloor displacements; and **d** geometry of the seafloor and fault, which help determine seafloor motion.

directivity. Since they travel at speeds much greater than the surface gravity wave speed, the tsunami emerges as a distinct feature by $t \approx 200$ s. The tsunami propagates dispersively until the end of the simulation. In this plot we also observe a tsunami wave caused by the horizontal motion of the nonphysical coastal "cliff" at $x = -220$ km,

mentioned in previous sections and visible in the bathymetry profile of Fig. 14d.

It is not possible to visualize a true initial condition on sea surface height; seafloor deformation occurs over tens of seconds and continuously excites surface gravity waves. Those waves begin to propagate even as the seafloor continues to deform. Furthermore, the dynamic nature of the tsunami

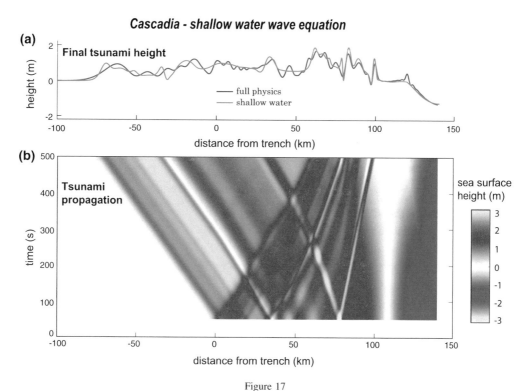

Figure 17
The Cascadia tsunami modeled using the shallow water wave equations, using η_{zg} as an initial condition on tsunami height. **a** A comparison with the final tsunami profile from the equivalent full-physics simulation. **b** A time-space plot of sea surface height, showing dispersion-free tsunami propagation using the shallow water wave equations

generation process means that the initial tsunami waveform is superimposed with ocean acoustic waves, guided Rayleigh waves, oceanic PL waves, and other seismic waves. In Fig. 14c we show two approaches to estimate the "initial" tsunami height. The first approach is to calculate the Tanioka and Satake initial condition on sea surface height, η_{ts}, from Eq. (1) using the static seafloor displacement. We plot the total $\eta_{ts}(x)$ as well as the separate contributions from vertical and horizontal displacement, $u_y(x)$ and $-m(x)u_x(x)$. The Tanioka and Satake initial condition considers neither nonhydrostatic effects nor the time-dependent rupture process, and therefore is a very simple interpretation of the initial tsunami. But it does yield an understanding of the relative importance of vertical and horizontal seafloor deformation.

The second approach involves running an identical simulation but with gravitational acceleration set to zero, i.e., $g = 0$. The zero gravity simulation results in a static sea surface perturbation, η_{zg}, that

captures contributions from both vertical and horizontal displacements of the seafloor as well as the nonhydrostatic response of the ocean (Kajiura 1963); it is as close as possible to an initial condition for the tsunami.

We see that vertical and horizontal displacements both play a substantial role in generating the tsunami at the Japan Trench. The horizontal contribution, $-mu_x$, increases gradually from $x \approx -100$ km to the trench, while the vertical contribution, u_y, is about level until $x \approx -20$ km, where it dips a bit before peaking at the trench due to the influence of the compliant prism. We find that η_{zg} is nearly equivalent to η_{ts} except, notably, in the vicinity of the trench. The differences can be attributed to nonhydrostatic effects, which effectively filter short wavelengths to make η_{zg} smoother than η_{ts}.

Figure 15 shows the life cycle of a tsunami in the Nankai Trough, for the simulation with $\lambda = 0.9$ and shallow $b - a = -0.002$. Figure 15a shows the final tsunami height, while Fig. 15b shows the tsunami's

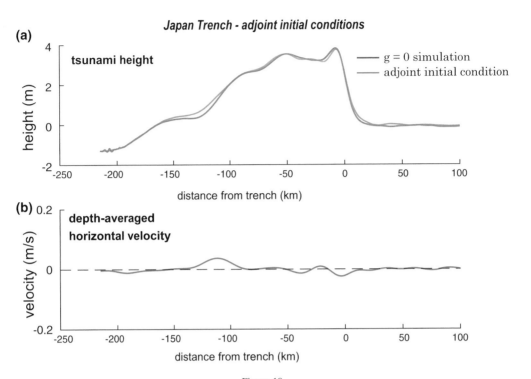

Figure 18
Initial conditions determined from the adjoint (time-reversed) wave propagation problem for the Japan Trench. The tsunami initial conditions occur at a time equivalent to $t = 61$ s. **a** Initial tsunami height compared to the final sea surface height of the zero-gravity simulation, η_{zg}. **b** Initial depth-averaged horizontal velocity

progression from start to finish. For this set of parameters, we observe that the splay ($x \sim -25$ km) is the largest contributor to tsunami height, though slip on the décollement produces waves that manifest in the final tsunami height as an extension of the first seaward peak and in the second landward peak. Dispersion is clearly visible in the final tsunami profile between $x = -50$ km and $x = 25$ km. As with the Japan Trench, we observe oceanic Rayleigh waves and other guided waves traveling seaward.

The zero-gravity sea surface profile, η_{zg}, shows the relative contributions of slip on the splay and décollement in the initial tsunami. (This plot would look substantially different for other Nankai simulations which have more or less slip on the décollement.) In the Nankai Trough, u_y plays a larger role in tsunami generation than $-mu_x$, except in the vicinity in the compliant prism and in the sloping area near the continental shelf, $-100 < x < 80$ km; much of the region between here and the surface expression of the splay is flat, i.e., $m \approx 0$. Also, as seen for the

Japan Trench, nonhydrostatic ocean response at short wavelengths smooths η_{zg} relative to η_{ts}.

The tsunami in the Cascadia Subduction Zone (Fig. 16) is more complex than that of either of the others, from start to finish. Rather than being dominated by one or two leading peaks traveling in opposite directions, the final tsunami profile is replete with many smaller peaks and troughs, arising from complexities in the seafloor bathymetry as well as dispersion during the subsequent propagation. The initial tsunami profile as determined from a zero-gravity simulation shows at least seven small crests. Closer to the coast, where the fault rupture is deeper, the Tanioka and Satake initial condition η_{ts} closely parallels η_{zg}, but that match breaks down closer to the deformation front, where $x < 45$ km. A closer look at the components of η_{ts} reveals that most of the variations in the tsunami perturbation are due to horizontal deformation, and that vertical seafloor displacement is quite even.

(a)

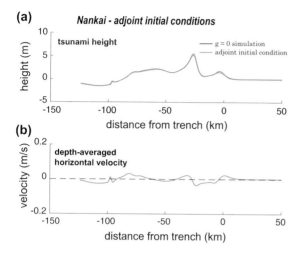

(b)

Figure 19
Initial conditions determined from the adjoint (time-reversed) wave propagation problem for the Nankai Trough. The tsunami initial conditions occur at a time equivalent to $t = 32$ s. **a** Initial tsunami height compared to the final sea surface height of the zero-gravity simulation, η_{zg}. **b** Initial depth-averaged horizontal velocity

(a)

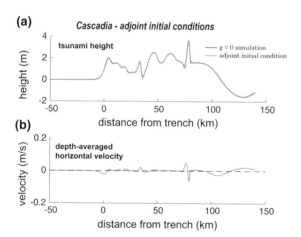

(b)

Figure 20
Initial conditions determined from the adjoint (time-reversed) wave propagation problem for the Cascadia Subduction Zone. The tsunami initial conditions occur at a time equivalent to $t = 30$ s. **a** Initial tsunami height compared to the final sea surface height of the zero-gravity simulation, η_{zg}. **b** Initial depth-averaged horizontal velocity

Had we interpreted a smoother, more simplified version of the seafloor bathymetry, we would not observe such a complex tsunami profile. Some of the peaks in Fig. 16a might not be present in a real Cascadia tsunami due to along-strike variations in slip and seafloor bathymetry. Those variations would produce even more complexity in the tsunami signal,

even though many of the features of our bathymetry profile are fairly coherent along the direction parallel to the coast.

In order to better understand the complexity of the Cascadia tsunami, we run an equivalent simulation using a simple form of the shallow water wave equations linearized about a state of rest, similar to our previous study (Lotto et al. 2017). The equations are written as $\partial\eta/\partial t + \partial(hv)/\partial x = 0$ and $\partial v/\partial t + g\partial\eta/\partial x = 0$, with ocean depth h, sea surface height η, and depth-averaged horizontal velocity v. The water is inviscid and incompressible; advection, nonlinearity, and bottom friction are neglected. The latter processes become relevant near shore and during inundation, but the approximations are justified given our focus on tsunami generation and offshore propagation. The shallow water equations require two initial conditions: one on initial sea surface height $\eta_0(x)$ and another on initial velocity $v_0(x)$. For the initial condition on sea surface height, we use $\eta_0(x) = \eta_{zg}$, the sea surface perturbation for the zero-gravity simulation. We set $v_0(x) = 0$, as is typically assumed, though we revisit the question of appropriate initial conditions in the next section.

Figure 17 shows the propagation of the tsunami using the shallow water wave equations, as well as a comparison of the final "full physics" tsunami of Fig. 16a to the best-matching time step of the shallow water solution. Due to the finite rupture process of the full-physics model, the shallow water solver is initiated at $t = 52$ s; this provides the best match in solutions at $t = 500$ s. We see in Fig. 17a that the two simulations have significant disagreements; even where peaks can be colocated they tend to differ from one another in amplitude. Comparing Figs. 16b and 17b reveals that the differences are largely due to the absence of dispersion in the shallow water model and its presence in the full physics simulation. The finite rupture duration is also responsible for some of the disparities, although its effect is small because the tsunami does not travel very far in the ~ 50 s source process.

9.3. Tsunami Initial Conditions

In this section we take a closer look at the initial conditions of the tsunamis in our study, using an adjoint wavefield procedure developed by Lotto et al. (2017). While there may be several advantages to using fully coupled numerical methods to model earthquakes and tsunamis, shallow water approaches tend to be far more computationally efficient and are therefore widely used by the tsunami modeling community. Still, it is important to justify our assumptions with regard to tsunami initial conditions. Song and collaborators (e.g., Song et al. 2008, 2017) have argued that horizontal momentum transfer from the Earth to the ocean requires the use of a nonzero initial condition on depth-averaged horizontal velocity, v_0, but virtually all other studies assume $v_0 = 0$. In our previous work (Lotto et al. 2017) we demonstrated that $v_0 = 0$ is the optimal initial condition for subduction zone earthquakes, though we largely studied a simplified subduction zone geometry. Here, we extend our previous work but with more realistic subduction zone geometries.

To generate self-consistent initial conditions for the Japan Trench, Nankai, and Cascadia, we begin by running a full-physics simulation to some final time where all seismic and ocean acoustic waves have left the computational domain, leaving only the slower-traveling tsunami. We then use the resulting stress and velocity fields as an initial condition on a time-reversed adjoint simulation, which involves simply flipping the sign of the particle velocity. We continue to use absorbing boundaries on the sides and bottom of the computational domain, and the waves that have left the domain are not re-injected in the adjoint problem. The governing equations are self-adjoint, so running it backward in time requires no additional changes to the code. The tsunami waves propagate backward in time toward the center of the domain where they converge above the earthquake source region, unperturbed by seismic or acoustic waves as they were in the forward simulation. We choose as the initial condition the point when sea surface height most closely matches η_{zg} from the zero-gravity simulation. To determine v_0, the initial condition on horizontal velocity, we take the average of horizontal particle velocity over the ocean depth.

For each of the three subduction zones (Figs. 18, 19 and 20), we can find a time step in the time-reversed simulation that produces a tsunami closely matching η_{zg}. We observe that at the determined

adjoint initial conditions, the tsunami from each subduction zone is associated with a small but nonzero horizontal velocity. The peaks in horizontal velocity may be loosely correlated with bathymetric features, although that relationship is difficult to distinguish. We used the results of the Cascadia adjoint simulation (with small, but nonzero v_0) as initial conditions for a shallow water simulation, as in Fig. 17 which has $v_0 = 0$, and found only negligible differences between the two shallow water simulations. Thus, we conclude that the contribution of horizontal velocity to tsunami height is inconsequential.

10. Conclusions

Though most earthquake rupture models consider subduction zone geometries and structures that are idealized to a greater or lesser extent, there is much to be gleaned from making realistic choices about material properties and seafloor bathymetry.

In the Japan Trench, we have a classic subduction zone geometry, with a plate boundary fault dipping from the seafloor through a layer of sediments and a relatively small sedimentary prism all below a deep ocean layer. The presence of the compliant prism leads to enhanced slip and seafloor deformation near the trench, but this causes only a slight increase in tsunami height, given the small width of the prism (~ 20 km) relative to wavelengths filtered by the nonhydrostatic response of the ocean ($< \sim 40$ km).

The presence of a prominent splay fault and a large and compliant sedimentary prism makes it hard to predict rupture pathway in the Nankai Trough. Some evidence suggests that rupture has occurred on the splay, and indeed all of our simulations show at least 10 m of slip at the splay fault tip. The magnitude of slip on the décollement is highly variable. Simulations with shallow highly velocity-strengthening friction result in almost zero décollement slip, whereas most others have slip on both fault segments. The splay poses the greater near-field hazard; motion on the splay more efficiently generates tsunami waves and those waves emanate from locations closer to shore, meaning that they will have the first and often greatest impact on human settlements.

The lack of a trench-breaking fault in our Cascadia model causes earthquakes and tsunamis to behave quite differently. Cascadia's thick sediment layer and gradually dipping plate boundary thrust, along with its rough bathymetry, lead to a complex tsunami signal. To the extent that its bathymetric ridges are coherent along strike we expect them to play a major role in tsunami generation, via the second term of Eq. (1). Even if they are not coherent along-strike, they will contribute to tsunami generation, but in an incoherent manner. Since the recent historic record includes no significant interplate seismicity in Cascadia, it is difficult to constrain the behavior of an tsunamigenic earthquake here. In a real event, thrust faults verging up from the décollement may be activated during a megathrust rupture, which could alter the pattern of shallow slip and thus the tsunami source.

In addition to focusing on individual subduction zones, we can make some general conclusions about the influence of friction, stress, and material structure on the earthquake rupture and tsunami generation process. Increases in prestress—including those due to decreases in pore pressure λ—consistently lead to increased fault displacement, seafloor motion, and tsunami height. The value of the rate-and-state parameter $b - a$ also has a major influence on shallow slip and tsunami heights. Not surprisingly, more velocity-strengthening friction usually leads to less slip, although in the Nankai case that relationship is complicated by the fact that changes in friction can lead the rupture to take entirely different pathways. Figure 10 demonstrates the effect of using realistic structural models on rupture pathway and tsunami height. Structural models with uniform material properties and models that only crudely employ compliant sediments tend to underestimate tsunami amplitudes.

Simulating realistic subduction models allows us to gain insights about tsunami physics and initial conditions for tsunami models decoupled from earthquake rupture. Running zero-gravity simulations allows us to get a sense of initial tsunami height without interference from acoustic and seismic waves. By comparing sea surface profiles from those zero-gravity simulations, η_{zg}, to the components of the Tanioka and Satake tsunami initial condition, η_{ts},

we observe that vertical and horizontal components of seafloor motion both contribute significantly to tsunami height. Neglecting contributions from horizontal seafloor displacement would cause one to underestimate tsunami height and qualitatively misrepresent the shape of the tsunami waveform.

But simply using η_{ts} as a tsunami initial condition would not be appropriate; the ocean's nonhydrostatic response smooths out the short wavelength variations in η_{ts}. This filtering effect, first described by Kajiura (1963), accounts for the difference between η_{ts} and η_{zg} in Figs. 14, 15, and 16. Ignoring the nonhydrostatic correction to η_{ts} would produce overly sharp peaks in the tsunami waveform.

What accounts for the differences between the tsunami of our full-physics model and that of the shallow water wave model in Fig. 17? Space–time plots of sea surface height reveal that dispersion is responsible for much of the mismatches in the two profiles. To a lesser extent, the time-dependent rupture process of the full-physics model contributes disagreement, although in the cases studied it is acceptable to assume the tsunami does not travel sufficiently far over the tsunami source duration to require a time-dependent initial condition. The compressibility of the ocean in the full-physics model also has a negligible effect on tsunami propagation, echoing conclusions of much earlier studies (Sells 1965). We also use an adjoint wave propagation method to run our full simulation backward in time, to produce initial tsunami conditions. The adjoint simulations demonstrate that, just as for the simplified geometry of our previous study (Lotto et al. 2017), depth-averaged horizontal velocity in the ocean has a very small amplitude and hence a negligible effect on tsunami height for realistic subduction zone geometries.

Fully coupled earthquake and tsunami codes can be difficult to implement and are more computationally expensive than methods that decouple the tsunami from its source process. However, coupled earthquake and tsunami simulations enable us to understand the details of subduction zone megathrust behavior and answer key questions about the physics of tsunami generation.

References

Aldam, M., Xu, S., Brener, E. A., Ben Zion, Y., & Bouchbinder, E. (2017). Nonmonotonicity of the frictional bimaterial effect. *Journal of Geophysical Research: Solid Earth*, *122*(10), 8270–8284.

Ando, M. (1975). Source mechanisms and tectonic significance of historical earthquakes along the Nankai Trough, Japan. *Tectonophysics*, *27*(2), 119–140.

Andrews, D. J., & Ben Zion, Y. (1997). Wrinkle like slip pulse on a fault between different materials. *Journal of Geophysical Research*, *102*, 553–571.

Audet, P., Bostock, M. G., Christensen, N. I., & Peacock, S. M. (2009). Seismic evidence for overpressured subducted oceanic crust and megathrust fault sealing. *Nature*, *457*(7225), 76.

Benson, B. E., Atwater, B. F., Yamaguchi, D. K., Amidon, L. J., Brown, S. L., & Lewis, R. C. (2001). Renewal of tidal forests in Washington State after a subduction earthquake in AD 1700. *Quaternary Research*, *56*(2), 139–147.

Bilek, S. L., & Lay, T. (1999). Rigidity variations with depth along interplate megathrust faults in subduction zones. *Nature*, *400*(6743), 443.

Blanpied, M. L., Lockner, D. A., & Byerlee, J. D. (1995). Frictional slip of granite at hydrothermal conditions. *Journal of Geophysical Research: Solid Earth*, *100*(B7), 13045–13064.

Briggs, R. W., Sieh, K., Meltzner, A. J., Natawidjaja, D., Galetzka, J., Suwargadi, B., et al. (2006). Deformation and slip along the Sunda megathrust in the great 2005 Nias-Simeulue earthquake. *Science*, *311*(5769), 1897–1901.

Chester, F. M., Rowe, C., Ujiie, K., Kirkpatrick, J., Regalla, C., Francesca Remitti, J., et al. (2013). Structure and composition of the plate-boundary slip zone for the 2011 Tohoku-Oki earthquake. *Science*, *342*(6163), 1208–1211.

Cocco, M., & Rice, J. R. (2002). Pore pressure and poroelasticity effects in Coulomb stress analysis of earthquake interactions. *Journal of Geophysical Research: Solid Earth*, *107*(B2), 2.

Cummins, P. R., & Kaneda, Y. (2000). Possible splay fault slip during the 1946 Nankai earthquake. *Geophysical Research Letters*, *27*(17), 2725–2728.

DeDontney, N., & Hubbard, J. (2012). Applying wedge theory to dynamic rupture modeling of fault junctions. *Bulletin of the Seismological Society of America*, *102*(4), 1693–1711.

Fleming, S. W., & Trehu, A. M. (1999). Crustal structure beneath the central Oregon convergent margin from potential field modeling: Evidence for a buried basement ridge in local contact with a seaward dipping backstop. *Journal of Geophysical Research: Solid Earth*, *104*(B9), 20431–20447.

Fluck, P., Hyndman, R. D., & Wang, K. (1997). Three dimensional dislocation model for great earthquakes of the Cascadia subduction zone. *Journal of Geophysical Research: Solid Earth*, *102*(B9), 20539–20550.

Flueh, E. R., Fisher, M. A., Bialas, J., Childs, J. R., Klaeschen, D., Kukowski, N., et al. (1998). New seismic images of the Cascadia subduction zone from cruise SO108–ORWELL. *Tectonophysics*, *293*(1), 69–84.

Fujii, Y., Satake, K., Sakai, S., Shinohara, M., & Kanazawa, T. (2011). Tsunami source of the 2011 off the Pacific coast of Tohoku Earthquake. *Earth, Planets and Space*, *63*(7), 55.

Fujiwara, T., Kodaira, S., Kaiho, Y., Takahashi, N., & Kaneda, Y. (2011). The 2011 Tohoku-Oki earthquake: Displacement reaching the trench axis. *Science*, *334*(6060), 1240–1240.

Fulton, P. M., Brodsky, E. E., Kano, Y., Mori, J., Chester, F., Ishikawa, T., et al. (2013). Low coseismic friction on the Tohoku-Oki fault determined from temperature measurements. *Science*, *342*(6163), 1214–1217.

Gettemy, G. L., & Tobin, H. J. (2003). Tectonic signatures in centimeter scale velocity-porosity relationships of Costa Rica convergent margin sediments. *Journal of Geophysical Research: Solid Earth*, *108*(B10), 2494.

Chris Goldfinger, C., Nelson, H., Morey, A. E., Johnson, J. E., Patton, J. R., Karabanov, E., et al. (2012). Turbidite event history: Methods and implications for Holocene paleoseismicity of the Cascadia subduction zone. *US Geological Survey Professional Paper*, *1661*, 170.

Guilbault, J.-P., Clague, J. J., & Lapointe, M. (1996). Foraminiferal evidence for the amount of coseismic subsidence during a late Holocene earthquake on Vancouver Island, west coast of Canada. *Quaternary Science Reviews*, *15*(8–9), 913–937.

Gulick, S. P., Austin, J. A, Jr., McNeill, L. C., Bangs, N. L., Martin, K. M., Henstock, T. J., et al. (2011). Updip rupture of the 2004 Sumatra earthquake extended by thick indurated sediments. *Nature Geoscience*, *4*(7), 453.

Hirono, T., Tsuda, K., Tanikawa, W., Ampuero, J. P., Shibazaki, B., Kinoshita, M., et al. (2016). Near-trench slip potential of megaquakes evaluated from fault properties and conditions. *Scientific reports*, *6*, 28184.

Holbrook, W. S., Kent, G., Keranen, K., Johnson, H. P., Trehu, A., Tobin, H., et al. (2012). Cascadia fore arc seismic survey: Open-access data available. *Eos, Transactions American Geophysical Union*, *93*(50), 521–522.

Hower, J., Eslinger, E. V., Hower, M. E., & Perry, E. A. (1976). Mechanism of burial metamorphism of argillaceous sediment: 1. Mineralogical and chemical evidence. *Geological Society of America Bulletin*, *87*(5), 725–737.

Hubbert, M. K., & Rubey, W. W. (1959). Role of fluid pressure in mechanics of overthrust faulting: I. Mechanics of fluid-filled porous solids and its application to overthrust faulting. *Geological Society of America Bulletin*, *70*(2), 115–166.

Hyndman, R. D., & Wang, K. (1993). Thermal constraints on the zone of major thrust earthquake failure: The Cascadia subduction zone. *Journal of Geophysical Research: Solid Earth*, *98*(B2), 2039–2060.

Hyndman, R. D., Yamano, M., & Oleskevich, D. A. (1997). The seismogenic zone of subduction thrust faults. *Island Arc*, *6*(3), 244–260.

Ikari, M. J., Kameda, J., Saffer, D. M., & Kopf, A. J. (2015). Strength characteristics of Japan Trench borehole samples in the high-slip region of the 2011 Tohoku–Oki earthquake. *Earth and Planetary Science Letters*, *412*, 35–41.

Ikari, M. J., & Saffer, D. M. (2011). Comparison of frictional strength and velocity dependence between fault zones in the Nankai accretionary complex. *Geochemistry, Geophysics, Geosystems, 12*(4), Q0AD11.

Jennings, S., & Thompson, G. R. (1986). Diagenesis of Plio–Pleistocene sediments of the Colorado River delta, southern California. *Journal of Sedimentary Research*, *56*(1), 89–98.

Jeppson, T. N., Tobin, H. J., & Hashimoto, Y. (2018). Laboratory measurements quantifying elastic properties of accretionary wedge sediments: Implications for slip to the trench during the

2011 Mw 9.0 Tohoku-Oki earthquake. *Geosphere, 14*(4), 1411–1424.

Kajiura, K. (1963). The leading wave of a tsunami. *Bulletin of the Earthequake reserch Institute, 43*, 535–571.

Kame, N., Rice, J. R., & Dmowska, R. (2003). Effects of prestress state and rupture velocity on dynamic fault branching. *Journal of Geophysical Research: Solid Earth, 108*(B5), 2265.

Kamei, R., Pratt, R. G., & Tsuji, T. (2012). Waveform tomography imaging of a megasplay fault system in the seismogenic Nankai subduction zone. *Earth and Planetary Science Letters, 317*, 343–353.

Kanamori, H., & Kikuchi, M. (1993). The 1992 Nicaragua earthquake: a slow tsunami earthquake associated with subducted sediments. *Nature, 361*(6414), 714–716.

Kemp, A. C., Cahill, N., Engelhart, S. E., Hawkes, A. D., & Wang, K. (2018). Revising estimates of spatially variable subsidence during the AD 1700 Cascadia earthquake using a Bayesian foraminiferal transfer function. *Bulletin of the Seismological Society of America, 108*(2), 654–673.

Kikuchi, M., Nakamura, M., & Yoshikawa, K. (2003). Source rupture processes of the 1944 Tonankai earthquake and the 1945 Mikawa earthquake derived from low-gain seismograms. *Earth, Planets and Space, 55*(4), 159–172.

Kimura, G., Hina, S., Hamada, Y., Kameda, J., Tsuji, T., Kinoshita, M., et al. (2012). Runaway slip to the trench due to rupture of highly pressurized megathrust beneath the middle trench slope: The tsunamigenesis of the 2011 Tohoku earthquake off the east coast of northern Japan. *Earth and Planetary Science Letters, 339*, 32–45.

Kodaira, S., No, T., Nakamura, Y., Fujiwara, T., Kaiho, Y., Miura, S., et al. (2012). Coseismic fault rupture at the trench axis during the 2011 Tohoku-Oki earthquake. *Nature Geoscience, 5*(9), 646–650.

Kopf, A., & Brown, K. M. (2003). Friction experiments on saturated sediments and their implications for the stress state of the Nankai and Barbados subduction thrusts. *Marine Geology, 202*(3), 193–210.

Kopp, H., & Kukowski, N. (2003). Backstop geometry and accretionary mechanics of the Sunda margin. *Tectonics, 22*(6), 1072.

Kozdon, J. E., & Dunham, E. M. (2013). Rupture to the trench: Dynamic rupture simulations of the 11 March 2011 Tohoku earthquake. *Bulletin of the Seismological Society of America, 103*(2B), 1275–1289.

Kozdon, J. E., & Dunham, E. M. (2014). Constraining shallow slip and tsunami excitation in megathrust ruptures using seismic and ocean acoustic waves recorded on ocean-bottom sensor networks. *Earth and Planetary Science Letters, 396*, 56–65.

Kozdon, J. E., Dunham, E. M., & Nordstrom, J. (2013). Simulation of dynamic earthquake ruptures in complex geometries using high-order finite difference methods. *Journal of Scientific Computing, 55*(1), 92–124.

Lay, T., Ammon, C. J., Kanamori, H., Yamazaki, Y., Cheung, K. F., & Hutko, A. R. (2011). The 25 October 2010 Mentawai tsunami earthquake (Mw 7.8) and the tsunami hazard presented by shallow megathrust ruptures. *Geophysical Research Letters, 38*(6).

Leonard, L. J., Currie, C. A., Mazzotti, S., & Hyndman, R. D. (2010). Rupture area and displacement of past Cascadia great earthquakes from coastal coseismic subsidence. *Bulletin, 122*(11–12), 2079–2096.

Lotto, G. C., & Dunham, E. M. (2015). High-order finite difference modeling of tsunami generation in a compressible ocean from offshore earthquakes. *Computational Geosciences, 19*(2), 327–340.

Lotto, G. C., Dunham, E. M., Jeppson, T. N., & Tobin, H. J. (2017). The effect of compliant prisms on subduction zone earthquakes and tsunamis. *Earth and Planetary Science Letters, 458*, 213–222.

Lotto, G. C., Nava, G., & Dunham, E. M. (2017). Should tsunami simulations include a nonzero initial horizontal velocity? *Earth, Planets and Space, 69*(1), 117.

Ma, S. (2012). A self-consistent mechanism for slow dynamic deformation and tsunami generation for earthquakes in the shallow subduction zone. *Geophysical Research Letters, 39*(11), L11310

Ma, S., & Beroza, G. C. (2008). Rupture dynamics on a bimaterial interface for dipping faults. *Bulletin of the Seismological Society of America, 98*(4), 1642–1658.

Maeda, T., & Furumura, T. (2013). FDM simulation of seismic waves, ocean acoustic waves, and tsunamis based on tsunami-coupled equations of motion. *Pure and Applied Geophysics, 170*(1–2), 109–127.

McCaffrey, R. (1997). Influences of recurrence times and fault zone temperatures on the age-rate dependence of subduction zone seismicity. *Journal of Geophysical Research: Solid Earth, 102*(B10), 22839–22854.

McCrory, P. A., Blair, J. L., Waldhauser, F., & Oppenheimer, D. H. (2012). Juan de Fuca slab geometry and its relation to Wadati–Benioff zone seismicity. *Journal of Geophysical Research: Solid Earth, 117*(B9), B09306.

Mitchell, E. K., Fialko, Y., & Brown, K. M. (2015). Frictional properties of gabbro at conditions corresponding to slow slip events in subduction zones. *Geochemistry, Geophysics, Geosystems, 16*(11), 4006–4020.

Miura, S., Takahashi, N., Nakanishi, A., Tsuru, T., Kodaira, S., & Kaneda, Y. (2005). Structural characteristics off Miyagi forearc region, the Japan Trench seismogenic zone, deduced from a wide-angle reflection and refraction study. *Tectonophysics, 407*(3), 165–188.

Moore, G. F., Bangs, N. L., Taira, A., Kuramoto, S., Pangborn, E., & Tobin, H. J. (2007). Three-dimensional splay fault geometry and implications for tsunami generation. *Science, 318*(5853), 1128–1131.

Moore, J. C., & Vrolijk, P. (1992). Fluids in accretionary prisms. *Reviews of Geophysics, 30*(2), 113–135.

Mori, J., Chester, F. M., Eguchi, N., & Toczko, S. (2012). Japan Trench Fast Earthquake Drilling Project (JFAST). IODP Sci. Prosp, 343(10.2204).

Nakamura, Y., Kodaira, S., Cook, B. J., Jeppson, T., Kasaya, T., Yamamoto, Y., et al. (2014). Seismic imaging and velocity structure around the JFAST drill site in the Japan Trench: low V p, high V p/V s in the transparent frontal prism. *Earth, Planets and Space, 66*(1), 121.

Nakanishi, A., Takahashi, N., Park, J. O., Miura, S., Kodaira, S., Kaneda, Y., et al. (2002). Crustal structure across the coseismic rupture zone of the 1944 Tonankai earthquake, the central Nankai Trough seismogenic zone. *Journal of Geophysical Research: Solid Earth, 107*(B1), EPM-2.

Nelson, A. R., Atwater, B. F., Bobrowsky, P. T., Bradley, L.-A., Clague, J. J., Carver, G. A., et al. (1995). Radiocarbon evidence for extensive plate-boundary rupture about 300 years ago at the Cascadia subduction zone. *Nature, 378*(6555), 371.

Oleskevich, D. A., Hyndman, R. D., & Wang, K. (1999). The updip and downdip limits to great subduction earthquakes: Thermal and structural models of Cascadia, south Alaska, SW Japan, and Chile. *Journal of Geophysical Research: Solid Earth, 104*(B7), 14965–14991.

Ozawa, S., Nishimura, T., Suito, H., Kobayashi, T., Tobita, M., & Imakiire, T. (2011). Coseismic and postseismic slip of the 2011 magnitude-9 Tohoku-Oki earthquake. *Nature, 475*(7356), 373.

Park, J. O., Tsuru, T., Kodaira, S., Cummins, P. R., & Kaneda, Y. (2002a). Splay fault branching along the Nankai subduction zone. *Science, 297*(5584), 1157–1160.

Park, J. O., Tsuru, T., Takahashi, N., Hori, T., Kodaira, S., Nakanishi, A., & Kaneda, Y. (2002b). A deep strong reflector in the Nankai accretionary wedge from multichannel seismic data: Implications for underplating and interseismic shear stress release. *Journal of Geophysical Research: Solid Earth, 107*(B4), 2061.

Peacock, S. M., Christensen, N. I., Bostock, M. G., & Audet, P. (2011). High pore pressures and porosity at 35 km depth in the Cascadia subduction zone. *Geology, 39*(5), 471–474.

Plafker, G. (1972). Alaskan earthquake of 1964 and Chilean earthquake of 1960: Implications for arc tectonics. *Journal of Geophysical Research, 77*(5), 901–925.

Polet, J., & Kanamori, H. (2000). Shallow subduction zone earthquakes and their tsunamigenic potential. *Geophysical Journal International, 142*(3), 684–702.

Raimbourg, H., Hamano, Y., Saito, S., Kinoshita, M., & Kopf, A. (2011). Acoustic and mechanical properties of Nankai accretionary prism core samples. *Geochemistry, Geophysics, Geosystems, 12*(4), Q0AD10.

Rice, J. R. (1992). Fault stress states, pore pressure distributions, and the weakness of the San Andreas fault. In *International geophysics* (Vol. 51, pp. 475–503). Academic Press.

Saffer, D. M., & Marone, C. (2003). Comparison of smectite-and illite-rich gouge frictional properties: Application to the updip limit of the seismogenic zone along subduction megathrusts. *Earth and Planetary Science Letters, 215*(1), 219–235.

Saffer, D. M., & Tobin, H. J. (2011). Hydrogeology and mechanics of subduction zone forearcs: Fluid flow and pore pressure. *Annual Review of Earth and Planetary Sciences, 39*, 157–186.

Sagiya, T., & Thatcher, W. (1999). Coseismic slip resolution along a plate boundary megathrust: The Nankai Trough, southwest Japan. *Journal of Geophysical Research: Solid Earth, 104*(B1), 1111–1129.

Saito, T., & Tsushima, H. (2016). Synthesizing ocean bottom pressure records including seismic wave and tsunami contributions: Toward realistic tests of monitoring systems. *Journal of Geophysical Research: Solid Earth, 121*(11), 8175–8195.

Sakaguchi, A., Chester, F., Curewitz, D., Fabbri, O., Goldsby, D., Kimura, G., et al. (2011). Seismic slip propagation to the updip end of plate boundary subduction interface faults: Vitrinite reflectance geothermometry on Integrated Ocean Drilling Program NanTro SEIZE cores. *Geology, 39*(4), 395–398.

Satake, K. (1994). Mechanism of the 1992 Nicaragua tsunami earthquake. *Geophysical Research Letters, 21*(23), 2519–2522.

Satake, K., Shimazaki, K., Tsuji, Y., & Ueda, K. (1996). Time and size of a giant earthquake in Cascadia inferred from Japanese tsunami records of January 1700. *Nature, 379*(6562), 246–249.

Satake, K., Wang, K., & Atwater, B. F. (2003). Fault slip and seismic moment of the 1700 Cascadia earthquake inferred from Japanese tsunami descriptions. *Journal of Geophysical Research: Solid Earth, 108*(B11), 2535.

Sato, M., Ishikawa, T., Ujihara, N., Yoshida, S., Fujita, M., Mochizuki, M., et al. (2011). Displacement above the hypocenter of the 2011 Tohoku–Oki earthquake. *Science, 332*(6036), 1395–1395.

Sawai, M., Niemeijer, A. R., Hirose, T., & Spiers, C. J. (2017). Frictional properties of JFAST core samples and implications for slow earthquakes at the Tohoku subduction zone. *Geophysical Research Letters, 44*(17), 8822–8831.

Sawai, M., Niemeijer, A. R., Plumper, O., Hirose, T., & Spiers, C. J. (2016). Nucleation of frictional instability caused by fluid pressurization in subducted blueschist. *Geophysical Research Letters, 43*(6), 2543–2551.

Sells, C. L. (1965). The effect of a sudden change of shape of the bottom of a slightly compressible ocean. *Philosophical Transactions of the Royal Society of London, 258*(1092), 495–528.

Seno, T. (2009). Determination of the pore fluid pressure ratio at seismogenic megathrusts in subduction zones: Implications for strength of asperities and Andean-type mountain building. *Journal of Geophysical Research: Solid Earth, 114*(B5), B05405.

Shennan, I., Long, A. J., Rutherford, M. M., Green, F. M., Innes, J. B., Lloyd, J. M., et al. (1996). Tidal marsh stratigraphy, sea-level change and large earthquakes, I: A 5000 year record in Washington, USA. *Quaternary Science Reviews, 15*(10), 1023–1059.

Shipley, T. H., McIntosh, K. D., Silver, E. A., & Stoffa, P. L. (1992). Three-dimensional seismic imaging of the Costa Rica accretionary prism: Structural diversity in a small volume of the lower slope. *Journal of Geophysical Research: Solid Earth, 97*(B4), 4439–4459.

Skarbek, R. M., & Saffer, D. M. (2009). Pore pressure development beneath the decollement at the Nankai subduction zone: Implications for plate boundary fault strength and sediment dewatering. *Journal of Geophysical Research: Solid Earth, 114*(B7), B07401.

Song, Y. T., Fu, L.-L., Zlotnicki, V., Ji, C., Hjorleifsdottir, V., Shum, C. K., et al. (2008). The role of horizontal impulses of the faulting continental slope in generating the 26 December 2004 tsunami. *Ocean Modelling, 20*(4), 362–379.

Song, Y. T., Mohtat, A., & Yim, S. C. (2017). New insights on tsunami genesis and energy source. *Journal of Geophysical Research: Oceans, 122*(5), 4238–4256.

Stein, S., & Okal, E. A. (2005). Seismology: Speed and size of the Sumatra earthquake. *Nature, 434*(7033), 581.

Tamura, S., & Ide, S. (2011). Numerical study of splay faults in subduction zones: The effects of bimaterial interface and free surface. *Journal of Geophysical Research: Solid Earth, 116*(B10), B10309.

Tanioka, Y., & Satake, K. (1996). Tsunami generation by horizontal displacement of ocean bottom. *Geophysical Research Letters, 23*(8), 861–864.

Tanioka, Y., & Satake, K. (1996). Fault parameters of the 1896 Sanriku tsunami earthquake estimated from tsunami numerical modeling. *Geophysical Research Letters, 23*(13), 1549–1552.

Tanioka, Y., & Satake, K. (2001). Coseismic slip distribution of the 1946 Nankai earthquake and aseismic slips caused by the earthquake. *Earth, Planets and Space, 53*(4), 235–241.

Tobin, H. J., & Moore, J. C. (1997). Variations in ultrasonic velocity and density with pore pressure in the decollement zone, northern Barbados Ridge accretionary prism. In *Proceedings of the Ocean Drilling Program. Scientific results* (pp. 125–136). National Science Foundation.

Tobin, H. J., Moore, J. C., & Moore, G. F. (1995). Laboratory measurement of velocity vs. effective stress in thrust faults of the Oregon accretionary prism: Implications for fault zone overpressure. In *Proceedings of the Ocean Drilling Program. Scientific results* (Vol. 146, pp. 349–358). Ocean Drilling Program.

Tobin, H. J., & Saffer, D. M. (2009). Elevated fluid pressure and extreme mechanical weakness of a plate boundary thrust. Nankai Trough subduction zone. *Geology, 37*(8), 679–682.

Tse, S. T., & Rice, J. R. (1986). Crustal earthquake instability in relation to the depth variation of frictional slip properties. *Journal of Geophysical Research: Solid Earth, 91*(B9), 9452–9472.

Tsuji, T., Kamei, R., & Pratt, R. G. (2014). Pore pressure distribution of a mega-splay fault system in the Nankai Trough subduction zone: Insight into up-dip extent of the seismogenic zone. *Earth and Planetary Science Letters, 396*, 165–178.

Ujiie, K., Tanaka, H., Saito, T., Tsutsumi, A., Mori, J. J., Kameda, J., et al. (2013). Low coseismic shear stress on the Tohoku–Oki megathrust determined from laboratory experiments. *Science, 342*(6163), 1211–1214.

Von Huene, R., Ranero, C. R., & Scholl, D. W. (2009). Convergent margin structure in high-quality geophysical images and current kinematic and dynamic models. In *Subduction Zone Geodynamics* (pp. 137–157). Springer.

Wang, C. (1980). Sediment subduction and frictional sliding in a subduction zone. *Geology, 8*(11), 530–533.

Wang, K., & Hu, Y. (2006). Accretionary prisms in subduction earthquake cycles: The theory of dynamic Coulomb wedge. *Journal of Geophysical Research: Solid Earth, 111*(B6), B06410.

Wang, K., Mulder, T., Rogers, G. C., & Hyndman, R. D. (1995). Case for very low coupling stress on the Cascadia Ssubduction Fault. *Journal of Geophysical Research: Solid Earth, 100*(B7), 12907–12918.

Wang, K., & Trehu, A. M. (2016). Invited review paper: Some outstanding issues in the study of great megathrust earthquakes—The Cascadia example. *Journal of Geodynamics, 98*, 1–18.

Wang, K., Wells, R., Mazzotti, S., Hyndman, R. D., & Sagiya, T. (2003). A revised dislocation model of interseismic deformation of the Cascadia subduction zone. *Journal of Geophysical Research: Solid Earth, 108*(B1).

Wang, P. L., Engelhart, S. E., Wang, K., Hawkes, A. D., Horton, B. P., Nelson, A. R., et al. (2013). Heterogeneous rupture in the great Cascadia earthquake of 1700 inferred from coastal subsidence estimates. *Journal of Geophysical Research: Solid Earth, 118*(5), 2460–2473.

Webb, S. I. (2017). *Interaction of structure and physical properties in accretionary wedges: examples from the Cascadia and Nankai Trough subduction zones.* PhD thesis, University of Wisconsin, Madison.

Weertman, J. (1980). Unstable slippage across a fault that separates elastic media of different elastic constants. *Journal of Geophysical Research: Solid Earth, 85*(B3), 1455–1461.

Wendt, J., Oglesby, D. D., & Geist, E. L. (2009). Tsunamis and splay fault dynamics. *Geophysical Research Letters, 36*(15), L15303.

Witter, R. C., Kelsey, H. M., & Hemphill-Haley, E. (2003). Great Cascadia earthquakes and tsunamis of the past 6700 years, Coquille River estuary, southern coastal Oregon. *Geological Society of America Bulletin, 115*(10), 1289–1306.

(Received May 11, 2018, revised August 17, 2018, accepted September 6, 2018, Published online September 27, 2018)

Pure Appl. Geophys. 176 (2019), 4043–4068
© 2019 The Author(s)
https://doi.org/10.1007/s00024-019-02250-z

A Secondary Zone of Uplift Due to Megathrust Earthquakes

Ylona van Dinther,[1,2] 🔟 Lukas E. Preiswerk,[1,3,4] and Taras V. Gerya[4]

Abstract—The 1960 M9.5 Valdivia and 1964 M9.2 Alaska earthquakes caused a decimeters high secondary zone of uplift a few hundred kilometers landward of the trench. We analyze GPS data from the 2010 M8.8 Maule and 2011 M9.0 Tohoku-Oki earthquakes to reveal the persistent existence of a secondary zone of uplift due to great earthquakes at the megathrust interface. This uplift varies in magnitude and location, but consistently occurs at a few hundred kilometers landward from the trench and is likely mainly coseismic in nature. This secondary zone of uplift is systematically predicted by our 2D visco-elasto-plastic seismo-thermo-mechanical numerical simulations, which model both geodynamic and seismic cycle timescales. Through testing hypotheses in both simple and realistic setups, we propose that a superposition of two physical mechanisms could be responsible for this phenomenon. First, a wavelength is introduced through elastic buckling of a visco-elastically layered fore-arc that is horizontally compressed in the interseismic period. The consequent secondary zone of interseismic subsidence is elastically rebound during the earthquake into a secondary zone of relative uplift. Second, absolute and broader uplift is ensured through a mass conservation-driven return flow following accelerated slab penetration due to the megathrust earthquake. The dip and width of the seismogenic zone and resulting (deep) coseismic slip seem to have the largest affect on location and amplitude of the secondary zone of uplift. These results imply that subduction and mantle flow do not occur at constant rates, but are rather modulated by earthquakes. This suggests a link between deep mantle and shallow surface displacements even at time scales of minutes.

Key words: Subduction zone processes, earthquakes, numerical modeling, crustal deformation, geodesy.

Electronic supplementary material The online version of this article (https://doi.org/10.1007/s00024-019-02250-z) contains supplementary material, which is available to authorized users.

[1] Seismology and Wave Physics, Institute of Geophysics, ETH Zürich, Zurich, Switzerland. E-mail: y.vandinther@uu.nl

[2] Department of Earth Sciences, Utrecht University, Utrecht, The Netherlands.

[3] Laboratory of Hydraulics, Hydrology and Glaciology (VAW), ETH Zürich, Zurich, Switzerland.

[4] Geophysical Fluid Dynamics, Institute of Geophysics, ETH Zürich, Zurich, Switzerland.

1. Introduction

The first-order interseismic and coseismic surface displacements of the overriding plate in subduction zones are reasonably well understood (e.g., Wang 2007; Govers et al. 2018). In the interseismic period the overriding plate is coupled to the subducting plate along the seismogenic zone. Subduction thus drags the overriding plate down and landward. This causes subsidence from the trench throughout a large part of the seismogenic zone and interseismic compression causes uplift beyond. This uplift slowly tappers to zero in the far field. In the coseismic period surface displacements typically show elastic rebound of these interseismically accumulated displacements (e.g., Reid 1910; Moreno et al. 2010). This leads to strong uplift of the wedge as the overriding plate slips seaward, while the coastal regions typically located above the hypocenter manifest subsidence. Again farther land inward, standard (visco-)elastic models models show zero vertical displacements (e.g., Wang 2007; Meltzner et al. 2006).

This classical conceptual model is contrasted by two great megathrust earthquakes (i.e., $M_w > 8.5$) in the 1960s in Chile and Alaska, where a distinct secondary zone of uplift was measured landward of the hypocenter (Plafker 1969; Plafker and Savage 1970). However, these static measurements were made long ago and many years after the earthquakes, such that seperating coseismic and early postseismic contributions was not possible. Up to now a secondary zone of uplift is thus not considered as a persistent feature. The classical interpretation of a very gradual tapering to zero uplift is also contrasted by more recent seismo-thermo-mechanical (STM) models, which predict the presence of a secondary zone of uplift (van Dinther et al. 2013b). These models self-consistently simulate both subduction dynamics and seismogenesis in a setup where visco-elastic structure is governed by conservation laws and a visco-elasto-plastic rheology based on laboratory experiments.

Following the two great megathrust earthquakes in the 1960s several papers identified specific settings and physical mechanisms that would allow for a secondary zone of uplift. For the 1960 M9.5 Valdivia earthquake, Plafker and Savage (1970) reproduced the secondary zone of uplift by introducing a down-ward curving fault that steepens suddenly. Linde and Silver (1989) reanalyzed the same dataset and suggested that slip must have also occurred until depths up to 65–80 km, while a strong kink in the interface was required below the peak of the secondary bulge to reproduce this feature. Vita-Finzi and Mann (1994) explained the deformation pattern in Valdivia by elastic flexure of a continuous elastic beam following displacements of mass and resulting buoyancy effects. For the 1964 M9.2 Alaska earthquake, Plafker (1969) speculated that it could be caused by a sudden increase in horizontal compressional strain and termed it a 'Poisson bulge', while noting this feature as a major unresolved problem. Alternatively, Plafker (1972) shortly postulated a hypothesis of transverse crustal buckling resulting from horizontal compression of the continental plate.

The occurrence of great megathrust earthquakes in the last decade allowed for major advances in understanding surface displacements (e.g., Kido et al. 2011; Fujiwara et al. 2011; Wang et al. 2012; Sun et al. 2014; Klein et al. 2016; Govers et al. 2018). However, models aiming to reproduce the displacements during the 2010 and 2011 megathrust earthquakes do typically not reproduce a secondary zone of uplift (e.g., Govers et al. 2018). Interestingly some models do reproduce a similar uplift, but secondary zones of uplift are not described as such (Miyashita 1987) or are produced by postseismic mechanisms only (Klein et al. 2016; Li et al. 2017).

This literature overview shows that there is no consensus on whether a secondary zone of uplift is a universal physical phenomenon. Additionally, there is no consensus on the physical mechanisms responsible for such a secondary zone of uplift. Through re-analyzing high-quality data from the last decade and dedicated numerical models (Sect. 2) we aim to understand whether the classical conceptual model of

surface displacements should be extended with a secondary zone of uplift. Our analysis of published data for four great megathrust earthquakes confirms the existence of a secondary zone of uplift (Sect. 3.1). We then study STM models of different degrees of complexity to propose two physical mechanisms working together to form a secondary zone of uplift (Sects. 3.2–3.4). Finally, we discuss the limitations, implications and predictions of our findings (Sect. 4).

2. Methods

2.1. Data Collection from Literature

A secondary zone of uplift in nature can be detected by surveying land elevations before and after an earthquake. Decades ago methods as described in e.g. Plafker (1965) and Plafker and Savage (1970) provided estimates with measurement uncertainties on the order of a few decimeters. Near the coast line relative sea level changes were mapped using local markers, such as high-tide lines or vertical growth limits up to which specific sessile marine organisms or plants can live. Inland elevation changes were obtained by differencing results from leveling methods obtained in two subsequent surveys. Nowadays, land-based GPS data provide widespread information on vertical displacements with an accuracy on the order of centimeters.

We analyze the megathrust earthquakes for which a decent amount of measurements exists at a few hundred kilometers landward from the trench. This requires a coastline for relative sea level change measurements or land for GPS measurements. Accordingly, we identified four megathrust earthquakes: the 1960 M9.5 Valdivia (Plafker and Savage 1970), 1964 M9.2 Alaska (Plafker 1969), 2010 M8.8 Maule (Vigny et al. 2011) and 2011 M9.0 Tohoku-Oki (Ozawa et al. 2011; Sato et al. 2011) earthquakes. For these earthquakes, we combined published data into trench-perpendicular profiles of vertical displacements. Relevant aspects regarding the origin of these datasets, specific values for the resulting secondary zone of uplift, and the corresponding tectonic parameters for each subduction zone are summarized in Table 1. The values and uncertainties in the data are adopted from the

Table 1

An overview of the differences between studied earthquakes and data acquisition methods. Data sources: [a]Plafker and Savage (1970),
[b]Plafker (1969), [c]Vigny et al. (2011), [d]Ozawa et al. (2011), [e]Sato et al. (2011), [f]Johnson et al. (1996), [g]Holdahl and Sauber (1994), [h]Yue and
Lay (2013), [i]Moreno et al. (2009), [j]Holdahl and Sauber (1994), [k]Moreno et al. (2014), [l]Hayes et al. (2012), [m]Heuret et al. (2011)

	1960 M9.5 Valdivia	1964 M9.2 Alaska	2010 M8.8 Maule	2011 M9.0 Tohoku-Oki
Data type	Growth limits, eyewitness accounts[a]	Growth limits, leveling survey[b]	GPS[c]	GPS[d], seafloor geodesy[e]
Time span between earthquake and survey (Δt)	8 years	1–2 years	2–20 days	4 h (GPS), 1–4 months (sea floor geodesy)
Uncertainty	20–100 cm	30 cm	1–10 cm	2 cm (GPS), 20–60 cm (geodesy)
Peak slip	20–40 m[a]	22–30 m[f,g]	15 m[c]	50–60 m[h]
Rupture width	~ 130 km[i]	~ 300 km[j]	~ 190 km[k]	~ 200 km[d]
Average interface dip	21° [l]	12° [l]	21° [l]	17° [l]
Seismogenic zone downdip limit	210 km[m]	243 km[m]	135 km[m]	210 km[m]
First hinge point (HP_1)	95 km	215 km	120 km	120 km
Second hinge point (HP_2)	200 km	350 km (E), 500 km (W)	240 km	335 km
Primary subsidence ($S_{1,max}$)	2.7 m	1.9 m	0.73 m	1.158 m
Secondary uplift ($U_{2,max}$)	1.1 m	0.3 m	0.12 m	0.04 m

The time span between earthquake and survey (Δt) provides an indication of the potential amount of postseismic data included, while the uncertainties amongst others depend on the measuring techniques. We include estimates of peak slip, rupture width (defined as downdip width of the zone with slip > 5 m), average interface dip (defined here as the average dip for 100 km depth), and downdip limit of the seismogenic zone from the literature. The lowermost block characterizes the surface displacements: HP_1 is the transition from primary uplift to subsidence (first hinge point), whereas HP_2 is the second hinge point (the transition from subsidence to secondary uplift). $S_{1,max}$ is the maximum subsidence of the primary zone of subsidence and $U_{2,max}$ denotes maximum uplift of the secondary zone of uplift

referenced studies (Plafker and Savage 1970; Plafker 1969; Vigny et al. 2011; Ozawa et al. 2011; Sato et al. 2011). In general, uncertainties decrease as measurements are done latter in time with more accurate acquisition methods. Furthermore, the time interval between the earthquake and survey (Δt) gives a rough indication for the amount of postseismic deformation that is potentially included in the displacement data (see discussion in Sect. 4.3).

2.2. Numerical Model

We use the seismo-thermo-mechanical (STM) numerical models developed and detailed in van Dinther et al. (2013a), van Dinther et al. (2013b), and van Dinther et al. (2014). These models are based on the continuum-mechanics framework of I2ELVIS, which is a 2-D implicit, conservative finite difference thermo-mechanical code (Gerya and Yuen 2007). The fully staggered Eulerian grid is combined with a Lagrangian marker-in-cell technique to allow for large deformation through advecting properties (e.g. lithology, material parameters and stress) along with

the particles they belong to. The code solves for the pressure as well as horizontal and vertical velocity assuming conservation of mass in an incompressible medium (Eq. 5) and conservation of momentum with gravity and inertia (see Appendix 1, Eqs. 6, 7). The large-scale models also solve for temperature using the heat equation including advection, conduction, and heat generation due to shear heating, radioactive heating and adiabatic (de-)compression (Appendix 1, Eq. 8). Additionally, the large-scale setup includes basic formulations of (de-)hydration, fluid flow, and erosion (van Dinther et al. 2013b).

The constitutive equations relate strain rates $\dot{\varepsilon}'_{ij}$ to deviatoric stresses σ'_{ij} using a non-linear visco-elasto-plastic rheology according to

$$\dot{\varepsilon}'_{ij} = \underbrace{\frac{1}{2\eta} \cdot \sigma'_{ij}}_{\dot{\varepsilon}'_{ij(viscous)}} + \underbrace{\frac{1}{2G} \cdot \frac{D\sigma'_{ij}}{Dt}}_{\dot{\varepsilon}'_{ij(elastic)}} + \underbrace{\begin{cases} 0 & \text{for } \sigma'_{II} < \sigma_{yield} \\ \chi \frac{\partial g_{plastic}}{\partial \sigma'_{ij}} & \text{for } \sigma'_{II} = \sigma_{yield} \end{cases}}_{\dot{\varepsilon}'_{ij(plastic)}}.$$

$$(1)$$

This represents a Maxwell visco-elastic body in series with a frictional plastic slider, where η is effective viscosity, G is shear modulus, $\frac{D\sigma'_{ij}}{Dt}$ is the objective co-rotational time derivative solved using a time explicit scheme, $g_{plastic}$ is the plastic flow potential, χ is the plastic multiplier connecting plastic strain rates and stresses, and σ_{yield} is the plastic yield strength. The amount of elastic versus viscous deformation is determined by the viscoelasticity factor $(G\Delta t)/(G\Delta t + \eta_{vp})$ (e.g., Moresi et al. 2003; Gerya 2010). The non-linear ductile viscosity η is based on experimentally derived dislocation creep flow laws as

$$\eta = \left(\frac{1}{\sigma'_{II}}\right)^{(n-1)} \cdot \frac{1}{2A_D} \cdot \exp\left(\frac{E_a + PV_a}{RT}\right), \quad (2)$$

where R is the gas constant (8.314 J/(mol $^{\circ}$C). Stress exponent n, pre-exponential factor A_D, activation energy E_a and activation volume V_a are experimentally determined parameters set for each lithology.

Brittle deformation is modeled using non-associative plasticity with a pore fluid pressure-effective pressure-dependent yield strength

$$\sigma_{yield} = C + \mu \cdot \left(1 - \frac{P_f}{P_s}\right) \cdot P. \quad (3)$$

Earthquake-like events result from a strongly slip rate-dependent frictional formulation with

$$\mu_{eff} = \mu_s(1 - \gamma) + \mu_s \frac{\gamma}{1 + \frac{V}{V_c}}, \quad (4)$$

where μ_s is the static friction coefficient, V_c is the characteristic velocity, and γ represents the amount of slip velocity-induced weakening (i.e., $1 - \frac{\mu_d}{\mu_s}$, where μ_d is the dynamic friction coefficient). Spontaneous ruptures represent the occurrence of rapid threshold-exceeding slip during which permanent displacement and stress drop occur along a localized interface (van Dinther et al. 2013b).

In summary, the resulting code handles both long-term subduction dynamics and short-term seismogenesis in a physically consistent manner. In the large-scale setup this means that the slab and seismogenic zone geometries together with its thermal and viscosity structures evolve autonomously. They influence the corresponding stress and strength distributions, which ultimately lead to the generation of spontaneous earthquake-akin events. Hence we model the interaction between the lithosphere, slab and mantle through spontaneously developing faults

and events. An important disadvantage of the current version is that events durations are very long (on the order of a $\sim$ 100 years) due to the constant time step of 5 years (van Dinther et al. 2013b). This makes inertial dynamics negligible and prevents us from uniquely distinguishing coseismic from postseismic processes. However, inertial dynamics in terms of shear wave propagation is resolved in the simplified analogue model setup, although waves are somewhat slow due to low scaled shear wave speeds of gelatin.

2.3. Model Setups

To better narrow down the physical mechanisms governing a secondary zone of uplift we use two model setups. These setups vary in degree of lithological, rheological and geometrical complexity.

The most realistic setup represents a 1500 $\times$ 400 km^2 trench-normal section of the Southern Chilean active continental margin, which is similar albeit deeper than the setup used in van Dinther et al. (2013b) and van Dinther et al. (2014) (Fig. 1a). The setup in the seismic cycle phase spontaneously evolved from about 5 million years of thermo-mechanical subduction of an oceanic slab of age 40 Ma due to a slab push at a steady rate of 7.5 cm/year. This lead to a large sedimentary wedge adjacent to a continental overriding plate beneath which an oceanic crust and lithospheric mantle subduct into a visco-elastic upper mantle. The visco-elasto-plastic parameters of each lithology are based on a range of laboratory experiments and are similar to van Dinther et al. (2014). These parameters within the governing and constitutive equations (Sect. 2.2) lead to a self-consistent thermo-mechanical structure in which viscosity is calculated according a temperature-, pressure-, and stress-dependent flow law dominated by dislocation creep (Eq. 2). The resulting smooth viscosity profile contains a lower continental crust with a viscosity of about 10^{21-22}Pa s and an asthenospheric mantle with a viscosity on the order of 10^{18-20} Pa s (Fig. 1a). The subduction channel is largely formed by the top 2 km of oceanic crust, which accommodates most deformation and spontaneously transits from brittle to ductile deformation between about 350 and 450 $^{\circ}$C (van Dinther et al.

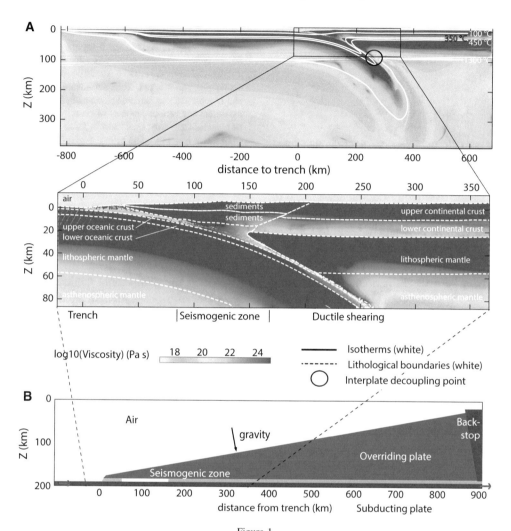

Figure 1
a Model setup of seismo-thermo-mechanical (STM) model with a large-scale setting resembling Southern Chile (van Dinther et al. 2013b). Top shows the entire model with relevant isotherms in white, whereas the lower panel shows a zoom on the region of interest and depicts the rock types and their boundaries in white. **b** Model setup of seismo-mechanical model (van Dinther et al. 2013a) simulating the analogue model of Corbi et al. (2013). Colors in all panels indicate the same viscosity scale

2013b). In terms of frictional parameters this megathrust interface is slip rate weakening (μ_s = 0.5, μ_d = 0.15, $\frac{P_f}{P_s}$ = 0.95), but for the shallowest section at which a transition to slip rate strengthening occurs from 150 to 100 °C.

The simplified setup is adapted from van Dinther et al. (2013a) and is based on the upscaled analogue modeling setup of Corbi et al. (2013). In this setup a rigid, straight slab subducts beneath a (visco-)elastic wedge, which is bounded by a rigid backstop (Fig. 1b). For ease of numerical computation we rotate the setup and gravity by a slab dip of 10° to align the megathrust interface and slab with the lower

boundary. The wedge-shaped fore-arc is confined by a backstop, which is moved further away from the trench to reduce its influence on simulated surface displacements to a minimum (see Sect. 3.3.1). The fore-arc wedge deforms elastically (99.7%). In this study we add lower crustal and upper mantle layers that largely deform viscously (98%). The megathrust interface additionally features plastic deformation, as controlled by a seismogenic zone with slip rate weakening friction bounded by slip rate strengthening friction regions.

In each setup a sticky air layer at the top deforms viscously at all time steps and approximates a free surface (Crameri et al. 2011; van Dinther et al. 2013b). This allows for unhampered evolution of both temporal and permanent topography.

3. Results and Analysis

We first compile published vertical displacement data to understand how universal a secondary zone of uplift is (Sect. 3.1). Second we study the universality, characteristics and evolution of a secondary zone of uplift in seismo-thermo-mechanical models with a realistic setup tailored to Southern Chile (Sect. 3.2). Section 3.3 analyzes the physical mechanisms responsible for such a secondary zone of uplift through studying both a realistic Southern Chile setup and a simplified wedge model. Finally, we discuss some parameters influencing a secondary zone of uplift (Sect. 3.4).

3.1. Natural Data

3.1.1 Coseismic Vertical Displacements

To understand whether a secondary zone of uplift exists in all great megathrust earthquakes we compile the available vertical surface displacements due to great megathrust earthquakes. These displacements aim to capture coseismic displacements, but post-seismic displacements are included when a larger time lag since the earthquake is involved (Table 1). The collected data for four out of four great earthquakes show a secondary zone of uplift (Figs. 2, 3). These secondary zones of uplift are remarkably spatially coherent with 164 out of 167 measurements indicating uplift (Fig. 2). Two measurements that show subsidence are obtained near the second hinge point of the 1960 Valdivia earthquake, which is the location where subsidence changes to uplift (HP_2). The location and magnitude of this secondary zone of uplift, however, vary significantly from one tectonic region to another. This can be appreciated quantitatively by studying Table 1), which contains the available data on the secondary zone of uplift and relating earthquake characteristics for each event.

Figure 2 ▶

Map view of the surface displacements in **a** southern Chile due to the M9.5 Valdivia earthquake, **b** central Chile due to the 2010 M8.8 Maule earthquake, **c** north-east Japan due to the 2011 M9.0 Tohoku-Oki earthquake, and **d** Alaska due to the 1964 M9.2 Alaska earthquake. Uplift is red, while subsidence is blue. Displacements aim to represent coseismic displacements, but due to a time lag in measurements various amounts of postseismic displacements can be included (see "time span" row in Table 1, where time lags vary from a few hours to a few years). Sources are given in the "data type" row of Table 1. Green stars denote epicenters and green shaded areas are approximate areas of coseismic slip. Thick black lines are the trenches (Coffin et al. 1998) and thin black lines indicate the horizontal distance to the trench with multiples of a 100 km. In all earthquakes studied, there is a secondary zone of uplift, but the second hinge point is at different distances from the trench

Within the comparable Chilean tectonic region, we observe a correlation in uplift magnitude with earthquake magnitude and slip. The M9.5 Valdivia earthquake shows a distinct secondary zone of uplift with a second hinge point at around 200 km from the trench and a maximum secondary uplift $U_{2,max}$ of ~ 1 m (Fig. 3a). The data points are more scattered due to large measurements errors (in the range of 20–60 cm) as well as local tectonic variations over a wide range during the long period of 8 years between earthquake and survey. The smaller 2010 M8.8 Maule earthquake ruptured the same subduction zone just north of the Valdivia earthquake. This earthquake also showed a secondary uplift with roughly 10 times smaller amplitude $U_{2,max}$ at roughly 50 km farther from the trench (HP_2 at 240 km, Fig. 3b).

The M9.0 Tohoku-Oki earthquake produced a minor secondary zone of uplift with uplift of about 4 cm beyond about 335 km from the trench (Fig. 3c). The map view confirms this minor uplift is widespread and spatially coherent uplift with all 64 stations landward of the main slip area measuring it (Fig. 2c).

The M9.2 Alaska earthquake caused a secondary uplift $U_{2,max}$ of maximum 0.3 m with a contrast between the eastern profile (recorded along the Richardson Highway) and the western profile (Alaska railroad, Fig. 3d). The second hinge point at the western transect was measured at ~ 500 km, while it occurred at ~ 350 km for the eastern transect (Fig. 2d). This difference could arise from the location with respect to the lateral limit of the rupture

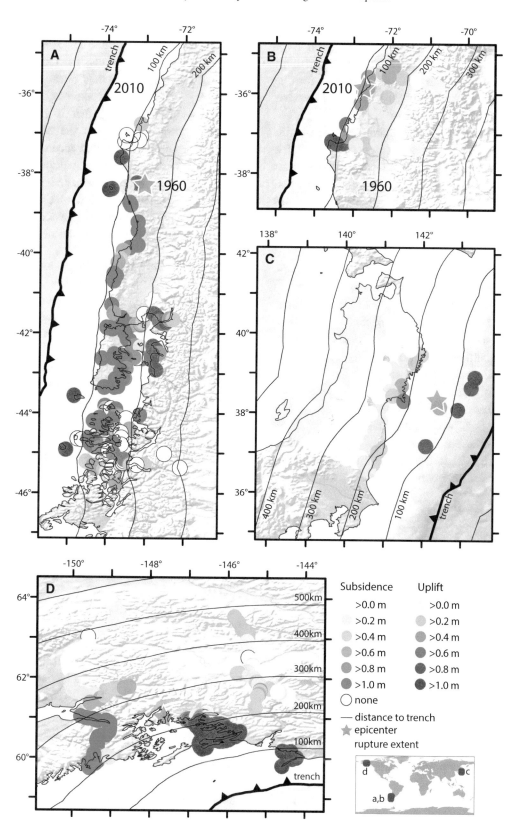

or from the sharp bend of the slab and trench in Alaska, which influences the slab dip. Such large differences in location of the secondary zone of uplift within one earthquake are not observed for the Chilean and Tohoku earthquakes, where the trenches and slabs are rather straight.

From the 2004 M9.2 Sumatra-Andaman earthquake, GPS stations in our region of interest that recorded the earthquake are limited to two stations (see Figs. 1b, 2n, o; Table 2 in Hashimoto et al. 2006). Station SAMP at the eastern side of Sumatra is located at roughly 400 km from the trench and recorded an uplift of 6.2 ± 8.5 mm. Additionally, an uplift of 12.5 ± 7.3 mm was recorded at 600–700 km from the trench in Phuket (Thailand). Interestingly, levelling data following the 1946 M8.2 Nankaido earthquake also shows three out of three locations with uplift in a secondary zone beyond 250 km from the trench (see Fig. 10 in Miyashita 1987). These two and three uplift measurements suggest a secondary zone of uplift could also be present for the 2004 M9.2 Sumatra-Andaman and 1946 M8.2 Nankaido earthquakes. Nonetheless, due the limited statistical meaning of both two and three data points in space, we exclude these two earthquakes from our formal analysis. We did not analyze the limited data in the area of interest for any other M < 8.5 earthquake.

3.1.2 Deciphering Tectonic Control

In an attempt to decipher which tectonic features influence this secondary zone of uplift, we compare values for this admittedly too low number of four earthquakes (Table 1). The amount of secondary uplift $U_{2,max}$ seems somewhat correlated to earthquake magnitude and thus the total amount of slip. Total slip can for this be approximated as amount of slip times slip area, which is derived from moment magnitude scaled to seismic moment (Blaser et al. 2010) and assumes shear moduli are roughly equal. More total slip or larger magnitude leads to a higher secondary zone of uplift with the exception of the M9.0 Tohoku earthquake. However, the amount of slip on the shallow portion of the Tohoku megathrust interface was exceptional (e.g., Fujiwara et al. 2011).

The distance from the second hinge point HP_2 to the trench seems to increase with the downdip width

of each rupture and decrease with the dip of the megathrust interface (Table 1). Alaska with the flattest subduction zone shows the most horizontally stretched pattern, whereas the Chilean slab dips most steeply and show a more compressed uplift pattern (Figs. 3, 4). This suggests that if earthquake slip penetrates farther away from the trench, the secondary zone of uplift is shifted accordingly.

In summary, all four megathrust earthquakes studied show a similar displacement pattern including a secondary zone of uplift. Differences in amplitude and hinge point position are considerable and are likely related to slab geometry and rupture size.

3.1.3 Postseismic Vertical Displacements

The data targeting coseismic displacements include different amounts of postseismic deformation mainly due to the different delay times when measuring the displacements (Table 1). To better constrain the coseismic or postseismic nature of secondary uplift, we shortly analyze the vertical displacements in the locations that showed a secondary zone of uplift for the 2011 and 2010 earthquakes (Fig. 2).

The stations that showed secondary coseismic uplift in the 2010 M8.8 Maule earthquake still show uplift up to 4 years after the event (e.g., Klein et al. 2016). However, the uplift rates are decreasing significantly. Averaged uplift rates up to 400 cm/year were observed in measurements best approximating the coseismic phase (Vigny et al. 2011). In the postseismic phase this uplift rate reduces to 5 cm/year in month 2–6 and 1.3 cm/year two to three years after (Klein et al. 2016). This significantly slower postseismic uplift corresponds to the mid-field zone of postseismic uplift as observed and potentially explained by Klein et al. (2016) and Li et al. (2017).

These findings of early postseismic uplift in the region of interest are contrasted by early postseismic observations following the 2011 M9.0 Tohoku earthquake. In Japan west coast stations showed coseismic uplift, as observed in the measurements done within 4 h after the earthquake [Figs. 2, 3 showing data from Ozawa et al. (2011)]. However, these west coast stations show subsidence at rates of about 1.7 cm/year in the first 1.5 and 2 years after the earthquake (e.g., Yamagiwa et al. 2015; Hu et al.

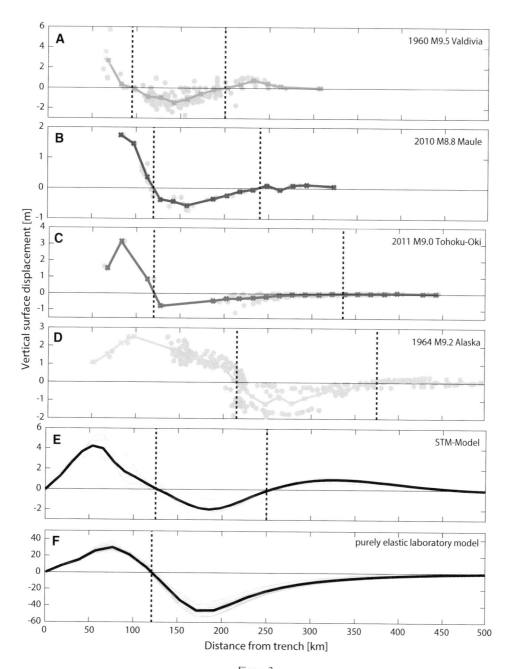

Figure 3
Cross section showing the coseismic subsidence and uplift for the **a** 1960 M9.5 Valdivia, **b** 2010 M8.8 Maule, **c** 2011 M9.0 Tohoku-Oki, **d** 1964 M9.2 Alaska earthquakes, **e** our seismo-thermo-mechanical (STM) models, and **f** a purely elastic laboratory-scale model. Various amounts of postseismic displacements can be included in these coseismic estimates (see Table 1; Sect. 4.1). Gray dots are data points located at each measurements minimum distance to the trench to account for along-strike variations of long trenches. Bold colored lines represent their means as binned over widths of 15 km. For the model gray lines represent individual events and black lines are means. The solid black line is zero vertical displacement and the dotted vertical lines indicate the two hinge points HP_1 and HP_2, where the surface displacement changes from uplift to subsidence and vice-versa. Note the different vertical scales. All earthquakes and the STM model show a secondary zone of uplift, but its amplitude and the position of HP_2 varies. The surface displacements of the purely elastic model asymptotically go to zero for large distances to the trench

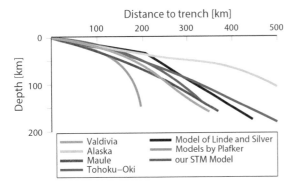

Figure 4
Cross section of slab depths from Slab1.0 (Hayes et al. 2012) and from models discussed in the text, showing how different the slabs are dipping in both nature and models

2016). This opposite direction of vertical displacements in Japan and Chile and the significantly decreasing uplift rates for Chile suggest that the zone of postseismic uplift in the midfield and the secondary zone of mostly coseismic uplift could be different features, which may overlap.

3.2. Seismo-thermo-mechanical models predict a secondary zone of uplift

To understand the universality and physical origin of the secondary zone of uplift we first analyze its evolution and characteristics in the realistic Southern Chile model setup (Fig. 1a). The model simulates 36 spontaneous, largely quasi-characteristic and quasi-periodic events during which on average the overriding plate displaces by about 18 m every 881 years. The surface displacements of the reference model reproduced the spatial pattern of vertical displacements for the 2010 M8.8 Maule earthquake unintentionally and without any tuning due to its physically-consistent basis (Fig. 5 in van Dinther et al. 2013b, which also corresponds to a grey line in Fig. 3e). They show a distinct secondary zone of uplift beyond ∼ 250 km from the trench with a peak around 330 km. Here, we analyze additional events of this same model (grey lines in Fig. 3e). Events are detected based on the slip velocity, i.e. the coseismic phase starts when the markers located just above the interface start to move significantly seaward and end when they return to pre-event levels. The resulting displacements and hinge points in Fig. 3e agree well

with the reference event of van Dinther et al. (2013b, Fig. 5 therein) and all show a secondary zone of uplift. The consistency of the location of the secondary zone of uplift for different events is notable and seems fairly independent of event details and rupture size. The limited amount of smaller events in the sequence causes minor variations in the pattern of uplift in the primary zone, while variations in the secondary zone of uplift are rather negligible.

The universal nature of the secondary zone of uplift is supported by extensive tests in which this realistic setup was more and less drastically changed in attempts to remove the secondary zone of uplift. In these tests we have taken care to ensure that the secondary zone of uplift is not influenced by numerical modeling parameters (e.g., domain size and boundary conditions). None of the tested numerical parameters influences the location and magnitude of the secondary zone of uplift in a noteworthy manner (e.g., compare red and blue lines with black lines in Fig. 5). Additionally, all models with a wide range of tectonic and material parameters reveal a secondary zone of uplift. These models will be discussed in more detail in Sects. 3.3.2 and 4.2 to better understand the physical mechanisms governing it.

To better understand what happens we analyze the spatiotemporal evolution of one quasi-characteristic event and one seismic cycle in detail (Fig. 6). Figure 6a portrays the vertical surface velocities as a function of distance to the trench (X) and time in time steps (Y). During the interseismic period the fore-arc within 100 km from the trench is dragged landward, since it is coupled to the landward subducting plate (also see spatial snapshots in Fig. 6b, d). This compression causes primary interseismic uplift of the overriding plate from about 100 to almost 300 km. Interestingly, a secondary zone of interseismic subsidence occurs at distances beyond 300 km from the trench. This model thus predicts very slow interseismic subsidence (i.e., less than mm/year) at a few hundred kilometers landward of the trench.

During the coseismic period accumulated displacements within the overriding plate are largely elastically rebound [see Fig. 5 in same models of van Dinther et al. (2013b), where on average more than 90% of interseismic displacements is rebound].

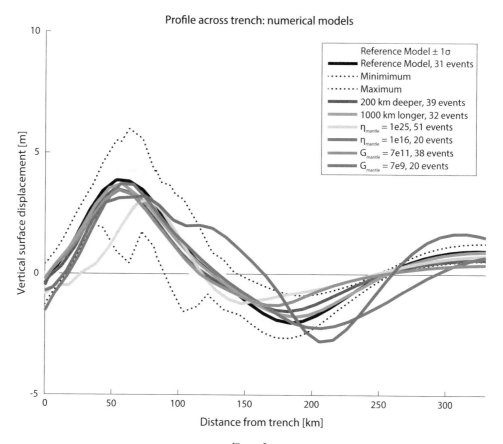

Figure 5

Vertical surface displacements accumulated over the coseismic period versus distance to the trench for different seismo-thermo-mechanical models. Black line (mean) and grey shading ($\pm 1\sigma$) represent the reference model with dashed lines showing the maximum and minimum of all events in the reference model. Colored solid lines represent the mean of different simulations, as calculated for the given number of events per simulation. Red and blue lines indicate simulations with a 200 km deeper (+ 100%) and a 1000 km longer model (+ 67%) respectively. Green and purple curves result from simulations with about five orders of magnitude larger and smaller viscosity in the mantle (the former only beneath the overriding plate). Orange and brown curves result from overriding mantle shear moduli respectively increased and decreased by one order of magnitude

Elastic rebound leads to a primary zone of uplift, which propagates seaward along with the rupture that nucleated just above the brittle-ductile transition. Behind this propagating uplift, the extending overriding plate experiences primary subsidence. Beyond about 250 km the secondary zone of uplift occurs (also see spatial snapshots in Fig. 6c, e). This uplift occurs over the same time period during which primary coseismic uplift occurs, although it lasts slightly longer. This suggests that the secondary zone of coseismic uplift at least has a distinct coseismic component, although exact distinctions are not allowed due to the low temporal resolution of the numerical model.

Analyzing velocities at depth shows that the secondary zone of uplift is connected to a broad pattern of uplift occurring throughout the whole lithosphere and mantle. Upward velocities occur landward of a dipping line connecting the secondary hinge point at the surface and the start of ductile deformation at the megathrust interface (T > 450 °C in Fig. 6e). The second hinge point at the surface is located roughly above the interplate decoupling point, which is located at the megathrust interface at depths of about 90–95 km. The interplate decoupling point marks the depth below which motions of the hanging wall decouple from those in the footwall, which from then on induce corner flow in the mantle

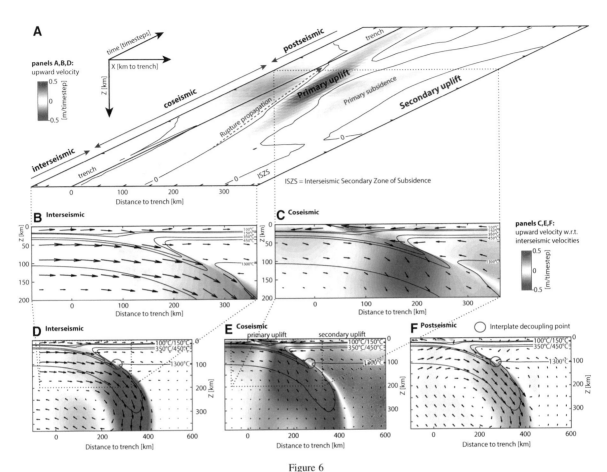

Figure 6

a Evolution of vertical surface velocities through space (Y) and time (X). Contours indicate separation of regions of uplift and subsidence. Note that the time axis is spaced in terms of time steps (five time steps per dash) with labels indicating duration of inter-, co- and postseismic periods. Values for time and coseismic velocities are not appropriate (resp. too long and slow) due to a constant time step of 5 years. The accumulated coseismic displacement due to the coseismic vertical velocities are shown in Fig. 5. Spatial cross-sections in the X–Z plane shown at inter- (**b, d**), co- (**c, e**) and postseismic (**f**) times. The middle panels (**b, c**) show zooms over the width shown in **a**, while lower panels (**d–f**) show larger cross sections revealing lithospheric and mantle flow patterns

wedge [definition updated from Furukawa (1993) starting with van Dinther et al. (2013b)]. This is evident from seaward flow in the astenosphere, as opposed to interseismic landward displacement of the lithosphere (Fig. 6d). This is facilitated by significantly decreased overriding plate viscosities due to ambient temperatures approaching 1300 °C. This roughly corresponds to the thermal definition of the lithosphere–astenosphere boundary within the overriding plate. At depth, corner flow is driven by slab penetration continuously, although at variable speeds (Fig. 6d–f). Subduction and slab penetration are slow within the interseismic period. As the megathrust interface decouples during an earthquake, subduction

and slab penetration are considerably accelerated. Displacements of the subducting plate are thus not rebound to their original position, but rather subduct faster to catch up with long-term subduction rates.

After the rupture arrests and the seismogenic zone relocks itself, postseismic velocities largely change back toward their interseismic pattern (also see spatial snapshot in Fig. 6f w.r.t. Fig.6d). This reversal is delayed within 50 km from the trench due to afterslip in the velocity-strengthening domain and beyond about 400 km from the trench due to viscous relaxation of the lithosphere–astenosphere. Additionally, the secondary zone of interseismic subsidence starts around 250 km, while it moves progressively

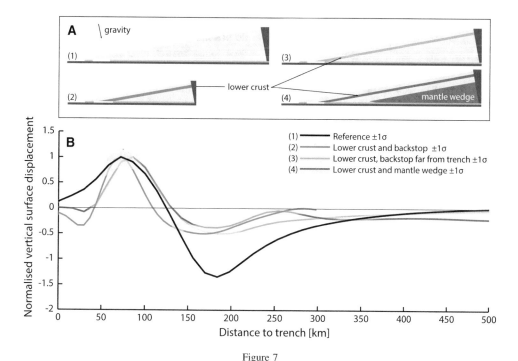

Figure 7

a Setups of laboratory-scale models that progressively increase in complexity: (1) dominantly elastic fore-arc, (2) add a visco-elastic lower crust with laboratory backstop, (3) move backstop to far-field, and (4) add a visco-elastic mantle wedge. The rheological layers are adopted from the large-scale thermo-mechanical model (Fig. 1a). **b** The resulting displacements of these models with uncertainties. To highlight the different shapes and not the amplitudes, the vertical displacements are normalized with respect to maximum primary uplift

downdip until it starts around 300 km in the late interseismic period.

3.3. Physical Mechanisms Governing a Secondary Zone of Uplift

Both observations from nature and results from our numerical models suggest that a secondary zone of uplift is a universal characteristic of great megathrust earthquakes. Here we attempt to identify possible physical mechanisms that are applicable to all subduction zones. To identify and directly test the proposed physical mechanisms we utilize two model setups (Sect. 2.3, Fig. 1). We aim to add a secondary zone of uplift to the simple elastic laboratory wedge setup, which simulates the classical model for surface displacements and thus does not include a secondary zone of uplift (Fig. 3f). Excluding an identified potential physical mechanism should also be able to remove the secondary zone of uplift from the realistic, albeit complex setup for Southern Chile (Fig. 3e). Satisfying both these criteria provides a

valid test for potential mechanisms, which discarded several potential mechanisms from acting at all or acting alone.

3.3.1 Elastic Rebound After Interseismic Buckling of Visco-elastically Layered Lithosphere

We start from the most simple setup, where a rigid slab subducts beneath a wedge-shaped fore-arc (van Dinther et al. 2013a) (Figs. 1b, 7a). When the fore-arc is homogeneous and virtually elastic, spontaneous cycles of megathrust events confirm results of elastic models with near trench uplift followed by subsidence (Fig. 7; a1 and black line in b). At large distances from the trench, the surface keeps subsiding, as it asymptotically approaches the zero level. Model sets 2, 3 and 4 shown in Fig. 7 demonstrate how this subsidence in the elastic model is modified by adding essential model complexities.

Model set 2 introduces a lower crustal layer with reduced viscosity in the original analogue model setup, which introduces a very small secondary zone of uplift around 250–300 km from the trench (cf. sets 1 and 2 in Fig. 7b). This modulation is related to the presence of a thin elastic beam (i.e., the upper crust) separated by a viscously deforming layer (i.e., the lower crust). The thin elastic beam is then free to buckle during the interseismic period in response to horizontal compression due to an end load (e.g., Turcotte and Schubert 2002). Elastic buckling due to horizontal forces typically introduces a secondary zone of subsidence at large distances from the trench (e.g., Fig. 6a). This zone of secondary subsidence is approximately rebound in the coseismic period due to reversal of elastic deformation (e.g., Reid 1910), thereby introducing a secondary zone of uplift. Buckling is not observed in the original model with the thick overriding plate (model 1 in Fig. 7b), since the horizontal compressional forces are not large enough to buckle a thick beam with a very large elastic thickness.

Model set 3 shows a gentle secondary zone of uplift reduces to a secondary zone of slight relative uplift, albeit absolute subsidence, when the backstop is moved further inland (Fig. 7a3, b). This suggests that, for buckling to be effective in introducing secondary uplift, a means to generate more localized compression is needed. A backstop corresponds to a region that is significantly stronger than the region just trenchward of it. This can result from a transition from sedimentary to magmatic/metamorphic rocks (Byrne et al. 1993) or a transition from thinner, warmer and weaker arc lithosphere to colder, thicker and stronger (e.g., cratonic) lithosphere (Sobolev and Babeyko 2005; Manea et al. 2012). However, it is arguable whether a backstop is present at the required location in all subduction settings. It is also only modestly present at a location closer to the trench in the large-scale model, which also simulates a secondary zone of uplift.

When a viscous mantle wedge is added in set 4, a minor kink in the reduced subsidence curve of set 3 is turned into a clear, albeit still subsiding, secondary bulge (Fig. 7a4, b). This is added to reproduce the mechanical structure of the overriding plate in the large-scale model (Fig. 1a). A weak lower crust and weak mantle wedge create a double-beam system made of the rigid upper crust (upper beam) and

mantle lithosphere (lower beam). This facilitates additional buckling of the lithospheric mantle and thus leads to a higher-order wavelength of surface displacements. Note here that the peak of the secondary bulge is roughly located above the mantle wedge tip, which colocates with the interplate decoupling point. This peak, however, remains below the zero vertical displacement level and the surface at 200–300 km is thus still subsiding.

3.3.2 Need for a Second Mechanism

Numerical experiments in the analogue model setup show elastic rebound following interseismic horizontal compression of a visco-elastically layered lithosphere is able to introduce a high-order buckling wavelength. However, in several dozens of attempts widely varying mechanical structure through geometries and parameters, we are not able to generate a consistent secondary zone of uplift with amplitudes above the zero level. Our summarized results in Fig. 7b show that a mechanism to generate broad uplift over larger wavelengths is missing. A rigid backstop in the near field (i.e., within 300 km) might be able to facilitate this for some events, although adding that to an analogue setup with complete visco-elastic layering still only partially ensures elevations above zero. However, the disputable presence of a backstop at those locations in both nature and our large-scale models suggests that a second mechanism is needed.

The need for a second mechanism is confirmed in dozens of experiments in a large-scale setup. A secondary zone of uplift can not be removed by eliminating the thin rigid beam structure that facilitates buckling of the overriding plate. Increasing viscosities of the lower crust and/or upper mantle on the landward side to 10^{25} Pa s shows that surface displacement patterns are affected to a minor extent, so that a secondary zone of uplift remains present and largely unchanged (green line in Fig. 5). This also suggests that, at least in the the analogue setup, visco-elastic relaxation within the hanging wall is not a key mechanism. Besides this experiment we ran numerous experiments in the large-scale model aimed to remove the secondary zone of uplift. None of these experiments lead to removing of the secondary zone

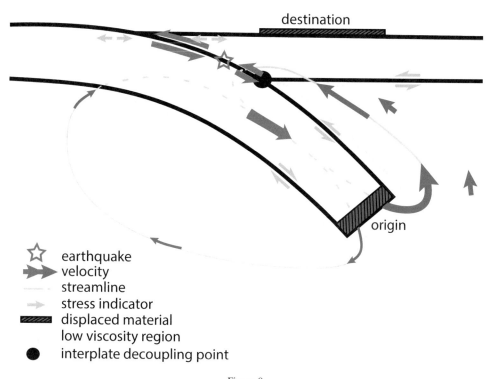

Figure 8
Schematic diagram illustrating how mass conservation due to accelerated slab penetration causes a secondary zone of uplift. Black lines represent lithosphere contours and other colors are described in the legend

of uplift, while generally its amplitude and location remained largely unchanged (a few selected experiments are shown in Fig. 5). This wide range of experiments rather shows that the second mechanism is a very basic feature of our model that can not be removed. One option that is difficult to quantify is the presence of a smoothly curved interface. However, since we do not observe a secondary zone of uplift in other smoothly curved models (e.g., Moreno et al. 2009) we suspect curvature is not the key component. If this holds, then the only other mechanism not present in our analogue model involves the slab and mantle kinematics and dynamics.

3.3.3 Mass Conservation Following Slab Penetration

To analyze the impact of subduction dynamics, we return to the spatial cross sections (Fig. 6d–f) and schematically represent our interpretation of the physical mechanism in Fig. 8. Interseismic velocities show relatively slow slab subsidence and motion towards the land, which drags along and compresses the overriding plate (Fig. 6b, d). During an event the seismogenic zone unlocks and accumulated overriding plate displacements rebound, whereas the slab instead accelerates downward (Fig. 6c, e). This penetration represents the footwall displacements of the thrust event and ensures that the slab at depth catches up with its long-term subduction rates, since subduction was partially stalled during the interseismic period due to locking at the megathrust interface (see velocity variations in oceanic domain in Appendix Fig. 10a, b). These coseismic displacements of the slab are comparable in size to those in the less constrained overriding plate (compare arrows in supplementary movie S1), which is likely due to catching up of interseismic loading and slab pull. Coseismic slab displacements are thus considerable and on the order of a few tens of meters. For a 150 km wide downgoing region in two dimensions this amounts to a displacement of mantle material of a few square kilometers (red dashed area in Fig. 8 indicated as "origin"), which needs to be displaced somewhere to conserve mass (and momentum). These large amounts of mantle are preferentially forced upward beneath the overriding plate (Figs. 6e, 8; Appendix Fig. 10b). This generates uplift on the landward side of the interplate decoupling point, which is mostly focused within 100–150 km landward of that (red dashed area in Fig. 8 indicated as "destination", i.e., the area of the secondary zone of uplift). These motions also uplift the lithosphere, where—at the same time—space is being created by the seaward displacements of the lithosphere. A simple calculation as suggested by Fig. 8 indicates that conserving the mass displaced as calculated above can lead to uplift on the order of tens of centimeters. In addition a smaller portion of mantle is displaced upward beneath the oceanic plate (Appendix Fig. 10b). These displacement patterns form two (tilted) convective cells similar to what is observed if mass is conserved due to a sinking object [see Fig. 1.2b in Gerya (2010)]. These two convective cells become more narrow and more clear in the postseismic phase (Fig. 6f). Finally, this mechanism can be better understood through an analogy in which you put your finger into a pot with honey and as a results of this the surface surrounding your finger is slightly uplifted.

In summary, both interseismic buckling of a double-beam overriding plate and upward mass-conservation driven flow in the mantle wedge are necessary to produce a secondary zone of uplift due to megathrust earthquakes.

3.4. Parameter Study

Lithosphere buckling and upward flow due to slab penetration suggest that the most influential parameters are related to the geometry of the fore-arc as well as slab and seismogenic zone geometry. This agrees with our preliminary findings based on the limited observations from nature, where the geometries of the slab interface and rupture seemed most important (Sect. 3.1).

To analyze the role of these geometries in a physically-consistent manner, we varied the slab age. This primarily affects the thermal structure within the slab and thus the megathrust geometry and seismogenic zone width. An older slab has more time to cool

and thus leads to lower temperatures at the megathrust interface and a deeper and slightly wider seismogenic zone (especially for the young slabs; see legend Fig. 9a). Ruptures in a wider seismogenic zone penetrate more landward and thus cause subsidence and extension further land inward, thereby leading to a landward shift of the spatial surface pattern and the second hinge point (Fig. 9a, b). Figure 9c shows that peak amplitude of the secondary zone of uplift hardly changes with slab age or downdip rupture limit. This is likely due to the fact that the seismogenic zone width only weakly increases with slab age (see legend Fig. 9a), such that maximum earthquake size increases only weakly.

We further analyze the role of event magnitude by also including other presented models to allow for a larger variation in earthquake size . We find that the location of the second hinge point is not correlated to event magnitude (Fig. 9d). The amplitude of the secondary zone of uplift does appear to increase with increasing earthquake size and slip (Fig. 9e).

Summarizing observations from nature and our models, the geometry of the secondary zone of uplift is mostly influenced by the geometry of the seismogenic zone (i.e., its dip and resulting downdip rupture limit).

4. Discussion

We first confirm the existence of a secondary zone of uplift both from an observational and numerical modeling perspective. We then hypothesize on the physical mechanisms governing this secondary zone of uplift and propose that two complementary mechanisms are needed. First, elastic rebound from interseismic buckling of a visco-elastic layered lithosphere due to horizontal compression to generate a wave length. Second, conservation of mass due to accelerated slab penetration to generate (focused) uplift. This section discusses the limitations, embedding of mechanisms within the literature, open challenges and implications of this.

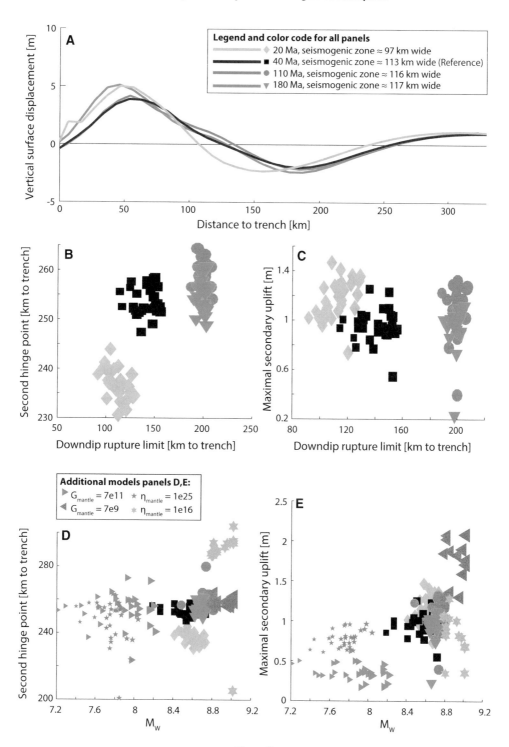

Figure 9

a Displacements of models with different slab ages showing a landward shift of the second hinge point for older slabs. **b** The downdip rupture limit and the second hinge point are clearly correlated. **c** The amplitude of the uplift is not related to downdip rupture limit and therefore slab age. **d** Models with distinctly different mantle parameters show that the second hinge point does not depend on the magnitude of the event. **e** The maximal secondary uplift however is related to the magnitude of the events

4.1. Limitations

The 2D STM models are able to model both long-term subduction dynamics and short-term seismogenesis. This innovation allows them to predict new features and processes, which previously went either not observed or unrecognized. One such example is this consistent presence of a secondary zone of uplift due to megathrust earthquakes. The challenge of bridging all relevant time scales of subduction zone processes has, however, not been fully completed and thus involves some limitations for the large-scale model (section 4.6, van Dinther et al. (2013b)). The main limitation is the exceptionally long coseismic duration of our events (years instead of seconds), which does not allow us to seperate postseismic and steady-state subduction contributions from the coseismic contribution. The same problem holds for natural data from the two earthquakes in the 1960s, since measurements were done years after the earthquakes. A second, relevant modeling limitation lies within the applied rheologies. These models use the assumption of incompressibility, which is a typical assumption made in geodynamic models. Compressibility could decrease the surface response due to accelerated slab penetration, although elimination of this characteristic response is not expected (as supported by similar motions in Sun and Wang (2015) for a Poisson's ratio of 0.25). These limitations have been overcome using adaptive time stepping, rate-and-state friction and compressibility in large-scale models of strike-slip settings (Herrendoerfer et al. 2018). However, bridging from millions of years of subduction to earthquake dynamics remains a challenge (Herrendoerfer 2018). This is thus far only partially accomplished by coupling two different models (van Zelst et al. 2019) or for resolving the postseismic phase (Sobolev and Muldashev 2017).

4.2. Embedding of Proposed Mechanisms

The proposed physical mechanisms form a new universal explanation for a secondary zone of uplift and for related modeled displacement patterns. In this section we compare our mechanisms and displacement patterns to other studies and start discussing open questions. Other studies with a (visco-)elastic rheology do typically not discuss a secondary zone of uplift (e.g., Wang 2007). Interestingly, also large-scale models of Miyashita (1987) reveal a secondary zone of coseismic uplift, which is also followed by subsidence that propagates landward in the postseismic phase.

In terms of physical mechanisms, the fact that they are universal and apply to all subduction zones is important, because we demonstrated its universal existence for various types of subduction zones. That suggests that a mechanism should not be valid only for one or two subduction zones, but rather for all. Hence the specific geometrical causes to locally uplift material due to abrupt curved steepening of this interface (Plafker and Savage 1970) or distinct slab interface kinks (Linde and Silver 1989) are not deemed relevant for all subduction zones. Moreover, a compilation of slab data rather shows smooth interfaces without such rapidly changing slab shapes (Fig. 4 based on data from Hayes et al. 2012).

The first mechanism of interseismic buckling due to horizontal compression falls in the same category as the transverse crustal buckling of a horizontally compressed continental plate mentioned by Plafker (1972). We, however, emphasize that compression occurred during the interseismic period (not coseismic) and resulted in a secondary zone of subsidence. This was subsequently rebound during the earthquake and hence showed secondary uplift. Second, we added that a realistic visco-elastic layering of the lithosphere is needed to have a thin enough elastic beam that can buckle (Turcotte and Schubert 2002). The need of a realistically layered visco-elastic lithosphere to show higher-order wavelengths in vertical displacements has also been shown by Pollitz (1997). Besides inducing lithospheric buckling through horizontal forces, Vita-Finzi and Mann (1994) modeled buckling and a primary and secondary zone of uplift of an elastic beam atop a viscous astenosphere due to vertical forces (i.e., flexure). These oscillations resulted largely from buoyancy effects following mass displacements within which accelerated subduction lead to overlap and a positive buoyancy force that uplifts the primary zone. In our model density displacements are different and it is rather elastic rebound from an overriding plate that is dragged down with the slab that causes primary uplift. Instead secondary uplift seems to result from a superposition of horizontal buckling and slab-induced return flow.

The second mechanism of accelerated slab penetration and resulting accelerated convection to conserve mass is new. However, the resulting convective displacement patterns throughout the lithosphere-mantle system are similar to those modeled in Sun et al. (2014); Sun and Wang (2015) [see Fig. 3A in Sun and Wang (2015)]. They, however, explain all displacements using the general term of visco-elastic relaxation and appoint an asymmetric rupture inducing greater tension in the upper plate as driving mechanism for these convective displacements with two cells. In our model the rupture is much more symmetric (Fig. 6e) as the slab moves more and thus induces stress changes that are comparable [see Fig. 3D in van Dinther et al. (2013b)]. Larger slab displacements are likely caused by more realistic cyclic loading of our self-consistent system [and presence of spontaneous low viscosity zones around the slab as in Figs. 1a, 8 and maybe off-fault plasticity (van Dinther et al. 2013b)]. Realistic interseismic loading ensures the relatively compressed slab catches up with slab pull in the coseismic period as subduction was partially inhibited in the interseismic period. Additionally, long-term subduction causes a pre-stress state with large extensional stresses in the slab and near neutral (locally compressional) stresses in the wedge. Nonetheless, we still interpret that rapid seaward motion of the wedge contributed, since these convective displacements fill up the space created by the seaward displaced wedge. Instead, as a primary control, slab penetration creates accelerated uplift as mass (and momentum) are conserved. Secondary, this uplift is tunneled towards the displaced wedge and creates a secondary zone of uplift (Fig. 8). This slab penetration mechanism might also be relevant for visco-elastic relaxation and should be considered and explored in physical explanations of it.

Our choice to not refer to this mechanism as visco-elastic relaxation per se is supported by experiments showing that a secondary zone of uplift also occurs when the overriding mantle deforms elastically (green line in Fig. 5). Therefore the ability of the visco-elastic mantle to delay and relax overriding plate displacements does not seem critical for a secondary zone of uplift. Instead letting the astenospheric mantle largely deform viscously does not remove its presence either, although it distinctly shifts the secondary zone of uplift in the direction of the land (purple line in Fig. 5). This could reflect a more distinct contribution from elastic rebound following interseismic buckling.

This second mechanism could be crudely related to a transient and coseismic version of the long-term thin viscous sheet model combined with corner flow that uplifts the volcanic arc (Wdowinski et al. 1989), as remarked upon by Vita-Finzi and Mann (1994). It namely relates to the corner flow of the mantle as induced by subduction to uplift a deformable lithosphere. We observe corner flow during the interseismic period (Appendix Fig. 10a). However, the displacement pattern of its accelerated version during coseismic slab penetration is distinctly skewed towards the seaside as the overriding plate is displaced in that direction (Appendix Fig. 10b). In addition our uplift is rapid and thus involves a larger elastic component.

It is possible that internal slab deformation and slab unbending due to the earthquake (around $X = 300$ km in Fig. 6c) could provide a minor contribution to this phenomenon. However, analysis of this phenomena demonstrated that it does not play an important role in our simulations. Additionally, these models can not exclude the role of deep afterslip in early postseismic displacements. It is implicitly included in the large-scale models in the form of dislocation creep (not frictional strengthening) in a spontaneous low viscosity channel from about $450°$ to $1300°$ (Fig. 1a). However, deep afterslip down to the interplate decoupling point rather seems to induce postseismic subsidence in the area of interest, just landward of the decoupling point (i.e., at 250–400 km in Fig. 6f). This sense of timing would support the dominance of two proposed mechanisms for a secondary zone of uplift during or shortly after megathrust earthquakes.

4.3. Coseismic or Postseismic Nature

The limited temporal resolution in the data and in the model make it challenging to decipher how much of the uplift in the secondary zone is coseismic and how much is postseismic in nature. However, the better resolved data during and following the 2010 M8.8 Maule and 2011 M9.0 Tohoku earthquakes suggest the secondary zone of uplift is mostly coseismic in nature. In the Maule region the coseismic data of Vigny et al. (2011) and postseismic data of Klein et al. (2016) combined show a strong decrease in uplift rates (Sect. 3.1.3). This decrease through time suggests that the small majority of the uplift occurred within the coseismic measurements of Vigny et al. (2011). The coseismic nature of the secondary zone of uplift is more evident for the Tohoku earthquake, where coseismic uplift on the west coast is followed by subsidence in the 2 weeks following the earthquake (Ozawa et al. 2011; Yamagiwa et al. 2015; Hu et al. 2016). Nonetheless, a relevant portion of the large amount of secondary uplift during the 1960 M9.5 Validivia earthquake was likely postseismic, since its tectonic setting is similar to the 2010 M8.8 Maule earthquake. In summary, we interpret that the secondary zone of uplift is coseismic in nature, whereas potential postseismic additions depend on the specifics of a subduction zone.

The numerical results can not unequivocally distinguish between coseismic and postseismic displacements, since the coseismic results likely contain a postseismic response due to the large time steps. However, the relative timing of surface displacements supports a mainly coseismic nature of the secondary zone uplift and the governing mechanisms. The relative timing is apparent from the simultaneous occurrence of a primary and secondary zone uplift (Fig. 6a). As the interface relocks and causes primary subsidence at around 50 km from the trench again, secondary subsidence starts to occur around 250–350 km from the trench as well.

It is open for discussion whether the second mechanism of mass conservation driven return flow (Fig. 8) occurs at coseismic and/or (early) postseismic time scales. Traditionally one might think that the response from the mantle will be too slow. However, we know that the mantle behaves elastically during and just after the earthquake, because of the prolonged propagation of seismic waves. Moreover, we know that mass must always be conserved, also when the large and heavy slab inevitably penetrates rapidly into the mantle. Together with the apparent occurrence of a mainly coseismic secondary zone of uplift, it seems this could occur at least during—or within the few days following—the earthquake. This might be observed on the Korean peninsula, where uplift is only observed for the five days following the 2011 M9.0 Tohoku earthquake (Kim and Bae 2012). Nonetheless, the penetration of the slab into the mantle will be at least partially delayed, since on both sides of the slab the mantle resists its penetration as it does for the overriding plate. The resulting shear stresses (Fig. 8) will need to be relaxed on the time scales mostly dictated by mantle viscosity. In our model these viscosities are low and on the order of 10^{18} Pa s (Figs. 1a, 8). This reduction with respect to the surrounding mantle with viscosities of about 10^{20} Pa s occurs due to increased strain rates around the slab, which feedback non-linearly via the stress dependence of dislocation creep viscosity (Eq. 2). Consequently, Maxwell relaxation times would be around one or a few years, unless accelerated slab penetration on time scales of minutes can increase them even further in the vicinity of the slab (e.g., to around 10^{15-16} Pa s as in Sobolev and Muldashev 2017). In summary, we estimate mass conservation following accelerated slab penetration operates on both coseismic and (early) postseismic time scales, where it may also affect viscoelastic relaxation. What contribution is coseismic and what is postseismic could also be affected by the specifics of a subduction zone.

4.4. Implications

These results imply that subduction is not a gradual process with subduction occurring at constant rates, as typically envisioned within the long-term communities. Subduction rather proceeds in shocks following the brittle stick-slip behaviour of the shallow seismogenic zone. During the interseismic period some subduction can occur. However, locking across a 100 km or 200 km portion of the megathrust interface can partially stall the penetration of the slab. When the whole megathrust unlocks during a great megathrust earthquake, subduction catches up and the earthquake-induced displacement of the slab induces a significant amount of mantle flow (Fig. 6e, Supplementary movie S1). This makes megathrust earthquakes an integral driver of mantle flow. Similar ideas exploring the interaction between earthquakes and mantle flow are explored in other numerical models, which feature modulation of astenospheric flow (Barbot 2018) and modulation of residual polar wander (Cambiotti et al. 2016). This thus suggests a link between deep mantle and shallow surface displacements on timescales from minutes to decades, which is shorter than previously considered.

These numerical results also demonstrate implications for geodetic-based source inversions. The inclusion of visco-elastic layering and their detailed geometrical implementation distinctly impacts the resulting coseismic surface displacements (e.g., Fig. 7). Conversely, when using these surface displacements to estimate slip at a fault within a homogeneous elastic medium typically used for source inversions, one would artificially adapt fault slip to compensate for the missed partially viscous features. This is similarly observed for fitting interseismic velocity data, where elastic models require the presence of a rigid micro plate (e.g., Simons et al. 2007) that is not required using viscoelastic models (Trubienko et al. 2013). Additionally, slip artifacts might be introduced by the secondary zone of uplift present in data, but not in an elastic forward model. This might make it difficult to fit model results to the data (e.g., Lin et al. 2013, for the Maule earthquake). This supports emerging results that it is important to include a realistic visco-elastic structure in inversion for interseismic crustal deformation and earthquake slip inversions (e.g., Wang et al. 2012; Trubienko et al. 2013; Sun and Wang 2015; Klein et al. 2016; Moore et al. 2017; Sun et al. 2018).

4.5. Predicting Future Observations

Based on our observational and numerical findings, we make several predictions that can be tested as more accurate data becomes available during and prior to future, large megathrust earthquakes.

We predict more secondary zones of uplift will be observed in future great megathrust earthquakes (and maybe also for M > 8 or smaller earthquakes). The location of the secondary hinge point, its wavelength,

its amplitude, and decay with time will vary with specifics of the various subduction zones (Figs. 3, 9). The location could move further inland for subduction zones with more shallowly dipping slabs, whose seismogenic zones are wider and earthquakes can thus penetrate more inland. More slip on the deep portion of the seismogenic megathrust likely also translates into higher amplitudes of secondary uplift. The contribution from interseismic buckling would be enhanced by the presence of more effective backstops (Fig. 7) and by compression of a thinner upper crust and/or mantle lithosphere [as eq. 3-124 in Turcotte and Schubert (2002) predicts lower amplitudes at larger distances]. The contribution of mass-conserving return flow due to slab penetration would be enhanced by increased uplift amplitudes when more slab material displaces more mantle and effectively tunnels it to just landward of the interplate decoupling point (Figs. 6e, 8). This is anticipated for subduction zones with slabs (and events) that have a larger lateral extent and/or larger thickness of the slab (as to some extent occurs for an older and cooler slab). Additionally, more slip also displaces more material that needs to be relocated.

Finally, we predict a secondary zone of interseismic subsidence to occur at similar distances of between 200 and 500 km from the trench (Fig. 6a). This interseismic subsidence will be very slow (Fig. 6d) and is the counterpart or cause that—through elastic rebound—leads to a secondary zone uplift due to megathrust earthquakes. It likely occurs just on the landward side of the interplate decoupling point, where mantle displacements beneath the overriding plate become dominant. This lithosphere - astenosphere transition facilitates both interseismic buckling and rapid upward displacements following slab penetration.

5. Conclusions

We propose to extend the classical earthquake vertical displacement pattern for great megathrust earthquakes from one to two zones of uplift that flank a zone of primary subsidence. A second, minor zone of uplift was first predicted by physically consistent models starting to bridge long- and short-term dynamics. Subsequently we observed it for all four great megathrust earthquakes studied. This secondary zone of uplift starts at distances between 200 km and 350 km (or 500 km) from the trench and varies in magnitude from a maximum of 0.4 to 11 decimeter.

Extensive numerical experiments in both realistic and simple setups could not identify a single physical mechanism that is able to respectively remove and add a secondary zone of uplift to the two setups. Instead we hypothesize that a superposition of at least two mechanisms is needed to generate a secondary zone of uplift. We need a visco-elastically layered fore-arc to form two thin rigid beams that can buckle elastically in response to horizontal compression due to end loading in the interseismic period. This introduces a higher-order wavelength with a secondary zone of very minor interseismic subsidence. Elastic rebound due to an earthquake then causes a secondary zone of relative uplift. This is uplifted above zero by displacements that conserve mass (and momentum) following the earthquake-triggered penetration of the slab into the mantle. These upward displacements particularly localize in the about 150 km's landward of the interplate decoupling point, which typically corresponds to the area of the secondary zone of uplift.

Recent postseismic data and coincident, albeit unresolved, timing in our numerical model point to a mainly coseismic nature of the secondary zone of uplift. Uplift in the region of interest can either be enlarged or decreased during the postseismic period depending on the specifics of a subduction zone. The exact coseismic and postseismic contributions in various subduction zones and corresponding processes remain to be confirmed.

We estimate that the most important parameters affecting the secondary zone of uplift are the seismogenic zone dip and (deep) coseismic slip magnitude and limit (or earthquake size). Predictions from our models in terms of in the future verifiable observations include more secondary zones of coseismic uplift (potentially also for smaller earthquakes) and a secondary zone of very minor subsidence for interseismic displacements. Additionally, we propose a suite of tectonic influences that could start to explain variations in its size and location. In any case a more accurate representation of the visco-elastic structure of the fore-arc helps to understand and invert for inter-, co- and postseismic displacements. Finally, our results imply that subduction is not a gradual processes, but that is rather accelerated and decelerated through seismic cycles following the slab penetration during great megathrust earthquakes. This suggests a link between deep mantle and shallow surface displacements at time scales as short as minutes.

Acknowledgements

All natural data used for this paper are properly cited and referred to in the reference list. The data from the numerical experiments are available from the authors upon request. We are grateful to Andreas Fichtner and Paul Tackley for additional funding to continue the MSc thesis project of Lukas Preiswerk. This work was also supported by a computing resources Grant from the Swiss National Supercomputing Centre (CSCS; s741). We thank Arnauld Heuret, Emilie Klein, Saulė Simutė and Christophe Vigny for providing us with data as well as Rob Govers for discussions. We also thank three anonymous reviewers, who helped to largely improve the readability of the manuscript, and guest editor Sylvain Barbot for his patience and understanding in accommodating our time restrictions. We acknowledge the use of the geolib package for Matlab (Karney 2013). Finally, we acknowledge our contributions for transparency. L.P. performed and analyzed the computations and wrote the first version of the manuscript with support from Y.D.. Y.D. designed the study, closely supervised the work, and rewrote the manuscript during major revisions. T.G. supervised the project. All authors discussed the results.

Appendix A: Conservation Equations

To obtain horizontal velocity v_x, vertical velocity v_z, and pressure P we solve the conservation of mass and momentum as

$$\frac{\partial v_x}{\partial x} + \frac{\partial v_z}{\partial z} = 0 \qquad (5)$$

$$\frac{\partial \sigma'_{xx}}{\partial x} + \frac{\partial \sigma'_{xz}}{\partial z} - \frac{\partial P}{\partial z} = \rho \frac{Dv_x}{Dt} \qquad (6)$$

$$\frac{\partial \sigma'_{zx}}{\partial x} + \frac{\partial \sigma'_{zz}}{\partial z} - \frac{\partial P}{\partial z} = \rho \frac{Dv_z}{Dt} - \rho g. \qquad (7)$$

Here σ'_{ij} represents the 2D deviatoric stress tensor. The conservation of momentum includes gravitational acceleration g and the inertial term, represented by density ρ times the Lagrangian time derivative of the respective velocity components $\frac{Dv}{Dt}$. The momentum equations include the inertial term to stabilise high coseismic slip rates at low time steps (van Dinther et al. 2013a). A time step of five years, however, reduces our formulation to a virtually quasi-static one.

In the large-scale model we also solve the heat equation

$$\rho C_p \left(\frac{DT}{Dt} \right) = -\frac{\partial q_x}{\partial x} - \frac{\partial q_z}{\partial z} + H_a + H_s + H_r, \qquad (8)$$

where C_p is isobaric heat capacity, DT/Dt is the Lagrangian time derivative of temperature, and q_x and q_z are the horizontal and vertical heat flux, respectively. The equation includes contributions from conductive heat transport and volumetric internal heat generation H due to adiabatic (de-)compression H_a, shear heating during non-elastic deformation H_s and lithology-specific radioactive heat production H_r (e.g., Gerya and Yuen 2003, 2007).

Appendix B: Large-scale velocities throughout the seismic cycle

Figure 10 shows how mantle flow is affected by slab penetration at each different stage in the seismic

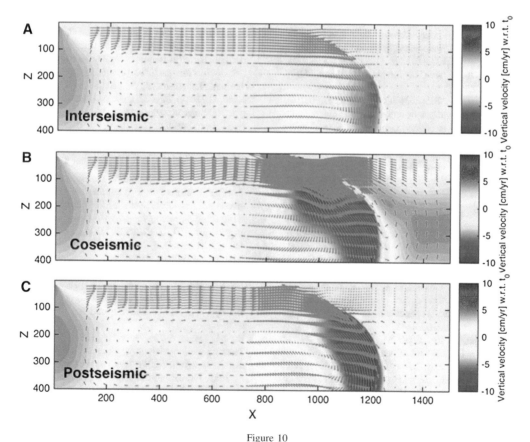

Figure 10
Snapshots of **a** interseismic, **b** coseismic, and **c** postseismic velocities throughout the whole model domain. This illustrates the reach and size of convective cells, which ensures mass is conserved as regular or accelerated subduction occurs. Colors indicate vertical velocities with respect to those at the start of the experiment

cycle. Accelerated slab penetration in the coseismic period leads to localized upward flow that is mainly returning on the landside of the slab. At the landside of the slab we also observe mantle flow is most focused within 150 km landward of the interplate decoupling point, where it could possibly fill up the space created by the displacement of the overriding plate.

Publisher's Note   Springer Nature remains neutral with regard to jurisdictional claims in published maps and institutional affiliations.

REFERENCES

Barbot, S. (2018). Asthenosphere flow modulated by megathrust earthquake cycles. *Geophysical Research Letters, 45*, 6018–6031.

Blaser, L., Kruger, F., Ohrnberger, M., & Scherbaum, F. (2010). Scaling relations of earthquake source parameter estimates with special focus on subduction environment. *Bulletin of the Seismological Society of America, 100*(6), 2914–2926.

Byrne, D. E., Wang, W. H, & Davis, D. M. (1993). Mechanical role of backstops in the growth of forearcs. *Tectonics, 12*(1), 123–144.

Cambiotti, G., Wang, X., Sabadini, R., & Yuen, D. (2016). Residual polar motion caused by coseismic and interseismic deformations from 1900 to present. *Geophysical Journal International, 205*(2), 1165–1179.

Coffin, M. F., Gahagan, L. M, & Lawler, L. A. (1998). *Present-day plate boundary digital data compilation.* Tech. Rep. 174, University of Texas Institute for Geophysics. https://doi.org/10.1029/2005GL022437/full.

Corbi, F., Funiciello, F., Moroni, M.b., van Dinther, Y., Mai, P.M., Dalguer, L.A., Faccenna, C. (2013). The seismic cycle at subduction thrusts: 1. Insights from laboratory models. *Journal of Geophysical Research 118*(4), 1483–1501. http://www.scopus.com/inward/record.url?eid=2-s2.0-84880710737&partnerID=40&md5=400ba823e2159242374c2b67a7d6639f.

Crameri, F., Schmeling, H., Golabek, G. J., Duretz, T., Orendt, R., Buiter, S., et al. (2011). A benchmark comparison of numerical topography—what are suitable sticky air parameters? *Geophysical Journal International, 200*, 1–12.

Fujiwara, T., Kodaira, S., No, T., Kaiho, Y., Takahashi, N., & Kaneda, Y. (2011). The 2011 tohoku-oki earthquake: Displacement reaching the trench axis. *Science, 334*(6060), 1240–1240.

Fukahata, Y. (2015). Afterslip and viscoelastic relaxation following the 2011 Tohoku-oki earthquake (Mw 9.0) inferred from inland GPS and seafloor GPS/Acoustic data. *Geophysical Research Letters, 42*(1), 66–73. https://doi.org/10.1002/2014GL061735.

Furukawa, Y. (1993). Depth of the decoupling plate interface and thermal structure under arcs. *Journal of Geophysical Research Solid Earth (1978–2012), 98*(B11), 20005–20013. https://doi.org/10.1029/93JB02020.

Gerya, T. (2010). *Introduction to numerical geodynamic modelling.* Cambridge: Cambridge University Press.

Gerya, T. V., & Yuen, D. (2003). Characteristics-based marker-in-cell method with conservative finite-differences schemes for modeling geological flows with strongly variable transport properties. *Physics of the Earth and Planetary Interiors, 140*(4), 293–318.

Gerya, T. V., & Yuen, D. A. (2007). Robust characteristics method for modelling multiphase visco-elasto-plastic thermo-mechanical problems. *Physics of the Earth and Planetary Interiors,163*(1–4), 83–105. https://doi.org/10.1016/j.pepi.2007.04.015.

Govers, R., Furlong, K. P., van de Wiel, L., Herman, M. W., & Broerse, T. (2018). The geodetic signature of the earthquake cycle at subduction zones: Model constraints on the deep processes. *Reviews of Geophysics, 56*(1), 6–49.

Hashimoto, M., Choosakul, N., Hashizume, M., Takemoto, S., Takiguchi, H., & Fukuda, Y., et al. (2006). Crustal deformations associated with the great Sumatra-Andaman earthquake deduced from continuous GPS observation. *Earth Planets Space,58*(2), 127–139. https://doi.org/10.1186/BF03353369.

Hayes, G. P., Wald, D. J., & Johnson, R. L. (2012). Slab1.0: A three-dimensional model of global subduction zone geometries. *Journal of Geophysical Research, 117*(B1), B01302. https://doi.org/10.1029/2011JB008524.

Herrendoerfer, R. (2018). *Modeling of the slip spectrum along mature and spontaneously forming faults in a visco-elasto-plastic continuum.* PhD thesis, ETH.

Herrendoerfer, R., Gerya, T., & van Dinther, Y. (2018). An invariant rate-and-state dependent friction formulation for viscoelastoplastic earthquake cycle simulations. *Journal of Geophysical Research Solid Earth, 123*(6), 5018–5051.

Heuret, A., Lallemand, S., Funiciello, F., Piromallo, C., & Faccenna, C. (2011). Physical characteristics of subduction interface type seismogenic zones revisited. *Geochemistry Geophysics Geosystems.* https://doi.org/10.1029/2010GC003230.

Holdahl, S.R., & Sauber, J. (1994). Coseismic slip in the 1964 Prince–William–Sound Earthquake—a new geodetic inversion. *Pure and Applied Geophysics, 142*(1), 55–82.

Hu, Y., Bürgmann, R., Uchida, N., Banerjee, P., & Freymueller, J. T. (2016). Stress-driven relaxation of heterogeneous upper mantle and time-dependent afterslip following the 2011 Tohoku earthquake. *Journal of Geophysical Research, 121*(1), 385–411. https://doi.org/10.1002/2015JB012508.

Johnson, J. M., Satake, K., Holdahl, S. R., & Sauber, J. (1996). The 1964 Prince William Sound earthquake: Joint inversion of tsunami and geodetic data. *Journal of Geophysical Research, 101*(B1), 523–532. https://doi.org/10.1029/95JB02806.

Karney, C. F. F. (2013). Algorithms for geodesics. *Journal of Geodesy, 87*(1), 43–55. https://doi.org/10.1007/s00190-012-0578-z.

Kido, M., Osada, Y., Fujimoto, H., Hino, R., & Ito, Y. (2011). Trench-normal variation in observed seafloor displacements associated with the 2011 Tohoku-Oki earthquake. *Geophysical Research Letters.* https://doi.org/10.1029/2011GL050057.

Kim, S. K., & Bae, T. S. (2012). Analysis of crustal deformation on the Korea Peninsula after the 2011 Tohoku Earthquake. *Journal of the Korean Society of Surveying Geodesy Photogrammetry and Cartography, 30*(1), 87–96.

Klein, E., Fleitout, L., Vigny, C., & Garaud, J. D. (2016). Afterslip and viscoelastic relaxation model inferred from the large-scale post-seismic deformation following the 2010 Mw8.8 Maule earthquake (Chile). *Geophysical Journal International, 205*(3), 1455–1472.

Li, S., Moreno, M., Bedford, J., Rosenau, M., Heidbach, O., Melnick, D., et al. (2017). Postseismic uplift of the Andes following the 2010 Maule earthquake: Implications for mantle rheology. *Geophysical Research Letters, 44*(4), 1768–1776. https://doi.org/10.1002/2016GL071995.

Lin YN, Sladen, A., Ortega-Culaciati, F., Simons, M., Avouac, J. P., Fielding, E. J., et al. (2013). Coseismic and postseismic slip associated with the 2010 Maule Earthquake, Chile: Characterizing the Arauco Peninsula barrier effect. *Journal of Geophysical Research, 118*(6), 3142–3159. https://doi.org/10.1002/jgrb.50207.

Linde, A. T., & Silver, P. G. (1989). Elevation changes and the Great 1960 Chilean Earthquake: Support for aseismic slip. *Geophysical Research Letters, 16*(11), 1305–1308. https://doi.org/10.1029/GL016i011p01305.

Manea, V., Perez-Gussinye, M., & Manea, M. (2012). Chilean flat slab subduction controlled by overriding plate thickness and trench rollback. *Geology, 40*(1), 35–38.

Meltzner, A. J., Sieh, K., Abrams, M., Agnew, D. C., Hudnut, K. W., Avouac, J. P., et al. (2006). Uplift and subsidence associated with the great Aceh-Andaman earthquake of 2004. *Journal of Geophysical Research, 111*(B2), B02407–n/a. https://doi.org/10.1029/2005JB003891.

Miyashita, K. (1987). A model of plate convergence in southwest Japan, inferred from leveling data associated with the 1946 Nankaido earthquake. *Journal of Physics of the Earth, 35*, 449–467.

Moore, J. D. P., Yu, H., Tang, C. H., Wang, T., Barbot, S., Peng, D., et al. (2017). Imaging the distribution of transient viscosity after the 2016 Mw 7.1 Kumamoto earthquake. *Science, 356*(6334), 163–167.

Moreno, M., Bolte, J., Klotz, J., & Melnick, D. (2009). Impact of megathrust geometry on inversion of coseismic slip from geodetic data: Application to the 1960 Chile earthquake. *Geophysical Research Letters, 36*(16), L16310. https://doi.org/10.1029/2009GL039276.

Moreno, M., Haberland, C., Oncken, O., Rietbrock, A., Angiboust, S., & Heidbach, O. (2014). Locking of the Chile subduction zone controlled by fluid pressure before the 2010 earthquake. *Nature Geoscience, 7*(4), 292–296. https://doi.org/10.1038/ngeo2102.

Moreno, M., Rosenau, M., & Oncken, O. (2010). Maule earthquake slip correlates with pre-seismic locking of Andean subduction zone. *Nature, 467*, 198–202.

Moresi, L., Dufour, F., & Mühlhaus, H. (2003). A Lagrangian integration point finite element method for large deformation modeling of viscoelastic geomaterials. *Journal of Computational Physics, 184*(2), 476–497.

Ozawa, S., Nishimura, T., Suito, H., Kobayashi, T., Tobita, M., & Imakiire, T. (2011). Coseismic and postseismic slip of the 2011 magnitude-9 Tohoku-Oki earthquake. *Nature, 475*(7356), 373–376. https://doi.org/10.1038/nature10227.

Plafker, G. (1965). Tectonic deformation associated with the 1964 Alaska earthquake. *Science (New York, NY), 148*(3678), 1675–1687. https://doi.org/10.1126/science.148.3678.1675.

Plafker, G. (1969). *Tectonics of the March 27, 1964 Alaska earthquake.* US Geological Survey Professional Paper 543-I, United States Geological Survey.

Plafker, G. (1972). Alaskan earthquake of 1964 and Chilean earthquake of 1960: Implications for arc tectonics. *Journal of Geophysical Research, 77*(5), 901–925. https://doi.org/10.1029/JB077i005p00901.

Plafker, G., & Savage, J. C. (1970). Mechanism of the Chilean Earthquakes of May 21 and 22, 1960. *Geological Society of America Bulletin, 81*(4), 1001–1030. https://doi.org/10.1130/0016-7606(1970)81[1001:MOTCEO]2.0.CO;2.

Pollitz, F. (1997). Gravitational viscoelastic postseismic relaxation on a layered spherical Earth. *Journal of Geophysical Research, 102*(B8), 17921–17941.

Reid, H. F. (1910). *The California earthquake of April 18, 1906. Report of the State earthquake investigation commission.* Tech. rep., Carnegie Institution of Washington, Washington, D.C.

Sato, M., Ishikawa, T., Ujihara, N., Yoshida, S., Fujita, M., & Mochizuki, M., et al. (2011). Displacement above the hypocenter of the 2011 Tohoku-Oki earthquake. *Science (New York, NY), 332*(6036), 1395–1395. https://doi.org/10.1126/science.1207401.

Simons, W. J. F., Socquet, A., Vigny, C., Ambrosius, B. A. C., Haji Abu, S., Promthong, C., et al. (2007). A decade of GPS in Southeast Asia: Resolving Sundaland motion and boundaries. *Journal of Geophysical Research, 112*(B6), 686–20.

Sobolev, S. V., & Babeyko, A. (2005). What drives orogeny in the Andes? *Geology, 33*(8), 617–620.

Sobolev, S. V., & Muldashev, I. A. (2017). Modeling seismic cycles of great megathrust earthquakes across the scales with focus at postseismic phase. *Geochemistry Geophysics Geosystems, 18*(12), 4387–4408.

Socquet, A., Peyrat, S., Ruegg, J. C., Métois, M., Madariaga, R., Morvan, S., et al. (2011). The 2010 Mw 8.8 Maule megathrust earthquake of Central Chile, monitored by GPS. *Science (New York, NY), 332*(6036):1417–1421. https://doi.org/10.1126/science.1204132.

Sun, T., & Wang, K. (2015). Viscoelastic relaxation following subduction earthquakes and its effects on afterslip determination. *Journal of Geophysical Research, 120*, 1329–1344.

Sun, T., Wang, K., & He, J. (2018). Crustal deformation following great subduction earthquakes controlled by earthquake size and mantle rheology. *Journal Of Geophysical Research Solid Earth, 123*(6), 5323–5345.

Sun, T., Wang, K., Iinuma, T., Hino, R., He, J., Fujimoto, H., et al. (2014). Prevalence of viscoelastic relaxation after the 2011 Tohoku-oki earthquake. *Nature, 514*, 84–87.

Trubienko, O., Fleitout, L., Garaud, J. D., & Vigny, C. (2013). Interpretation of interseismic deformations and the seismic cycle associated with large subduction earthquakes. *Tectonophysics, 589*, 126–141.

Turcotte, D. L., & Schubert, G. (2002). *Geodynamics*, 2nd edn. Cambridge University Press, Cambridge.

van Dinther, Y., Gerya, T. V., Dalguer, L. A., Corbi, F., Funiciello, F., & Mai, P. M. (2013a). The seismic cycle at subduction thrusts: 2. Dynamic implications of geodynamic simulations validated with laboratory models. *Journal of Geophysical Research, 118*(4), 1502–1525. https://doi.org/10.1029/2012JB009479.

van Dinther, Y., Gerya, T. V., Dalguer, L. A., Mai, P. M., Morra, G., & Giardini, D. (2013b). The seismic cycle at subduction thrusts: Insights from seismo-thermo-mechanical models. *Journal of Geophysical Research, 118*(12), 6183–6202. https://doi.org/10.1002/2013JB010380.

van Dinther, Y., Mai, P. M., Dalguer, L. A., & Gerya, T. V. (2014). Modeling the seismic cycle in subduction zones: The role and spatiotemporal occurrence of off-megathrust earthquakes.

Geophysical Research Letters, *41*(4), 1194–1201. https://doi.org/10.1002/2013GL058886.

van Zelst, I., Wollherr, S., Gabriel, A. A., Madden, E., & van Dinther, Y. (2019). *Modelling coupled subduction and earthquake dynamics*. https://doi.org/10.31223/osf.io/f6ng5,eartharxiv.org/f6ng5.

Vita-Finzi, C., & Mann, C. D. (1994). Seismic folding in coastal south central Chile. *Journal of Geophysical Research*, *99*(B6), 12289–12299. https://doi.org/10.1029/93JB03061.

Wang, K. (2007). Elastic and viscoelastic models of crustal deformation in subduction earthquake cycles. *The seismogenic zone of subduction thrust faults* (pp. 1–36). Columbia: Columbia University Press.

Wang, K., Hu, Y., & He, J. (2012). Deformation cycles of subduction earthquakes in a viscoelastic Earth. *Nature*, *484*, 327–332.

Wdowinski, S., O'Connell, R. J., & England, P. (1989). A continuum model of continental deformation above subduction zones: Application to the Andes and the Aegean. *Journal of Geophysical Research*, *94*(B8), 10331–10346. https://doi.org/10.1029/JB094iB08p10331.

Yue, H., & Lay, T. (2013). Source rupture models for the Mw 9.0. (2011). Tohoku earthquake from joint inversions of high-rate geodetic and seismic data. *Bulletin of the Seismological Society of America,103*(2B), 1242–1255. https://doi.org/10.1785/0120120119.

(Received April 26, 2018, revised May 15, 2019, accepted June 7, 2019, Published online July 2, 2019)